Essential Relativistic Celestial Mechanics

Essential Relativistic Celestial Mechanics

V A Brumberg

Institute of Applied Astronomy of the Academy of Sciences of the USSR

CRC Press
Taylor & Francis Group
Boca Raton London New York

CRC Press is an imprint of the
Taylor & Francis Group, an **informa** business

CRC Press
Taylor & Francis Group
6000 Broken Sound Parkway NW, Suite 300
Boca Raton, FL 33487-2742

First issued in paperback 2019

ISBN-13: 978-0-367-40306-5

Library of Congress Cataloging-in-Publication Data

Catalog record is available from the Library of Congress

Visit the Taylor & Francis Web site at
http://www.taylorandfrancis.com

and the CRC Press Web site at
http://www.crcpress.com

Contents

Preface

In recent years the general theory of relativity (GRT) in its simplest applications in celestial mechanics and astrometry is no longer seen as a theory still to be proved, but rather serves as a necessary framework for the construction of accurate dynamical ephemerides (10^{-8} to 10^{-9} with respect to the main Newtonian terms) and in the discussion of high-precision observations ($0.001''$ in angular distance, 1 microsecond in time, 10^{-14} in frequency). It is of interest to note that in 1974 at IAU Colloquium no 26, devoted to the problems of reference systems for astronomy and geodynamics, the word 'relativity' was used only occasionally (Kolaczek and Weiffenbach 1975). However, by 1980 at IAU Colloquium no 56 devoted to the same problems several authors had already presented papers within the framework of GRT (Gaposchkin and Kolaczek 1981), and in 1984 the Japanese Astronomical Almanac for 1985 appeared, containing a detailed exposition of principles of GRT as applied to ephemeris astronomy (Japanese Ephemeris 1985). In fact, this was the start of broadly utilizing GRT in ephemeris astronomy. GRT was coming into use in the exposition of traditional questions of astrometry and spherical astronomy (Murray 1983, Green 1985). Finally, 1985 saw IAU Symposium no 114, 'Relativity in Celestial Mechanics and Astrometry', the first IAU meeting devoted exclusively to the problems of relativistic celestial mechanics and astrometry (*RCMA* 1986). As great an interest in the practical application of GRT is expressed by the International Association of Geodesy.

GRT is the physical foundation of celestial mechanics. Disregarding this meaning of GRT, and from a purely operational point of view, the distinction between Newtonian mechanics and GRT is displayed, mathematically, by the structure of the field equations and the equations of motion and, physically, by the way we compare observational and theoretical data. The first question is the subject of relativistic celestial mechanics, the second is the subject of relativistic astrometry. Only a consistent simultaneous treatment of both questions leads to physically meaningful results. The specific feature of the second question is that, in contrast to the inertial coordinates of Newtonian mechanics, the GRT coordinates have no physical meaning and cannot be considered to be physically measurable quantities. Therefore, the results of the relativistic dynamical theories expressed in terms of the coordinates are not unique; they depend on the type of the coordinate conditions used and so cannot be directly confirmed or refuted by observations. Only in terms of measurable quantities do the conclusions of the dynamical theories become unique and comparison with observation become possible.

These logically very simple statements of GRT present difficulties in the practical application of GRT by astronomers (and not only by them!) accustomed to Newtonian conceptions of space and time. In spite of the extensive penetration of GRT into celestial mechanics and astrometry there still exists the danger of superficially using GRT as a theory for only adding small corrections to Newtonian mechanics. This is a completely false idea leading to improper interpretation of GRT solutions and observational data. For complete utilization of the high-precision data of modern observational tools, such as space astrometry, VLBI, pulsar timing, etc, celestial mechanicians and astrometrists should accept the GRT framework in such fundamental domains as reference frames, timescales, reduction of observations, etc.

The aim of this book is to describe the essential questions of relativistic celestial mechanics in relation to relativistic astrometry. This book was preceded by *Relativistic Celestial Mechanics* by the same author, published in Russian (Moscow, 1972). In spite of the close interrelation between these books the present one is in fact a new monograph rather than a new edition of the preceding work. The main emphasis here is given to results obtained in relativistic celestial mechanics in recent years. Similar questions are partly considered in the recent book by Soffel (1989) but the present monograph contains different topics and has its own manner of treatment.

The book consists of seven chapters. The first chapter discusses the elements of Newtonian celestial mechanics, tensor analysis and the special theory of relativity. This chapter does not claim to be a comprehensive treatment of these topics and is aimed at avoiding too much reference to the appropriate specialized textbooks.

The second chapter is devoted to a description of the fundamentals of GRT. It deals with the post-Newtonian metric for an isolated system of bodies such as the Solar System, the post-Newtonian equations of test particle motion and light propagation, and the problem of measurement of infinitesimal time intervals and distances within the framework of the tetrad formalism. Not all the results of this chapter are used in later chapters but it seems reasonable to reflect here on the various mathematical techniques of current papers on GRT and relativistic astronomy.

The third chapter is concerned with one-body GRT problems. In deriving the Schwarzschild metric a more general approach has been used as compared with the techniques more commonly used. Much attention is focused on those aspects which are of importance in application to celestial mechanics and astrometry. The post-Newtonian solution of the equations of motion of a test particle is given both in coordinate form and by means of expressions for the orbital osculating elements. The equations of light propagation are solved in the post-post-Newtonian approximation. The contributions to the secular evolution of the orbit of a test particle caused by the rotation of an attracting body and its quadrupole moment are also discussed.

Approximate solutions of the field equations and approximate equations of motion of the N-body problem are examined in Chapter 4. The post-Newtonian equations of motion of the N-body problem are formulated in the barycentric reference system (BRS) after which the transformation to the planetocentric, and in particular the geocentric, reference system (GRS) is performed. Along with the traditional expansion in powers of v/c (v is the characteristic velocity of the bodies, c is the velocity of light), a technique based on expansions in powers of the ratio of the planetary masses to the mass of the Sun is also developed. The problem of two bodies of comparable masses is treated here, as well as the motion of a binary system, taking into account gravitational radiation.

The methods discussed in Chapter 4 are applied in Chapter 5 to derive the equations of motion of Solar System bodies: Earth's artificial satellites, the major planets and the Moon. The Earth satellite equations of motion are derived first in BRS and are then transformed to GRS equations. The equations of planetary motion are described in BRS, which is the most convenient for this case. Although relativistic perturbations for the GRS lunar motion are two orders of magnitude less than the BRS perturbations, BRS consideration of the lunar motion presented here does not lose its importance in providing a uniform treatment of planetary and lunar motion.

The relativistic reduction of astrometric measurements is analyzed in Chapter 6. The hierarchy of astronomical reference systems relating to the Solar System barycentre, the geocentre, a ground station on the surface of the Earth and a satellite station in orbit around the Earth, respectively, are constructed. This enables one to perform, in an uniform manner, typical reduction of astrometric measurements including optical, radio technical, laser and VLBI observations.

Geodynamical effects in time measurement on the Earth and in circumterrestrial space are briefly discussed in Chapter 7. Most attention is paid here to the timescales for the Earth and to the problem of clock synchronization via satellite. Relativistic effects in geodynamics form a comparatively new domain of investigation with a promising future.

This book contains few numerical data. Experience shows that such data quickly become obsolete. Therefore, emphasis is given here to practically useful relations and methodological questions which become particularly important in the extensive application of GRT in celestial mechanics and astrometry. Unsolved or not completely investigated problems are often indicated in the text. Examples of this type are the possibility of relativistic resonance in the critical inclination case in the theory of Earth satellite motion (section 3.3.3), the interaction of Schwarzschild terms and Newtonian planetary perturbations in the theory of motion of the major planets (section 5.2.3), the development of the expansion technique in powers of the ratio of the planetary masses to the mass of the Sun avoiding expansions in powers of v/c as well as motion within the cosmological background (section 4.3), the general structure and statement of the initial value problem of the motion taking into account gravitational radiation (section 4.4.6), the consistent application of the relativistic theory of astronomical reference systems to the problems of astrometry (section 6.2), etc.

A general remark may be made here. Relativistic celestial mechanics is meant here as celestial mechanics based on Einstein's general theory of relativity. Alternative theories of gravitation are not touched on at all (except for using sometimes the main parameters characterizing post-Newtonian metric theories for the sake of mathematical generalization). This is for two reasons. Firstly, the extensive programme of parametrized post-Newtonian (PPN) formalism performed during the last two decades has resulted in the experimental foundation of GRT as the theory best fitted to observation (Will 1985, 1986). Nowadays there is no reliable observational result inconsistent with GRT and calling for any other theory. Secondly, it is not unusual that some authors of current alternative theories accompany the development of their theories by unjustified criticism with respect to GRT, attributing fictitious shortcomings to it. This evokes nothing but regret and prevents citation of the corresponding papers.

The author is indebted to Dr A M Nobili and Professor A E Roy for their encouragement and stimulation of this work. Long and close collaboration with Dr S M Kopejkin from Sternberg Astronomical Institute (Moscow) affected many parts of this book and is gratefully acknowledged. Particular thanks is given for the invaluable help received in preparing and checking the computer version of the manuscript to Drs E Z Chotimskaya, S A Klioner and A V Voinov from the Institute of Applied Astronomy (Leningrad). The author is also immensely grateful to the staff of Adam Hilger, and most particularly to Mr Richard Fidczuk, the Desk Editor, for the high-quality work of improving the author's English and converting the computer form of the manuscript from Chiwriter to TEX, resulting in better reproduction of the book.

V A Brumberg
Leningrad

1

Mathematical Tools

1.1 ELEMENTS OF NEWTONIAN CELESTIAL MECHANICS

1.1.1 Two-body problem

Newtonian celestial mechanics is based on Newton's law of universal gravitation (theory of the Newtonian potential) and the laws of Newtonian mechanics (theory of motion). The simplest problem of Newtonian celestial mechanics is the problem of two bodies (point masses). The differential equations of motion of this problem in some fixed rectangular reference system x, y, z have the form

$$\begin{aligned} \ddot{\boldsymbol{r}}_1 &= -\,GM_2\frac{\boldsymbol{r}_1-\boldsymbol{r}_2}{r^3} \\ \ddot{\boldsymbol{r}}_2 &= -\,GM_1\frac{\boldsymbol{r}_2-\boldsymbol{r}_1}{r^3} \end{aligned} \tag{1.1.1}$$

where M_i are the masses, $\boldsymbol{r}_i = (x_i, y_i, z_i)$ are the position vectors of the two bodies, $r = |\boldsymbol{r}_2 - \boldsymbol{r}_1|$ is the mutual distance between the bodies and G is the gravitational constant. Replacing $\boldsymbol{r}_1$ and $\boldsymbol{r}_2$ by the position vector of the Newtonian centre of mass

$$\boldsymbol{r}_0 = \frac{1}{M}(M_1\boldsymbol{r}_1 + M_2\boldsymbol{r}_2) \qquad M = M_1 + M_2$$

and the relative position vector of the second body with respect to the first one

$$\boldsymbol{r} = \boldsymbol{r}_2 - \boldsymbol{r}_1$$

system (1.1.1) is split into two systems:

$$\ddot{\boldsymbol{r}}_0 = 0 \tag{1.1.2}$$

and

$$\ddot{\boldsymbol{r}} = -GM\frac{\boldsymbol{r}}{r^3}. \tag{1.1.3}$$

From (1.1.2) it follows that the centre of mass of the two-body system is in uniform rectilinear motion (a reference system is called barycentric provided that the centre of mass is at rest at the origin of this system). Equations (1.1.3) determine the relative motion of M_2 with respect to M_1.

The trajectories of the two-body problem are conic sections (straight lines in the degenerate case). In what follows only the case of elliptic motion is of importance, i.e. the hyperbolic or parabolic motion will not be considered. The size and the shape of an elliptic orbit are characterized by the semi-major axis a and the eccentricity e ($0 \leq e < 1$). The orientation in space of the plane of motion is defined by the inclination i (the angle between the xy plane and the orbital plane) and the longitude of the ascending node Ω (the angle in the xy plane between the x axis and the line of nodes, i.e. the line of intersection of the xy plane and the orbital plane; it is assumed here that in passing through the ascending node, the coordinate z of the moving body changes from negative values to positive ones). For orientation of the orbit in the plane of motion one uses the argument of the pericentre ω (the angular orbital distance between the pericentre and the ascending node). These five Keplerian elements may be replaced by the vectorial elements, for example, by the mutually orthogonal area vector $\boldsymbol{c}$ and Laplace vector $\boldsymbol{f}$:

$$\boldsymbol{c} = \boldsymbol{r} \times \dot{\boldsymbol{r}} \qquad \boldsymbol{f} = \left(\dot{\boldsymbol{r}}^2 - \frac{GM}{r}\right)\boldsymbol{r} - (\boldsymbol{r}\dot{\boldsymbol{r}})\dot{\boldsymbol{r}}. \tag{1.1.4}$$

Expressions (1.1.4) are the first integrals of equations (1.1.3). Along with them there is the energy integral

$$\dot{\boldsymbol{r}}^2 = GM\left(\frac{2}{r} - \frac{1}{a}\right). \tag{1.1.5}$$

The arbitrary constants occurring in (1.1.4) are related by two equations:

$$\boldsymbol{c}\boldsymbol{f} = 0 \qquad GM\boldsymbol{c}^2 + a\boldsymbol{f}^2 = (GM)^2 a.$$

It is convenient to introduce the triad of unit vectors $\boldsymbol{l}$, $\boldsymbol{m}$, $\boldsymbol{k}$ directed along the line of the nodes toward the ascending node of the orbit, transversely to the line of nodes in the orbital plane and transversely to the orbital plane, respectively:

$$\boldsymbol{l} = \begin{pmatrix} \cos\Omega \\ \sin\Omega \\ 0 \end{pmatrix} \qquad \boldsymbol{m} = \begin{pmatrix} -\cos i \sin\Omega \\ \cos i \cos\Omega \\ \sin i \end{pmatrix} \qquad \boldsymbol{k} = \begin{pmatrix} \sin i \sin\Omega \\ -\sin i \cos\Omega \\ \cos i \end{pmatrix}. \tag{1.1.6}$$

Let $\boldsymbol{P}$, $\boldsymbol{Q}$ be the unit vectors directed along the line of apsides towards the pericentre and transversely to the line of apsides in the orbital plane, respectively:

$$\boldsymbol{P} = \boldsymbol{l}\cos\omega + \boldsymbol{m}\sin\omega \qquad \boldsymbol{Q} = -\boldsymbol{l}\sin\omega + \boldsymbol{m}\cos\omega. \tag{1.1.7}$$

Then

$$\boldsymbol{c} = (GMp)^{1/2}\boldsymbol{k} \qquad \boldsymbol{f} = GMe\boldsymbol{P} \tag{1.1.8}$$

with $p = a(1-e^2)$ being the parameter of orbit. The coordinates of the body in the orbital plane are the radius vector r and the argument of latitude u (the angular orbital distance reckoned from the ascending node). In terms of r and u one has

$$\begin{aligned} \boldsymbol{r} &= r(\boldsymbol{l}\cos u + \boldsymbol{m}\sin u) \\ \dot{\boldsymbol{r}} &= \left(\frac{GM}{p}\right)^{1/2}[-\boldsymbol{l}(\sin u + e\sin\omega) + \boldsymbol{m}(\cos u + e\cos\omega)]. \end{aligned} \tag{1.1.9}$$

Instead of u one may use the true anomaly f, i.e. the angular orbital distance between the pericentre and the moving body. Evidently,

$$u = f + \omega. \tag{1.1.10}$$

In terms of the true anomaly there results

$$r = \frac{p}{1 + e\cos f} \tag{1.1.11}$$

and

$$\begin{aligned} \boldsymbol{r} &= r(\boldsymbol{P}\cos f + \boldsymbol{Q}\sin f) \\ \dot{\boldsymbol{r}} &= \left(\frac{GM}{p}\right)^{1/2}[-\boldsymbol{P}\sin f + \boldsymbol{Q}(\cos f + e)]. \end{aligned} \tag{1.1.12}$$

The dependence on time t is determined by the transcendental Kepler equation

$$E - e\sin E = l. \tag{1.1.13}$$

The eccentric anomaly E is related to the true anomaly f by means of

$$\tan\frac{f}{2} = \left(\frac{1+e}{1-e}\right)^{1/2}\tan\frac{E}{2}. \tag{1.1.14}$$

Therefore,

$$r\cos f = a(\cos E - e) \tag{1.1.15}$$

$$\begin{aligned} r \sin f &= a(1-e^2)^{1/2} \sin E \\ r &= a(1 - e \cos E). \end{aligned} \tag{1.1.16}$$

The mean anomaly l represents a linear function of time

$$l = l_0 + n(t - t_0). \tag{1.1.17}$$

l_0, the mean anomaly at the epoch, serves as the sixth, dynamical, element of the elliptic motion. The mean motion n is related to the semi-major axis a by Kepler's third law:

$$n^2 a^3 = GM. \tag{1.1.18}$$

The set of these formulae defines completely the general solution of the elliptic two-body problem in terms of time t and six integration constants, the Keplerian elements a, e, i, Ω, ω, and l_0. The explicit dependence on time may be given by the Fourier series in multiples of the mean anomaly:

$$\left(\frac{r}{a}\right)^n \exp(imf) = \sum_{k=-\infty}^{\infty} X_k^{n,m}(e) \exp(ikl) \tag{1.1.19}$$

with arbitrary integers m and n. The Hansen coefficients $X_k^{n,m}$ depend only on e and may be easily calculated by means of expansions in powers of e^2. These expansions begin with terms of order $|m-k|$ with respect to e. The Hansen coefficients with the zero lower index are of particular importance yielding the mean value of functions of the left-hand side of (1.1.19). These coefficients are expressed by the hypergeometric polynomials

$$X_0^{n,m}(e) = \left(\frac{e}{2}\right)^{|m|} \frac{(-n-1-|m|)_{|m|}}{(1)_{|m|}} \times \begin{cases} F\left(\frac{|m|-n-1}{2}, \frac{|m|-n}{2}, 1+|m|, e^2\right) & n \geq |m|-1 \\ (1+\beta^2)^{|m|-n-1} F(|m|-n-1, -n-1, 1+|m|, \beta^2) & |m|-1 > n \geq -1 \\ 0 & -1 > n \geq -|m|-1 \\ (1-e^2)^{n+3/2} F\left(\frac{n+|m|+2}{2}, \frac{n+|m|+3}{2}, 1+|m|, e^2\right) & -|m|-1 > n \end{cases} \tag{1.1.20}$$

with

$$\beta = \frac{e}{1 + (1-e^2)^{1/2}}$$

and generalized factorials

$$(\alpha)_s = \alpha(\alpha+1)\ldots(\alpha+s-1) \qquad (\alpha)_0 = 1.$$

$F(\alpha, \beta, \gamma, x)$ is the hypergeometric series (a polynomial for the integral negative values of α or β)

$$F(\alpha, \beta, \gamma, x) = \sum_{s=0}^{\infty} \frac{(\alpha)_s (\beta)_s}{(\gamma)_s (1)_s} x^s.$$

Using (1.1.12) and (1.1.19) one may describe the general solution of the elliptic two-body problem as the trigonometric series

$$x + \mathrm{i}y = a \sum_{k=-\infty}^{\infty} X_k^{1,1}(e)\left(\cos^2\frac{i}{2}\exp[\mathrm{i}(kl+\omega+\Omega)] + \sin^2\frac{i}{2}\exp[\mathrm{i}(-kl-\omega+\Omega)]\right) \tag{1.1.21}$$

$$z = a\sin i \sum_{k=-\infty}^{\infty} X_k^{1,1}(e)\sin(kl+\omega). \tag{1.1.22}$$

The angular elements Ω, ω, l_0 enter into these series as trigonometric arguments. The coefficients of the series are functions of the linear elements a, e, i analogous to the action variables. The anomaly l is often used instead of l_0. In the typical case of direct motion ($0 \le i \le 90°$) ω and l are often replaced by the longitude of pericentre $\pi = \Omega + \omega$ and the mean longitude $\lambda = \pi + l$. In accordance with (1.1.17)

$$\lambda = n(t - t_0) + \epsilon \tag{1.1.23}$$

with $\epsilon = \pi + l_0$ being the mean longitude at the epoch. With small values of eccentricity and inclination one often employs

$$h = e\cos\pi \quad k = e\sin\pi \quad p = \sin i\cos\Omega \quad q = \sin i\sin\Omega \tag{1.1.24}$$

or analogous combinations. From the structure of the series (1.1.21) and (1.1.22) and the Hansen coefficients it follows that the coordinates x, y, z are holomorphic functions of these quantities.

1.1.2 Perturbations of osculating elements

Most practical problems of celestial mechanics are to a certain extent close to the two-body problem, enabling them to be solved by the perturbation theory techniques. The standard equations of the perturbed motion have the form

$$\ddot{\boldsymbol{r}} = -\frac{GM}{r^3}\boldsymbol{r} + \boldsymbol{F} \tag{1.1.25}$$

with the perturbing force $\boldsymbol{F}$ being generally dependent on $\boldsymbol{r}$, $\dot{\boldsymbol{r}}$, t and possibly $\ddot{\boldsymbol{r}}$. This force is proportional to some small parameter specified by a given problem. The general principle of the perturbation theory is to solve equations (1.1.25) with the aid of series in powers of this parameter (or by a sequence of iterations with respect to the parameter). With $\boldsymbol{F} = 0$ equations (1.1.25) reduce to equations (1.1.3) of the two-body problem and then $\boldsymbol{r}$, $\dot{\boldsymbol{r}}$ are expressed by (1.1.11)–(1.1.17) in terms of time and six arbitrary constants, for example, the Keplerian elements a, e, i, Ω, ω, l_0. Using the method of the variation of arbitrary constants these formulae may be retained in the disturbed motion with $\boldsymbol{F} \neq 0$. The quantities a, e, i, Ω, ω, l_0 are thereby some functions of time to be determined by definite system of differential equations. Such functions are called osculating elements. The equations for their determination are as follows:

$$
\begin{aligned}
\frac{\mathrm{d}a}{\mathrm{d}t} &= \frac{2}{n(1-e^2)^{1/2}}\left(Se\sin f + T\frac{p}{r}\right)\\
\frac{\mathrm{d}e}{\mathrm{d}t} &= \frac{(1-e^2)^{1/2}}{na}[S\sin f + T(\cos f + \cos E)]\\
\frac{\mathrm{d}i}{\mathrm{d}t} &= \frac{r\cos u}{na^2(1-e^2)^{1/2}}W\\
\frac{\mathrm{d}\Omega}{\mathrm{d}t} &= \frac{r\sin u}{na^2(1-e^2)^{1/2}\sin i}W\\
\frac{\mathrm{d}\omega}{\mathrm{d}t} &= -\cos i\frac{\mathrm{d}\Omega}{\mathrm{d}t} + \frac{(1-e^2)^{1/2}}{nae}\left[-S\cos f + T\left(1+\frac{r}{p}\right)\sin f\right]\\
\frac{\mathrm{d}l_0}{\mathrm{d}t} &= -(1-e^2)^{1/2}\left(\frac{\mathrm{d}\omega}{\mathrm{d}t} + \cos i\frac{\mathrm{d}\Omega}{\mathrm{d}t}\right) - S\frac{2r}{na^2}
\end{aligned}
\qquad (1.1.26)
$$

where the polar coordinates r and u are to be expressed in terms of elements by the formulae of the two-body problem. The two last equations may be replaced by the equivalent equations

$$
\begin{aligned}
\frac{\mathrm{d}\pi}{\mathrm{d}t} &= 2\sin^2\frac{i}{2}\frac{\mathrm{d}\Omega}{\mathrm{d}t} + \frac{(1-e^2)^{1/2}}{nae}\left[-S\cos f + T\left(1+\frac{r}{p}\right)\sin f\right]\\
\frac{\mathrm{d}\epsilon}{\mathrm{d}t} &= \frac{e^2}{1+(1-e^2)^{1/2}}\frac{\mathrm{d}\pi}{\mathrm{d}t} + 2(1-e^2)^{1/2}\sin^2\frac{i}{2}\frac{\mathrm{d}\Omega}{\mathrm{d}t} - S\frac{2r}{na^2}.
\end{aligned}
$$

S, T, W are the components of the perturbing acceleration $\boldsymbol{F}$ on the vector $\boldsymbol{r}$, the transversal and the normal to the orbital plane. Therefore

$$
S = \frac{1}{r}\boldsymbol{r}\boldsymbol{F} \qquad T = \frac{1}{r}(\boldsymbol{k}\times\boldsymbol{r})\boldsymbol{F} \qquad W = \boldsymbol{k}\boldsymbol{F}. \qquad (1.1.27)
$$

In composing equations for l_0 or ϵ it is assumed that in contrast with (1.1.17) and (1.1.23) the mean anomaly and the mean longitude for the disturbed motion are defined by the formulae

$$
l = l_0 + \int_{t_0}^{t} n\,\mathrm{d}t \qquad \lambda = \epsilon + \int_{t_0}^{t} n\,\mathrm{d}t. \qquad (1.1.28)
$$

This modification enables one to avoid the occurrence of terms proportional to t on the right-hand side of the equations for l_0 and ϵ. Instead of these equations one may use directly the equations for l and λ

$$\frac{\mathrm{d}l}{\mathrm{d}t} = n + \frac{\mathrm{d}l_0}{\mathrm{d}t} \qquad \frac{\mathrm{d}\lambda}{\mathrm{d}t} = n + \frac{\mathrm{d}\epsilon}{\mathrm{d}t}.$$

For the osculating vectorial elements $\boldsymbol{c}$ and $\boldsymbol{f}$ one has

$$\dot{\boldsymbol{c}} = \boldsymbol{r} \times \boldsymbol{F} \qquad \dot{\boldsymbol{f}} = 2(\dot{\boldsymbol{r}}\boldsymbol{F})\boldsymbol{r} - (\boldsymbol{r}\boldsymbol{F})\dot{\boldsymbol{r}} - (\boldsymbol{r}\dot{\boldsymbol{r}})\boldsymbol{F}. \tag{1.1.29}$$

Equations (1.1.29) are equivalent to the first five equations (1.1.26). Indeed, in virtue of (1.1.8) one obtains

$$\dot{\boldsymbol{c}} = \frac{na}{2(1-e^2)^{1/2}}\frac{\mathrm{d}p}{\mathrm{d}t}\boldsymbol{k} + na^2(1-e^2)^{1/2}\left(\boldsymbol{l}\sin i\frac{\mathrm{d}\Omega}{\mathrm{d}t} - \boldsymbol{m}\frac{\mathrm{d}i}{\mathrm{d}t}\right) \tag{1.1.30}$$

$$\begin{aligned}\dot{\boldsymbol{f}} &= n^2a^3\frac{\mathrm{d}e}{\mathrm{d}t}\boldsymbol{P} + n^2a^3e \\ &\times \left[\left(\frac{\mathrm{d}\omega}{\mathrm{d}t} + \cos i\frac{\mathrm{d}\Omega}{\mathrm{d}t}\right)\boldsymbol{Q} + \left(\sin\omega\frac{\mathrm{d}i}{\mathrm{d}t} - \cos\omega\sin i\frac{\mathrm{d}\Omega}{\mathrm{d}t}\right)\boldsymbol{k}\right].\end{aligned} \tag{1.1.31}$$

The scalar product of $\dot{\boldsymbol{c}}$ and $\dot{\boldsymbol{f}}$ by $\boldsymbol{k}$, $\boldsymbol{l}$, $\boldsymbol{m}$ and $\boldsymbol{P}$, $\boldsymbol{Q}$ respectively results again in the first five equations (1.1.26).

Sometimes it may be convenient to take for the independent argument in the equations of the osculating elements the anomalies f or E. This substitution is performed by means of the relations

$$\frac{\mathrm{d}f}{\mathrm{d}t} = \frac{na^2}{r^2}(1-e^2)^{1/2} - \left(\frac{\mathrm{d}\omega}{\mathrm{d}t} + \cos i\frac{\mathrm{d}\Omega}{\mathrm{d}t}\right) \tag{1.1.32}$$

$$\frac{\mathrm{d}E}{\mathrm{d}t} = \frac{na}{r} - \frac{r}{a(1-e^2)^{1/2}}\left(\frac{\mathrm{d}\omega}{\mathrm{d}t} + \cos i\frac{\mathrm{d}\Omega}{\mathrm{d}t} + \frac{\sin f}{1-e^2}\frac{\mathrm{d}e}{\mathrm{d}t}\right) \tag{1.1.33}$$

where the derivatives of the osculating elements on the right-hand side should by taken from equations (1.1.26). If in solving equations (1.1.26) accuracy of first order with respect to perturbations is sufficient then one may reject these derivatives on the right-hand sides of (1.1.32) and (1.1.33).

Integration of equations (1.1.26) introduces six arbitrary constants. As these constants one may adopt either the values of the osculating elements at an initial moment or some constant quantities to be chosen from specific conditions. Such constant quantities are called mean elements. They should be distinguished from the osculating elements.

1.1.3 Perturbations of contact elements

Equations (1.1.26) are valid for any disturbing force. If the equations of the disturbing motion are represented in Lagrangian form then equations (1.1.26) may by replaced by more convenient ones. Indeed, if there exists a disturbing function $R = R(\boldsymbol{r}, \dot{\boldsymbol{r}}, t)$ such that

$$\boldsymbol{F} = \frac{\partial R}{\partial \boldsymbol{r}} - \frac{\mathrm{d}}{\mathrm{d}t}\frac{\partial R}{\partial \dot{\boldsymbol{r}}} \tag{1.1.34}$$

then equations (1.1.25) may be put in Lagrangian form with the Lagrangian

$$L = \frac{1}{2}\dot{\boldsymbol{r}}^2 + \frac{GM}{r} + R. \tag{1.1.35}$$

Introducing impulses

$$\boldsymbol{p} = \boldsymbol{r} + \frac{\partial R}{\partial \dot{\boldsymbol{r}}} \tag{1.1.36}$$

one may rewrite the equations of motion in terms of $\boldsymbol{r}$, $\boldsymbol{p}$ in canonical form with Hamiltonian

$$H = \frac{1}{2}\boldsymbol{p}^2 - \frac{GM}{r} - V \tag{1.1.37}$$

with

$$V = R + \frac{1}{2}\left(\frac{\partial R}{\partial \dot{\boldsymbol{r}}}\right)^2. \tag{1.1.38}$$

If for $R = 0$ the canonical variables $\boldsymbol{r}$ and $\boldsymbol{p}$ are expressed by the formulae of the elliptic motion in terms of certain elements $\tilde{a}$, $\tilde{e}$, $\tilde{i}$, $\tilde{\Omega}$, $\tilde{\omega}$, $\tilde{l}_0$, called the contact elements, then according to the variation of arbitrary constants method the same formulae remain valid for the disturbed motion provided that the contact elements as functions of time satisfy the well-known Lagrange equations with disturbing function V. These equations are of the form

$$\begin{aligned}
\frac{\mathrm{d}\tilde{a}}{\mathrm{d}t} &= \frac{2}{\tilde{n}\tilde{a}}\frac{\partial V}{\partial \tilde{l}_0}\\
\frac{\mathrm{d}\tilde{e}}{\mathrm{d}t} &= \frac{1-\tilde{e}^2}{\tilde{n}\tilde{a}^2\tilde{e}}\frac{\partial V}{\partial \tilde{l}_0} - \frac{(1-\tilde{e}^2)^{1/2}}{\tilde{n}\tilde{a}^2\tilde{e}}\frac{\partial V}{\partial \tilde{\omega}}\\
\frac{\mathrm{d}\tilde{i}}{\mathrm{d}t} &= \frac{\cot\tilde{i}}{\tilde{n}\tilde{a}^2(1-\tilde{e}^2)^{1/2}}\frac{\partial V}{\partial \tilde{\omega}} - \frac{\csc\tilde{i}}{\tilde{n}\tilde{a}^2(1-\tilde{e}^2)^{1/2}}\frac{\partial V}{\partial \tilde{\Omega}}\\
\frac{\mathrm{d}\tilde{\Omega}}{\mathrm{d}t} &= \frac{\csc\tilde{i}}{\tilde{n}\tilde{a}^2(1-\tilde{e}^2)^{1/2}}\frac{\partial V}{\partial \tilde{i}}\\
\frac{\mathrm{d}\tilde{\omega}}{\mathrm{d}t} &= -\frac{\cot\tilde{i}}{\tilde{n}\tilde{a}^2(1-\tilde{e}^2)^{1/2}}\frac{\partial V}{\partial \tilde{i}} + \frac{(1-\tilde{e}^2)^{1/2}}{\tilde{n}\tilde{a}^2\tilde{e}}\frac{\partial V}{\partial \tilde{e}}\\
\frac{\mathrm{d}\tilde{l}_0}{\mathrm{d}t} &= -\frac{1-\tilde{e}^2}{\tilde{n}\tilde{a}^2\tilde{e}}\frac{\partial V}{\partial \tilde{e}} - \frac{2}{\tilde{n}\tilde{a}}\frac{\partial V}{\partial \tilde{a}}.
\end{aligned} \tag{1.1.39}$$

Compared with equations (1.1.26) these equations have the advantage of involving one function V instead of three functions S, T, W. In the classical problems of celestial mechanics such as the planetary or satellite three-body problem R does not depend on $\dot{\boldsymbol{r}}$. Hence $\boldsymbol{p} = \dot{\boldsymbol{r}}$ and the contact elements coincide with the osculating elements. For the general case when R depends on $\dot{\boldsymbol{r}}$ the contact elements differ from the osculating elements by quantities of the order of perturbations. Transformation from one set of elements to another set is performed by analytic formulae. In fact, since the contact elements $\tilde{\boldsymbol{c}}$, $\tilde{\boldsymbol{f}}$ satisfy the equations analogous to (1.1.4) by replacing $\dot{\boldsymbol{r}}$ by $\boldsymbol{p}$, then

$$\boldsymbol{c} - \tilde{\boldsymbol{c}} = \frac{\partial R}{\partial \dot{\boldsymbol{r}}} \times \boldsymbol{r} \tag{1.1.40}$$

$$\boldsymbol{f} - \tilde{\boldsymbol{f}} = -2\left(\dot{\boldsymbol{r}}\frac{\partial R}{\partial \dot{\boldsymbol{r}}}\right)\boldsymbol{r} + \left(\boldsymbol{r}\frac{\partial R}{\partial \dot{\boldsymbol{r}}}\right)\dot{\boldsymbol{r}} + (\boldsymbol{r}\dot{\boldsymbol{r}})\frac{\partial R}{\partial \dot{\boldsymbol{r}}} + \frac{\partial R}{\partial \dot{\boldsymbol{r}}} \times \left(\frac{\partial R}{\partial \dot{\boldsymbol{r}}} \times \boldsymbol{r}\right). \tag{1.1.41}$$

Having obtained the differences $\boldsymbol{c} - \tilde{\boldsymbol{c}}$, $\boldsymbol{f} - \tilde{\boldsymbol{f}}$ it is easy to derive the differences for the Keplerian elements $a - \tilde{a}$, $e - \tilde{e}$, $i - \tilde{i}$, $\Omega - \tilde{\Omega}$, $\omega - \tilde{\omega}$. As for the difference $l - \tilde{l}_0$ one has from the Kepler equation (1.1.13) with (1.1.28) and a similar equation for the contact elements,

$$l_0 - \tilde{l}_0 = -\int_{t_0}^{t} (n - \tilde{n})\, dt + E - \tilde{E} - (e \sin E - \tilde{e} \sin \tilde{E}) \tag{1.1.42}$$

where $\tilde{E}$ is expressed in terms of $\boldsymbol{r}$ and $\boldsymbol{p}$ in the same way that E is expressed in terms of $\boldsymbol{r}$ and $\dot{\boldsymbol{r}}$, for example

$$E = \tan^{-1}\left[\frac{\boldsymbol{r}\dot{\boldsymbol{r}}}{r\dot{r}^2 - GM}\left(\frac{2GM}{r} - \dot{r}^2\right)^{1/2}\right].$$

Expressions (1.1.40)–(1.1.42) are rigorous. Within the first-order accuracy obtained from the comparison of (1.1.40) and (1.1.42) with (1.1.29) the expressions for $a - \tilde{a}$, $e - \tilde{e}$, $i - \tilde{i}$, $\Omega - \tilde{\Omega}$, $\omega - \tilde{\omega}$ coincide with the corresponding right-hand sides of equations (1.1.26) to be calculated with the 'force' $\boldsymbol{F} = -\partial R/\partial \dot{\boldsymbol{r}}$. As seen from (1.1.42) the difference $l_0 - \tilde{l}_0$ within the first-order accuracy will contain in addition to (1.1.26) one extra term

$$l_0 - \tilde{l}_0 = -(1 - e^2)^{1/2}[(\omega - \tilde{\omega}) + (\Omega - \tilde{\Omega}) \cos i] + \frac{2}{na^2}\boldsymbol{r}\frac{\partial R}{\partial \dot{\boldsymbol{r}}} - \frac{3}{na^2}\int_{t_0}^{t} \dot{\boldsymbol{r}}\frac{\partial R}{\partial \dot{\boldsymbol{r}}}\, dt. \tag{1.1.43}$$

1.1.4 Secular perturbations

It is to be noted that in celestial mechanics there are many efficient methods enabling one to derive in analytical form the approximate solutions of the equations of the disturbed motion, like (1.1.25). For the purposes of relativistic celestial mechanics it is generally not necessary to apply rather complicated methods designed for calculation of the high-order perturbations. Presently, in virtue of the actual smallness of the relativistic parameter entering into the equations of relativistic celestial mechanics for Solar System bodies the terms of high order are not needed in practice. Therefore, in relativistic problems one commonly has to do only with the first-order perturbations. The most efficient tool for this is the method of variation of arbitrary constants. Of course, one may apply different methods such as, for example, methods in rectangular coordinates. But the method of variation of arbitrary constants seems to be the most universal one.

In most applications equations (1.1.26) or (1.1.39) cannot be solved rigorously. As a consequence approximate solution techniques are needed. The most simple and widespread technique is to substitute into the right-hand sides of these equations the constant values for the elements resulting after integration in the first-order perturbations. In doing so secular terms proportional to t may occur in e and i. These terms lead to the secular terms in coordinates and velocity components. This results from the adopted integration technique and is in no way characteristic of the actual evolution of motion. The approximate solution derived in such a manner is valid for a limited interval of time permitting no conclusion to be drawn about the actual evolution of motion. Such a technique is often used in the problem of motion of the major planets. As the motions of the perihelia and nodes of the planetary orbits are extremely small compared with the planetary mean motions, it is possible initially to substitute into the right-hand sides of the equations in osculating elements the constant values for Ω and ω. In satellite problems, such as the problem of motion of the Moon, with much faster secular motions of perigee and node, such a 'planetary' method of integration turns out to be too rough. In such problems it is suitable to apply a 'satellite' method deriving initially the secular terms in Ω, ω, l_0 and substituting afterwards into the right-hand sides the constant values for a, e, i and the linear functions of time for Ω, ω, l_0. Then e, i and, consequently, coordinates and velocities will not contain secular terms. The solution obtained in this manner has a purely trigonometric form with respect to time and is formally valid for an unlimited interval of time. But using this formal solution one cannot again draw any conclusion regarding the evolutionary character.

In many problems the secular terms are of particular interest. For small eccentricities and inclinations, in particular restricting to the first degree terms, it may be possible to derive the rigorous solution of the differential equations for the secular perturbations. It is customary to determine secular perturbations in the osculating elements. If R does not depend on t outside the trigonometric arguments then, as seen from (1.1.40) and (1.1.41), the secular perturbations for geometric osculating and contact elements are the same. As for the mean anomaly at the epoch one should take into account the last term in (1.1.43). Finally, the equations for the first-order secular perturbations in the osculating elements have the form

$$\begin{aligned}
\frac{da}{dt} &= \frac{2}{na}\frac{\partial[R]}{\partial l_0} \\
\frac{de}{dt} &= \frac{1-e^2}{na^2e}\frac{\partial[R]}{\partial l_0} - \frac{(1-e^2)^{1/2}}{na^2e}\frac{\partial[R]}{\partial\omega} \\
\frac{di}{dt} &= \frac{\cot i}{na^2(1-e^2)^{1/2}}\frac{\partial[R]}{\partial\omega} - \frac{\csc i}{na^2(1-e^2)^{1/2}}\frac{\partial[R]}{\partial\Omega} \\
\frac{d\Omega}{dt} &= \frac{\csc i}{na^2(1-e^2)^{1/2}}\frac{\partial[R]}{\partial i} \\
\frac{d\omega}{dt} &= -\frac{\cot i}{na^2(1-e^2)^{1/2}}\frac{\partial[R]}{\partial i} + \frac{(1-e^2)^{1/2}}{na^2e}\frac{\partial[R]}{\partial e} \\
\frac{dl_0}{dt} &= -\frac{1-e^2}{na^2e}\frac{\partial[R]}{\partial e} - \frac{2}{na}\frac{\partial[R]}{\partial a} - \frac{3}{na^2}\left[\dot{r}\frac{\partial R}{\partial \dot{r}}\right]
\end{aligned} \tag{1.1.44}$$

with brackets denoting averaging over the fast changing angular variables (like the mean anomaly l).

1.2 ELEMENTS OF RIEMANNIAN GEOMETRY AND TENSOR ANALYSIS

1.2.1 Euclidean space

The affine space represents the set of points and vectors governed by the laws of ordinary vector algebra. This space is homogeneous and isotropic. The dimension of the space is determined by the maximal possible number of linearly independent vectors. In the affine n-dimensional space A_n a set of n linearly independent vectors $\boldsymbol{e}_1, \ldots, \boldsymbol{e}_n$ emanating from an arbitrary point O constitutes an affine reference basis. An arbitrary vector $\boldsymbol{x}$ emanating from point O may be decomposed on the basis vectors

$$\boldsymbol{x} = x^i \boldsymbol{e}_i \tag{1.2.1}$$

where the coefficients x^i are called affine coordinates of the vector. Here and everywhere below the Einstein summation rule is used. This means the summation from 1 to n is taken over every index occurring twice in any expression. Under the affine (linear) transformation

$$e_{i'} = A^{i}_{i'} e_i \qquad i, i' = 1, 2, \ldots, n \tag{1.2.2}$$

from one basis $e_1, \ldots, e_n$ to another basis $e_{1'}, \ldots, e_{n'}$ the affine coordinates of a vector are changed as follows:

$$x^{i'} = A^{i'}_{i} x^{i} \tag{1.2.3}$$

i.e. with the aid of the transposed inverse matrix of the starting transformation (1.2.2)

$$A^{k'}_{i} A^{j}_{k'} = \delta^{j}_{i} \tag{1.2.4}$$

Here δ^j_i denotes the Kronecker symbol with the values

$$\delta^j_i = \begin{cases} 0 & i \neq j \\ 1 & i = j \end{cases} . \tag{1.2.5}$$

In expressions like (1.2.2) and (1.2.3) it should be remembered that the index i' is not related to the index i. In fact an accent does not refer to the index itself but is characteristic of the new reference basis. As seen from (1.2.3) the coordinates of a vector are distinct in the different reference systems. However, all relations described in the vectorial form retain their form in any reference system. Generalization of the vectorial form may be carried out with the aid of tensors. Tensors represent sets of quantities changing by a simple linear law in transforming from one reference system to another. Tensors enable one to define operations remaining invariant under the transformation of coordinates. All relations expressed in the tensorial form retain their form in any reference system.

A tensor covariant of rank k and contravariant of rank m is defined as a collection of n^{k+m} quantities $a^{j_1 \ldots j_m}_{i_1 \ldots i_k}$ given in any reference system and transforming with (1.2.2) in accordance with

$$a^{j'_1 \ldots j'_m}_{i'_1 \ldots i'_k} = A^{j'_1}_{j_1} \ldots A^{j'_m}_{j_m} A^{i_1}_{i'_1} \ldots A^{i_k}_{i'_k} a^{j_1 \ldots j_m}_{i_1 \ldots i_k} . \tag{1.2.6}$$

The lower indices obeying the transformation law (1.2.2) are called covariant. The upper indices transforming in accordance with (1.2.3) are called contravariant. The basic tensor algebra operations are addition, multiplication, contraction and change of indices. The first two operations generalize the corresponding operations of vector algebra. Contraction is to choose all components of a given tensor for which a definite upper index is equal to a definite lower index. There results a new tensor whose rank is two units less than the rank of the starting tensor. Changing indices enables us to derive new tensors by altering the order of arrangement of indices of a given tensor.

Euclidean n-dimensional space R_n represents the affine space A_n with the operation of scalar (inner) products of vectors introduced. This defines all metric properties of the space, enabling us, in particular, to measure the lengths of vectors and curve arcs. If the scalar product of the vectors of some affine reference basis is denoted as

$$g_{ij} = \boldsymbol{e}_i \boldsymbol{e}_j \tag{1.2.7}$$

then in virtue of (1.2.1) the scalar product of vectors $\boldsymbol{x}$ and $\boldsymbol{y}$ will be

$$\boldsymbol{x}\boldsymbol{y} = g_{ij} x^i y^j . \tag{1.2.8}$$

The scalar square of vector $\boldsymbol{x}$ is expressed by the quadratic form with respect to its affine coordinates

$$\boldsymbol{x}^2 = g_{ij} x^i x^j . \tag{1.2.9}$$

The scalar product of vectors in R_n has properties of symmetry and indegeneracy. Therefore

$$g_{ij} = g_{ji} \qquad g = \det \| g_{ij} \| \neq 0. \tag{1.2.10}$$

Under transformation (1.2.2) the quantities g_{ij} act as the components of a covariant tensor of rank two. This tensor, called the metric tensor,determines all structure of R_n at hand. For a new reference system

$$g_{i'j'} = A^i_{i'} A^j_{j'} g_{ij} . \tag{1.2.11}$$

Therefore

$$g' = (\det \| A^i_{i'} \|)^2 g \tag{1.2.12}$$

i.e. the determinant g formed from the components g_{ij} is the relative invariant of weight 2. Elements g^{ij} of the inverse matrix of $\| g_{ij} \|$ are the components of the contravariant metric tensor of rank two. With the aid of g_{ij} and g^{ij} one can perform in R_n the operations of index raising and lowering. For example, the covariant coordinates of the vector $\boldsymbol{x}$ are

$$x_i = g_{ij} x^j$$

having the simple meaning

$$x_i = \boldsymbol{x}\boldsymbol{e}_i . \tag{1.2.13}$$

In turn, the contravariant coordinates of the vector result from the covariant ones by index raising,

$$x^i = g^{ij} x_j .$$

Vectors $\boldsymbol{x}$ and $\boldsymbol{y}$ are orthogonal if their scalar product vanishes. The length of vector is determined as $|\boldsymbol{x}| = (\boldsymbol{x}^2)^{1/2}$.

In real R_n admitting no complex numbers the scalar squares of vectors may take only real values. The real Euclidean space can be classified as proper Euclidean with strictly positive scalar squares for all non-zero vectors or pseudo-Euclidean where there exist the non-zero vectors with positive, negative and zero scalar squares. Hence, in the pseudo-Euclidean spaces the length of a non-zero vector may be positive or purely imaginary or zero. A non-zero vector with the zero length is called an isotropic vector.

A curve in A_n may be represented analytically as a one-parametric set of the coordinates $x^i = x^i(t)$. The position vector of any point of the curve may be decomposed along the basis vectors

$$\boldsymbol{x}(t) = x^i(t)\boldsymbol{e}_i$$

with the tangent vector

$$\frac{\mathrm{d}\boldsymbol{x}}{\mathrm{d}t} = \frac{\mathrm{d}x^i}{\mathrm{d}t}\boldsymbol{e}_i.$$

The arc length from M_1 to M_2 is determined in R_n as

$$s = \int_{M_1}^{M_2} |\mathrm{d}\boldsymbol{x}| = \int_{t_1}^{t_2} \left|\frac{\mathrm{d}\boldsymbol{x}}{\mathrm{d}t}\right| \mathrm{d}t$$

with the differential

$$\mathrm{d}s = |\mathrm{d}\boldsymbol{x}| = \left|\frac{\mathrm{d}\boldsymbol{x}}{\mathrm{d}t}\right| \mathrm{d}t.$$

Depending on the sign of the square of the differential $\mathrm{d}s^2 = \mathrm{d}\boldsymbol{x}^2$ the curve may have real length ($\mathrm{d}s^2 > 0$) or purely imaginary length ($\mathrm{d}s^2 < 0$). For $\mathrm{d}s^2 = 0$ the curve of the zero length is called isotropic. If the arc has real length then s may be taken as the parameter t along the curve and $\boldsymbol{u} = \mathrm{d}\boldsymbol{x}/\mathrm{d}s$ is the unit tangent vector. If the arc has purely imaginary length then the real variable $\sigma = s/\mathrm{i}$ may be taken as a parameter t and the vector $\boldsymbol{u} = \mathrm{d}\boldsymbol{x}/\mathrm{d}\sigma$ is the imaginary unit tangent vector.

An arbitrary reference basis in A_n has $n(n+1)$ independent parameters. In R_n one can construct an orthonormal basis characterized by $n(n+1)/2$ independent parameters. Such a basis for the pseudo-Euclidean space consists of k unit and $n-k$ imaginary unit vectors ($0 < k < n$). For given R_n the number k, called the space index, is the same for any reference basis. The spaces with the indices k and $n-k$ ($0 \le k \le n$) are not different since all the lengths of these spaces are distinguished by the common factor i. The simplest example of pseudo-Euclidean space is the pseudo-Euclidean plane ($n = 2$, $k = 1$). In this case (and generally for all pseudo-Euclidean spaces of index 1) it is convenient to enumerate coordinates from 0. In the orthonormal basis $e_0^2 = 1, e_1^2 = -1$ one has

$$x^2 = x^{0^2} - x^{1^2} \qquad xy = x^0y^0 - x^1y^1. \tag{1.2.14}$$

Transformation from e_0, e_1 to a new basis $e_{0'}, e_{1'}$ is given by (1.2.2). The relations of orthonormality of $e_{0'}, e_{1'}$ imply the explicit form of this transformation

$$e_{0'} = \frac{e_0 + \beta e_1}{(1-\beta^2)^{1/2}} \qquad e_{1'} = \frac{\beta e_0 + e_1}{(1-\beta^2)^{1/2}} \tag{1.2.15}$$

with $-1 < \beta < 1$. In general, the square roots in (1.2.15) may be taken with arbitrary signs resulting in four types of rotation of reference basis in the pseudo-Euclidean plane. For application to the theory of relativity the case of proper rotation is of importance characterized by both positive signs in (1.2.15). The determinant of such transformation is equal to +1.

The properties of the pseudo-Euclidean plane may be extended to the n-dimensional pseudo-Euclidean space of index 1. For the case $n = 4$ of the prime importance for the theory of relativity one has (putting in fact $k = 3$)

$$e_0^2 = 1 \qquad e_1^2 = e_2^2 = e_3^2 = -1. \tag{1.2.16}$$

The scalar square of a vector is then

$$x^2 = x^{0^2} - x^{1^2} - x^{2^2} - x^{3^2}. \tag{1.2.17}$$

Equation $x^2 = 0$ determines the isotropic hypercone along the x^0 axis. The curves of the real length are passing inside the hypercone. The curves of the imaginary length lie outside it. Transformation of the reference basis in this space reduces to rotation (1.2.15) of the pseudo-Euclidean plane and trivial rotation of the three-dimensional basis e_1, e_2, e_3 in the proper Euclidean space R_3. With the appropriate choice of the axes one has

$$e_{0'} = \frac{e_0 + \beta e_1}{(1-\beta^2)^{1/2}} \quad e_{1'} = \frac{\beta e_0 + e_1}{(1-\beta^2)^{1/2}} \quad e_{2'} = e_2 \quad e_{3'} = e_3. \tag{1.2.18}$$

The contravariant coordinates of a vector are transformed thereby as

$$x^{0'} = \frac{x^0 - \beta x^1}{(1-\beta^2)^{1/2}} \quad x^{1'} = \frac{-\beta x^0 + x^1}{(1-\beta^2)^{1/2}} \quad x^{2'} = x^2 \quad x^{3'} = x^3. \tag{1.2.19}$$

All preceding formulae involve only affine coordinates. This is quite natural since in doing so all relations in A_n and R_n take the most simple form. But, in principle, one may introduce in A_n the curvilinear coordinates related with the affine ones by means of a non-singular, non-linear transformation. Let x^i be such coordinates. At any point M of A_n one constructs n coordinate lines. Along any coordinate line x^i the position vector of a point is a function of this coordinate alone, i.e. $\boldsymbol{x} = \boldsymbol{x}(x^i)$. The derivative $\boldsymbol{x}_i = \mathrm{d}\boldsymbol{x}/\mathrm{d}x^i$ at point M yields the tangent vector to this line at the given point. The set of vectors $\boldsymbol{x}_i (i = 1, 2, \ldots, n)$ together with point M form the local reference basis in A_n. Under the non-singular, non-linear transformation relating the curvilinear coordinates x^i and $x^{i'}$

$$x^{i'} = x^{i'}(x^1, \ldots, x^n) \qquad x^i = x^i(x^{1'}, \ldots, x^{n'}) \tag{1.2.20}$$

there results a new local reference basis

$$\boldsymbol{x}_{i'} = \frac{\partial x^i}{\partial x^{i'}} \boldsymbol{x}_i. \tag{1.2.21}$$

As an extension of (1.2.6) the components of a tensor with respect to the local basis are transformed under (1.2.20) as follows:

$$a^{j'_1 \ldots j'_m}_{i'_1 \ldots i'_k}(M) = \frac{\partial x^{j'_1}}{\partial x^{j_1}}(M) \ldots \frac{\partial x^{j'_m}}{\partial x^{j_m}}(M) \frac{\partial x^{i_1}}{\partial x^{i'_1}}(M) \ldots \frac{\partial x^{i_k}}{\partial x^{i'_k}}(M) a^{j_1 \ldots j_m}_{i_1 \ldots i_k}(M). \tag{1.2.22}$$

Algebraic operations are valid in curvilinear coordinates as well. But the absence of a unified affine reference basis manifests itself, first of all, in parallel transporting the vectors. Consider at some point $M(x^i)$ of the curve $x^i = x^i(t)$ a constant vector $\boldsymbol{\xi}$ with coordinates ξ^k relative to the local basis at M

$$\boldsymbol{\xi} = \xi^k(t) \boldsymbol{x}_k. \tag{1.2.23}$$

Parallel transport of this vector to the neighbouring point $\tilde{M}(x^i + \mathrm{d}x^i)$ results due to the change of the local basis in new coordinates $\xi^k + \mathrm{d}\xi^k$ of this vector. The condition of parallel transport, $\mathrm{d}\boldsymbol{\xi} = 0$, enables us to deduce the law of the change of coordinates under this transport

$$\mathrm{d}\xi^k = -\Gamma^k_{ij} \xi^j \mathrm{d}x^i. \tag{1.2.24}$$

Quantities Γ^k_{ij} symmetrical in their lower indices and called the connection coefficients are the coefficients of the expansion

$$\boldsymbol{x}_{ij} = \frac{\partial^2 \boldsymbol{x}}{\partial x^i \partial x^j} = \Gamma^k_{ij} \boldsymbol{x}_k. \tag{1.2.25}$$

For affine coordinates $\boldsymbol{x}_i = \boldsymbol{e}_i, \boldsymbol{x}_{ij} = 0$, and $\Gamma^k_{ij} = 0$. Vanishing of Γ^k_{ij} is necessary and sufficient for curvilinear coordinates x^i in A_n to be affine. From (1.2.21) and (1.2.25) there results

$$\Gamma^{k'}_{i'j'} = \frac{\partial^2 x^k}{\partial x^{i'} \partial x^{j'}} \frac{\partial x^{k'}}{\partial x^k} + \frac{\partial x^i}{\partial x^{i'}} \frac{\partial x^j}{\partial x^{j'}} \frac{\partial x^{k'}}{\partial x^k} \Gamma^k_{ij}. \tag{1.2.26}$$

Therefore, the connection coefficients are not tensors and only under the linear transformation act as tensors.

In using curvilinear coordinates the metric tensor of R_n is determined as the scalar product of the local basis vectors,

$$g_{ij}(M) = \boldsymbol{x}_i(M)\boldsymbol{x}_j(M) \tag{1.2.27}$$

being the function of a point. Let $x^i = x^i(t)$ and $\boldsymbol{x} = \boldsymbol{x}(x^1, \ldots, x^n)$ be some curve in R_n and a position vector of any point of the curve. Then the differential of the position vector is $\mathrm{d}\boldsymbol{x} = \boldsymbol{x}_i \mathrm{d}x^i$ and the square of the differential of arc $\mathrm{d}s^2 = \mathrm{d}\boldsymbol{x}^2$ is expressed by the metric quadratic form

$$\mathrm{d}s^2 = g_{ij}\mathrm{d}x^i\mathrm{d}x^j. \tag{1.2.28}$$

Given metric tensor g_{ij} or metric form $\mathrm{d}s^2$, all the geometry of R_n may be determined. In particular, it is easy to find the connection coefficients. In fact, from (1.2.25) there results

$$\boldsymbol{x}_{ij}\boldsymbol{x}_m = \Gamma_{mij} \tag{1.2.29}$$

with

$$\Gamma_{mij} = g_{km}\Gamma^k_{ij}. \tag{1.2.30}$$

From (1.2.30) it follows that

$$\Gamma^k_{ij} = g^{km}\Gamma_{mij}. \tag{1.2.31}$$

Hence, the formal operations (1.2.30) and (1.2.31) of raising and lowering indices are valid although the connection coefficients are not tensors. Differentiation of (1.2.27) yields

$$\frac{\partial g_{ij}}{\partial x^k} = \Gamma_{ijk} + \Gamma_{jik} \tag{1.2.32}$$

or after circularly changing the indices

$$\Gamma_{mij} = \frac{1}{2}\left(\frac{\partial g_{im}}{\partial x^j} + \frac{\partial g_{jm}}{\partial x^i} - \frac{\partial g_{ij}}{\partial x^m}\right). \tag{1.2.33}$$

The connection coefficients Γ_{kij} and Γ^k_{ij} are also called the Christoffel symbols of the first and second kind respectively.

1.2.2 Riemannian space

An elementary n-dimensional manifold of class C^N is meant as a set M_n with a given one-to-one map onto a domain of n coordinates x^i defined up to the N-differentiable transformation (1.2.20). Any open connected domain in A_n furnishes an example of M_n admitting, however, peculiar, affine, coordinates defined up to the linear transformations. A tensor at point M of M_n is defined as a set of quantities given in any reference system x^i and transformed in changing to new coordinates with (1.2.22). But in contrast to A_n the starting transformation is here the coordinate transformation (1.2.20) rather than the transformation of the reference basis (1.2.21). There is no local reference basis in M_n. However, it may be introduced in the tangent affine space A_n. The components of tensors in M_n in coordinates x^i may be interpreted as the tensor coordinates in A_n relative to the local reference basis appropriate to the coordinates x^i. The manifold M_n is called a Riemannian space V_n if for each point M there exists a covariant symmetrical and non-singular metric tensor of rank 2 $g_{ij}(M) = g_{ij}(x^1, \ldots, x^n)$. With the aid of $g_{ij}(M)$ the tangent affine space A_n at point M reduces to the Euclidean space R_n with the basic relation (1.2.8) for the scalar product of any vectors $\boldsymbol{x}$ and $\boldsymbol{y}$. V_n is the proper or pseudo-Riemannian space in the same sense as the tangent space R_n is the proper or pseudo-Euclidean. An infinitesimal displacement along a curve $x^i = x^i(t)$ in V_n generates an infinitesimal vector $\mathrm{d}x^i(t)$ in R_n and its length $|\mathrm{d}\boldsymbol{x}|$ is taken as the differential of arc $\mathrm{d}s$ retaining the valid original formula (1.2.28). In the pseudo-Riemannian space there exist curves of real ($\mathrm{d}s^2 > 0$), imaginary ($\mathrm{d}s^2 < 0$) and zero ($\mathrm{d}s^2 = 0$) length. Euclidean space R_n may be regarded as the particular case of V_n admitting such coordinates (affine) for which components g_{ij} are constant for the whole space (by normalization these constants may be put equal to $0, \pm 1$).

The determinant $g = \det \|g_{ij}\|$ is again the relative invariant of weight 2 transforming under (1.2.20) as

$$g' = \left(\det \left\| \frac{\partial x^i}{\partial x^{i'}} \right\| \right)^2 g. \tag{1.2.34}$$

Hence, the n-dimensional integral taken over some domain Ω

$$W = \int_\Omega \sqrt{|g|}\, \mathrm{d}x^1 \ldots \mathrm{d}x^n \tag{1.2.35}$$

is invariant under the transformation (1.2.20) determining the volume of domain Ω in curvilinear coordinates in R_n and in V_n.

As stated above, it is possible to introduce in R_n such coordinates (affine) that the connection coefficients Γ^k_{ij} vanish in the whole domain at hand. For V_n it is possible for Γ^k_{ij} to vanish at any given point M. Coordinates satisfying this condition are called geodesic. All operations may be significantly simplified thereby. Final results expressed in tensorial form remain valid for arbitrary coordinates. Transformation from the arbitrary coordinates x^i with $\Gamma^k_{ij}(M) \neq 0$ to the coordinates $x^{i'}$ geodesic at point M is easily performed. Equations determining $x^{i'}$ in terms of x^i follow immediately from (1.2.26):

$$\frac{\partial^2 x^{m'}}{\partial x^i \partial x^j}(M) = \Gamma^k_{ij}(M)\frac{\partial x^{m'}}{\partial x^k}(M). \tag{1.2.36}$$

These equations may be satisfied by putting

$$x^{m'} = a^{m'}_i(x^i - x^i_M) + \tfrac{1}{2}a^{m'}_k\Gamma^k_{ij}(M)(x^i - x^i_M)(x^j - x^j_M) \tag{1.2.37}$$

with non-singular constant matrix $||a^{m'}_i||$, x^i_M being the coordinates of point M.

1.2.3 Parallel transport and absolute differentiation

Parallel transport of vectors in V_n formally coincides with the corresponding operation (1.2.24) in curvilinear coordinates in R_n. In simultaneous parallel transport of two vectors along some curve their scalar product does not change, retaining the metric properties (lengths and angles between vectors) in parallel transport. But in contrast to R_n the parallel transport in V_n depends generally on the path.

Parallel transport is closely related to absolute differentiation, the most important operation of tensor analysis. Let $x^i = x^i(t)$ be a curve in V_n. Consider a tensor $a^{j_1\ldots j_m}_{i_1\ldots i_k}(t)$ at some point of this curve. At an infinitesimally close point this tensor has value $a^{j_1\ldots j_m}_{i_1\ldots i_k}(t+\mathrm{d}t)$. However, these tensors are related to different local reference bases and cannot be compared directly. The difference in these quantities is approximately equal to the non-tensorial differential $\mathrm{d}a^{j_1\ldots j_m}_{i_1\ldots i_k}(t)$. Denoting by $\tilde{a}^{j_1\ldots j_m}_{i_1\ldots i_k}(t)$ the tensor resulting from the parallel transport of $a^{j_1\ldots j_m}_{i_1\ldots i_k}(t+\mathrm{d}t)$ to point t one has approximately

$$\tilde{a}^{j_1\ldots j_m}_{i_1\ldots i_k}(t) - a^{j_1\ldots j_m}_{i_1\ldots i_k}(t) \approx \mathrm{D}a^{j_1\ldots j_m}_{i_1\ldots i_k}(t) \tag{1.2.38}$$

determining the absolute differential as a tensor of the same structure as the initial one.

For vectors there results from (1.2.24) and (1.2.38)

$$\mathrm{D}a^k = \mathrm{d}a^k + \Gamma^k_{ij}a^j\mathrm{d}x^i. \tag{1.2.39}$$

For a covariant vector a_k parallel transport is characterized by invariability of its contraction $\mathrm{d}(a_k\xi^k) = 0$ with an arbitrary contravariant vector ξ^k. From this equation it follows that

$$\mathrm{D}a_j = \mathrm{d}a_j - \Gamma^k_{ij}a_k\mathrm{d}x^i. \tag{1.2.40}$$

Generalizing (1.2.39) and (1.2.40) for a tensor of arbitrary structure one has

$$\begin{aligned} \mathrm{D}a^{j_1\ldots j_m}_{i_1\ldots i_k} = \mathrm{d}a^{j_1\ldots j_m}_{i_1\ldots i_k} + (\Gamma^{j_1}_{rs}a^{sj_2\ldots j_m}_{i_1\ldots i_k} + \ldots + \Gamma^{j_m}_{rs}a^{j_1\ldots j_{m-1}s}_{i_1\ldots i_k} \\ - \Gamma^{s}_{ri_1}a^{j_1\ldots j_m}_{si_2\ldots i_k} - \ldots - \Gamma^{s}_{ri_k}a^{j_1\ldots j_m}_{i_1\ldots i_{k-1}s})\mathrm{d}x^r . \end{aligned} \qquad (1.2.41)$$

Hence, under parallel transport of a tensor its absolute differential vanishes. Formula (1.2.41) may be rewritten in the form

$$\mathrm{D}a^{j_1\ldots j_m}_{i_1\ldots i_k} = \nabla_r a^{j_1\ldots j_m}_{i_1\ldots i_k}\mathrm{d}x^r . \qquad (1.2.42)$$

The quantity $\nabla_r a^{j_1\ldots j_m}_{i_1\ldots i_k}(t)$, denoted often as $a^{j_1\ldots j_m}_{i_1\ldots i_k;r}(t)$, represents the tensor with an extra lower index and is called the absolute or covariant derivative. Its explicit expression follows from (1.2.41) and (1.2.42). For scalars, and contravariant and covariant vectors one has respectively

$$\nabla_r a = \frac{\partial a}{\partial x_r} \qquad (1.2.43)$$

$$\nabla_r a^k = \frac{\partial a^k}{\partial x_r} + \Gamma^k_{rs}a^s \qquad (1.2.44)$$

$$\nabla_r a_k = \frac{\partial a_k}{\partial x_r} - \Gamma^s_{rk}a_s . \qquad (1.2.45)$$

The usual rules to differentiate sum, product, etc, are extended to absolute differentiation. Tensor contraction is permutable with absolute differentiation. By virtue of the definition of the Christoffel symbols one obtains

$$\nabla_k g_{ij} = 0 \qquad \nabla_k \delta^i_j = 0 \qquad \nabla_k g^{ij} = 0$$

i.e. the absolute differentials $\mathrm{D}g_{ij}$ and $\mathrm{D}g^{ij}$ vanish. Therefore, the operations of raising and lowering indices are permutable with absolute differentiation.

1.2.4 Geodesics and Fermi–Walker transport

A curve in V_n is called geodesic if any vector tangent to it at some point remains tangent in parallel transport along it. Differential geodesic equations result from (1.2.24). Indeed, let $\xi^k = \xi^k(t)$ be a tangent vector parallel transported along the geodesic $x^i = x^i(t)$. By virtue of the collinearity of tangent vectors one has

$$\frac{\mathrm{d}x^k}{\mathrm{d}t} = \alpha(t)\xi^k(t)$$

with the scalar factor $\alpha(t)$ dependent on a point of the curve. Introducing a parameter λ such that $\mathrm{d}\lambda = \alpha(t)\,\mathrm{d}t$ one obtains $x^i = x^i(\lambda)$ for the geodesic and $\xi^k = \mathrm{d}x^k/\mathrm{d}\lambda$ for the parallel transported tangent vector. Such a parameter λ defined up to a linear transformation is called canonical. In virtue of (1.2.24) the parallel transport of vector ξ^k will be determined by

$$\mathrm{d}\frac{\mathrm{d}x^k}{\mathrm{d}\lambda} = -\Gamma^k_{ij}\frac{\mathrm{d}x^j}{\mathrm{d}\lambda}\mathrm{d}x^i$$

yielding the differential equations of the geodesic line referred to the canonical parameter λ

$$\frac{\mathrm{d}^2x^k}{\mathrm{d}\lambda^2} + \Gamma^k_{ij}\frac{\mathrm{d}x^i}{\mathrm{d}\lambda}\frac{\mathrm{d}x^j}{\mathrm{d}\lambda} = 0 \tag{1.2.46}$$

or simply $\mathrm{D}(\mathrm{d}x^i/\mathrm{d}\lambda) = 0$. Since a parallel transported vector retains its length one has along the geodesic

$$g_{ij}\frac{\mathrm{d}x^i}{\mathrm{d}\lambda}\frac{\mathrm{d}x^j}{\mathrm{d}\lambda} = C \tag{1.2.47}$$

C being a constant. One may take as the canonical parameter λ either s (for a geodesic of real length) or $\sigma = s/\mathrm{i}$ (for a geodesic of imaginary length), resulting in $C = 1$ or $C = -1$ respectively. For a non-isotropic geodesic relation (1.2.47) is the first integral of equations (1.2.46). For an isotropic geodesic $C = 0$ and the relation (1.2.47) determining the canonical parameter should be considered in combination with equations (1.2.46). The isotropic geodesic referred to an arbitrary parameter $t = t(\lambda)$ is determined in accordance with (1.2.46) and (1.2.47) by the equations

$$\frac{\mathrm{d}^2x^k}{\mathrm{d}t^2} + \Gamma^k_{ij}\frac{\mathrm{d}x^i}{\mathrm{d}t}\frac{\mathrm{d}x^j}{\mathrm{d}t} = -\frac{\mathrm{d}^2t}{\mathrm{d}\lambda^2}\left(\frac{\mathrm{d}t}{d\lambda}\right)^{-2}\frac{\mathrm{d}x^k}{\mathrm{d}t} \tag{1.2.48}$$

$$g_{ij}\frac{\mathrm{d}x^i}{\mathrm{d}t}\frac{\mathrm{d}x^j}{\mathrm{d}t} = 0. \tag{1.2.49}$$

The geodesic equations may be obtained from the variational principle

$$\delta\int \mathrm{d}s = 0. \tag{1.2.50}$$

From (1.2.28) there results

$$ds = \sqrt{f}\mathrm{d}\lambda \tag{1.2.51}$$

with

$$f = g_{ij}\frac{\mathrm{d}x^i}{d\lambda}\frac{\mathrm{d}x^j}{d\lambda}. \tag{1.2.52}$$

The Euler–Lagrange equations of (1.2.50)

$$\frac{\mathrm{d}}{\mathrm{d}\lambda}\frac{\partial\sqrt{f}}{\partial(\mathrm{d}x^i/\mathrm{d}\lambda)} - \frac{\partial\sqrt{f}}{\partial x^i} = 0 \tag{1.2.53}$$

take the form

$$\frac{\mathrm{d}^2x^k}{\mathrm{d}\lambda^2} + \Gamma^k_{ij}\frac{\mathrm{d}x^i}{\mathrm{d}\lambda}\frac{\mathrm{d}x^j}{\mathrm{d}\lambda} = \frac{1}{2}\frac{\mathrm{d}\ln f}{\mathrm{d}\lambda}\frac{\mathrm{d}x^k}{\mathrm{d}\lambda}. \tag{1.2.54}$$

If λ is a canonical parameter then by (1.2.47) f = constant and equations (1.2.54) with vanishing right-hand sides become identical to the canonical equations of the geodesic (1.2.46). Equations (1.2.54) generalize equations (1.2.46) for an arbitrary parameter λ. In deriving these equations from (1.2.53) it is assumed that $f \neq 0$. Hence, the variational principle (1.2.50) is not valid for an isotropic geodesic. This principle is particularly convenient in choosing λ among coordinates x^i. Then it is suitable from the very beginning to use in the Lagrangian $\sqrt{f}$ this new independent variable. A non-isotropic geodesic will be determined thereby by $n-1$ equations of the second order involving $2n-2$ parameters.

Principle (1.2.50) gives the Lagrangian $\sqrt{f}$ in irrational form. But there exists the variational principle in rational form

$$\delta\int\left(\frac{\mathrm{d}s}{\mathrm{d}\lambda}\right)^2\mathrm{d}\lambda = 0 \tag{1.2.55}$$

containing the parameter λ explicitly. The variational principle in the form (1.2.55) is valid for the isotropic geodesic as well. The quantity f serves as the Lagrangian of principle (1.2.55) and appropriate equations coincide with (1.2.46). Therefore, λ should be a canonical parameter. f being explicitly independent of λ there exists the integral f = constant, i.e. (1.2.47). System (1.2.46) consists of n equations of the second order. Among $2n$ arbitrary constants there are two superfluous constants related with introducing λ. For an isotropic geodesic the relation $f = 0$ should be joined as an independent equation to equations (1.2.46) or to the variational principle.

A vector tangent to the geodesic remains tangent in parallel transport. If a curve is not geodesic then a vector tangent to it at some point does not remain tangent in parallel transport along this curve. Transport which enables a vector to remain tangent for any real curve is called Fermi–Walker transport. For a vector ξ^k and curve $x^i = x^i(s)$ the Fermi–Walker transport is determined by the equations

$$\frac{\mathrm{D}\xi^i}{\mathrm{d}s} = W^{ij}\xi_j \tag{1.2.56}$$

with

$$W^{ij} = A^iu^j - A^ju^i \qquad u^i = \frac{\mathrm{d}x^i}{\mathrm{d}s} \qquad A^i = \frac{\mathrm{D}u^i}{\mathrm{d}s}. \tag{1.2.57}$$

Since $u_i u^i = 1$ vector u^i is the unit tangent vector. The same relation implies that A^i and u^i are orthogonal vectors, i.e. $u_i A^i = 0$. It is easy to verify that the vector u^i satisfies the Fermi–Walker transport (1.2.56). If ξ^i, η^i are two vectors satisfying the Fermi–Walker transport then

$$\frac{\mathrm{D}}{\mathrm{d}s}(\xi_i \eta^i) = 0$$

demonstrating that similar to the parallel transport the Fermi–Walker transport preserves the scalar products of vectors and hence the angles between vectors and their lengths.

1.2.5 Curvature tensor

All operations with absolute differentials and derivatives of the first order are performed in the same manner as with ordinary differentials and derivatives. But even for the absolute differentials and derivatives of the second order this is not true because the commutative law is not valid anymore. If D, d are operators of absolute and ordinary differentiation in displacement from a given point x^i along some definite direction and derivatives $\tilde{\mathrm{D}}$, $\tilde{\mathrm{d}}$ are analogous operators in displacement along some other direction, then from (1.2.39) and (1.2.40) there results

$$\tilde{\mathrm{D}}\mathrm{D}a^j - \mathrm{D}\tilde{\mathrm{D}}a^j = -R_{kmi.}^{\cdot\cdot\cdot j} a^i \tilde{\mathrm{d}}x^k \mathrm{d}x^m \tag{1.2.58}$$

$$\tilde{\mathrm{D}}\mathrm{D}a_i - \mathrm{D}\tilde{\mathrm{D}}a_i = R_{kmi.}^{\cdot\cdot\cdot j} a_j \tilde{\mathrm{d}}x^k \mathrm{d}x^m \tag{1.2.59}$$

where

$$R_{kmi.}^{\cdot\cdot\cdot j} = \frac{\partial \Gamma^j_{ki}}{\partial x^m} + \Gamma^j_{mn}\Gamma^n_{ki} - \frac{\partial \Gamma^j_{mi}}{\partial x^k} - \Gamma^j_{kn}\Gamma^n_{mi}. \tag{1.2.60}$$

The tensor (1.2.60) is called the curvature tensor or Riemann–Christoffel tensor. In terms of absolute derivatives (1.2.58) and (1.2.59) may be rewritten

$$(\nabla_k \nabla_m - \nabla_m \nabla_k) a^j = -R_{kmi.}^{\cdot\cdot\cdot j} a^i \tag{1.2.61}$$

$$(\nabla_k \nabla_m - \nabla_m \nabla_k) a_i = R_{kmi.}^{\cdot\cdot\cdot j} a_j. \tag{1.2.62}$$

For a tensor of arbitrary structure the laws (1.2.58), (1.2.59), (1.2.61) and (1.2.62) are applied to each contravariant or covariant index separately. From (1.2.60) it is seen that the curvature tensor is antisymmetric in two first lower indices

$$R_{mki.}^{\cdot\cdot\cdot j} = -R_{kmi.}^{\cdot\cdot\cdot j} \tag{1.2.63}$$

and satisfies the Ricci identities

$$R_{kmi.}^{\cdot\cdot\cdot j} + R_{mik.}^{\cdot\cdot\cdot j} + R_{ikm.}^{\cdot\cdot\cdot j} = 0 \tag{1.2.64}$$

and the Bianchi identities

$$\nabla_k R_{mni.}^{\cdot\cdot\cdot j} + \nabla_m R_{nki.}^{\cdot\cdot\cdot j} + \nabla_n R_{kmi.}^{\cdot\cdot\cdot j} = 0. \tag{1.2.65}$$

By lowering an upper index one obtains the covariant curvature tensor

$$R_{kmij} = g_{jn} R_{kmi.}^{\cdot\cdot\cdot n} \tag{1.2.66}$$

or after some transformation based on (1.2.31), (1.2.33) and (1.2.60)

$$\begin{aligned} R_{kmij} = {} & \frac{1}{2}\left(\frac{\partial^2 g_{kj}}{\partial x^m \partial x^i} + \frac{\partial^2 g_{mi}}{\partial x^k \partial x^j} - \frac{\partial^2 g_{mj}}{\partial x^k \partial x^i} - \frac{\partial^2 g_{ki}}{\partial x^m \partial x^j}\right) \\ & + g_{pq}(\Gamma^p_{kj}\Gamma^q_{mi} - \Gamma^p_{mj}\Gamma^q_{ki}). \end{aligned} \tag{1.2.67}$$

The covariant curvature tensor satisfies the relations

$$R_{ijkm} = R_{kmij}, \quad R_{mkij} = -R_{kmij}, \quad R_{kmji} = -R_{kmij} \tag{1.2.68}$$

and Ricci identities take the form

$$R_{kmij} + R_{mikj} + R_{ikmj} = 0 \tag{1.2.69}$$

admitting cyclic permutation of any three indices. Due to all these identities the number of the significantly different components of the curvature tensor in V_n reduces from n^4 to $n^2(n^2-1)/12$. Along with tensor (1.2.66) one obtains on the basis of the curvature tensor the symmetrical Ricci tensor

$$R_{ij} = R_{kij.}^{\cdot\cdot\cdot k} \tag{1.2.70}$$

or

$$R_{ij} = g^{km} R_{kijm} \tag{1.2.71}$$

and the scalar curvature

$$R = g^{ij} R_{ij}. \tag{1.2.72}$$

In particular, the curvature tensor determines the behaviour of close geodesics in V_n. Let $x^i = x^i(t,p)$ be a one-parametric set of curves with parameter p. From (1.2.39) one has

$$\frac{\mathrm{D}u^i}{\mathrm{d}p} = \frac{\mathrm{D}q^i}{\mathrm{d}t} \tag{1.2.73}$$

where $u^i = \partial x^i/\partial t$ is a vector tangent to the reference curve of the set and $q^i = \partial x^i/\partial p$ represents a vector characterizing the deviation from the reference curve in the changing parameter p. Deviation of curves is determined by the infinitesimal vector

$$\eta^i = q^i \mathrm{d}p. \tag{1.2.74}$$

Differentiating (1.2.73) and using (1.2.58) yields

$$\frac{\mathrm{D}^2 q^i}{\mathrm{d}t^2} = \frac{\mathrm{D}}{\mathrm{d}p}\frac{\mathrm{D}u^i}{\mathrm{d}t} + R_{kmj.}^{\cdot\cdot\cdot i} u^m u^j q^k. \tag{1.2.75}$$

For the set of the geodesic lines with canonical parameter $t = \lambda$ the first term on the right-hand side of (1.2.75) vanishes and the final equations of the deviation of the geodesics take the form

$$\frac{\mathrm{D}^2 \eta^i}{\mathrm{d}\lambda^2} = R_{kmj.}^{\cdot\cdot\cdot i} u^m u^j \eta^k. \tag{1.2.76}$$

Completing here the exposition of the elements of Riemannian geometry and tensor analysis it is to be noted that regretfully there is no standard definition of the curvature tensor (no standard in writing indices) involving some non-uniqueness in further definitions. Definitions and generally all exposition techniques employed here are based on the textbook by Rashevsky (1953).

Let us note further that operations of tensor analysis are rather cumbersome. At present, performing these operations may be significantly facilitated by using specialized tensor systems of computer algebra. An example of such a systems is the FORTRAN based system GRATOS (Tarasevich *et al* 1987) offering the advantages of combination of operations of tensor analysis and Taylor expansion in small perturbations of the metric tensor g_{ij}.

1.3 ELEMENTS OF THE SPECIAL THEORY OF RELATIVITY

1.3.1 Lorentz transformations

Numerous experimental data obtained by the beginning of the 20th century resulted in four statements:

(1) all points of space and all moments of time are alike (homogeneity of space and time);
(2) all directions in space are alike (isotropy of space);
(3) all laws of nature are the same in all inertial reference frames (special principle of relativity);

(4) the velocity of light in vacuum c is the same constant in all inertial reference frames (postulate of the constancy of the velocity of light).

The first two statements are common both for Newtonian mechanics and the special theory of relativity. The latter two statements are specific for the special theory of relativity. The first basic law of Newtonian mechanics is the law of inertia. A reference frame providing the validity of this law is called an inertial frame. Strictly speaking, there exists no inertial system in nature and it may be realized by material bodies only to a greater or lesser degree of accuracy. Each reference frame moving uniformly and rectilinearly relative to a given inertial system is inertial as well. From the very beginning of Newtonian mechanics its laws were known to be valid in any inertial system in accordance with Galileo's principle of relativity. Mathematically this principle manifests itself as the invariability of equations of Newtonian mechanics under the Galilean transformations relating two inertial systems.

Let $S(t, x, y, z)$ and $S'(t', x', y', z')$ be two inertial systems. S' is assumed to move relative to S with constant velocity $\boldsymbol{v}$. If the position vector of some point is $\boldsymbol{r}$ for S and $\boldsymbol{r}'$ for S' then $\boldsymbol{r}' = \boldsymbol{r} - \boldsymbol{v}t$.

In Newtonian mechanics the time is absolute. In particular the time interval between two events is the same independent of the reference frame of its measurement. Then $t' = t$. These formulae characterize the Galilean transformations. Invariability of the equations of Newtonian mechanics under the Galilean transformations is achieved because neither the vector of acceleration of the left-hand side of these equations nor the vector of force of their right-hand member depend on the uniform rectilinear motion of the system. The Galilean transformations imply also that the space intervals between the points measured in different inertial systems are the same. This is due to the fact that the space of Newtonian mechanics is represented by the three-dimensional proper Euclidean space.

Adoption of the special principle of relativity and the postulate of the constancy of the velocity of light has radically changed Newtonian conceptions on space and time. The set of four statements indicated above results in the invariability for any two events $M_1(ct_1, x_1, y_1, z_1)$ and $M_2(ct_2, x_2, y_2, z_2)$ the square of length of the 4-vector M_1M_2 (the square of the space-time interval between the events)

$$(M_1M_2)^2 = c^2(t_2 - t_1)^2 - (x_2 - x_1)^2 - (y_2 - y_1)^2 - (z_2 - z_1)^2. \quad (1.3.1)$$

For infinitesimal events this means the invariability of

$$\mathrm{d}s^2 = c^2\mathrm{d}t^2 - \mathrm{d}x^2 - \mathrm{d}y^2 - \mathrm{d}z^2. \quad (1.3.2)$$

From this it follows that the space-time of the special theory of relativity is described by geometry of the four- dimensional pseudo-Euclidean space of index 1. Transformation between different inertial systems thus reduces to the transformation of the orthonormal reference bases. In the general case this transformation depends on $n(n+1)/2 = 10$ parameters. Four of them are due to the possibility of arbitrary choice of the initial point (homogeneity of space and time). Three parameters enable one to perform arbitrary rotation of the space axes (isotropy of space). The remaining three parameters are the components of the translatory velocity of one inertial system with respect to another one. Within the trivial transformation of translation and rotation this change from one inertial system S to another system S' is described by (1.2.19) putting

$$x^0 = ct \qquad x^1 = x \qquad x^2 = y \qquad x^3 = z. \tag{1.3.3}$$

From these formulae it follows that S' is moving relative to S along the x-axis with constant velocity $v = \beta c$ $(-c < v < c)$. Finally, for the particular case of coinciding axes of S and S' and motion along the x-axis this change reduces to the famous Lorentz transformations

$$t' = \frac{t - c^{-2}vx}{(1 - v^2/c^2)^{1/2}} \quad x' = \frac{-vt + x}{(1 - v^2/c^2)^{1/2}} \quad y' = y \quad z' = z. \tag{1.3.4}$$

A more general Lorentz transformation valid for arbitrary direction of the translatory velocity $\boldsymbol{v}$ and including the rotation of the space axes has the form

$$t' = \frac{t - c^{-2}(\boldsymbol{vr})}{(1 - v^2/c^2)^{1/2}} \tag{1.3.5}$$

$$T\boldsymbol{r}' = \boldsymbol{r} + \left[\left(\frac{1}{(1 - v^2/c^2)^{1/2}} - 1\right)\frac{\boldsymbol{vr}}{v^2} - \frac{t}{(1 - v^2/c^2)^{1/2}}\right]\boldsymbol{v}. \tag{1.3.6}$$

These formulae contain 6 parameters, i.e. three components of $\boldsymbol{v}$ and three angles of the spatial rotation specified by the orthogonal 3×3 matrix T. The lacking four parameters may be added with the trivial translation of the origin of the reference system. In the absence of rotation, T is replaced by the unit matrix E and may be simply omitted in (1.3.6). In the limit $c \to \infty$ the Lorentz transformations reduce to the Galilean transformations. The inverse transformation from S' to S is determined by the same formulae (1.3.5) and (1.3.6) replacing T by T^{-1} and interchanging the role of primed and unprimed quantities. Instead of $\boldsymbol{v}$ this inverse transformation involves velocity $\boldsymbol{v}'$ of S relative to S'

$$T\boldsymbol{v}' = -\boldsymbol{v} \tag{1.3.7}$$

and by absolute magnitude $v' = |\boldsymbol{v}'| = v = |\boldsymbol{v}|$.

1.3.2 Kinematical consequences of the Lorentz transformations

We now consider briefly the principal kinematical consequences of Lorentz transformations starting with the classical examples of moving material rods and clocks.

1. Contraction of the linear sizes of the moving bodies. If a material rod is at rest along the x'-axis of system S' then its proper length is equal to the differences of the abscissae of its end points $d = x'_2 - x'_1$. With respect to system S this rod is moving along the x-axis with velocity v. Its length $d' = x_2 - x_1$ in system S is the distance between the abscissae of its end points taken at the same moment t. Therefore, from (1.3.4) there results

$$d' = d(1 - v^2/c^2)^{1/2}. \tag{1.3.8}$$

2. Relativity of simultaneity. For two events M_1 and M_2 in accordance with (1.3.4) one has

$$t'_2 - t'_1 = \frac{t_2 - t_1 - c^{-2}v(x_2 - x_1)}{(1 - v^2/c^2)^{1/2}}. \tag{1.3.9}$$

If $t_2 = t_1$ and $x_2 \neq x_1$, i.e. if M_1 and M_2 are two simultaneous events in system S occurring in its different spatial points, then $t'_2 \neq t'_1$ demonstrating that these events are not simultaneous in system S'.

If the time interval between the events M_1 and M_2 is greater than the time needed for light to pass the distance between them then such events are called consecutive. The quantity (1.3.1) is positive for such events. If $t_1 < t_2$ then in any other system M_1 precedes M_2, i.e. $t'_1 < t'_2$. The four-dimensional interval between such events is called timelike. It is possible thereby to introduce such a reference frame where events M_1 and M_2 occur in the same space point. If M_1 and M_2 are such that the time interval between them is less than the time needed for light to pass the distance between them then the quantity (1.3.1) is negative. Such events are called quasi-simultaneous. Their succession in different systems may be different (i.e. it is possible to have $t_2 > t_1$ and $t'_2 < t'_1$). There is no causality relation between them because no interaction can propagate faster than light. The four-dimensional interval between such events is called spacelike. It is always possible to introduce a reference system where these events occur at the same moment of time.

3. Retardation of a moving clock. Consider a clock at rest in system S' (situated, for example, at point $x' = \text{constant}, y' = z' = 0$) reading time t'. Denoting by $\tau = t'_2 - t'_1$ the interval of proper time of the system (i.e. time reading of the clock at rest in the given system) with the aid of the inverse Lorentz transformation one has

$$\tau = (t_2 - t_1)(1 - v^2/c^2)^{1/2}. \tag{1.3.10}$$

The corresponding time interval $t_2 - t_1$ of the moving clock is greater than τ implying its retardation.

4. Addition of velocities. Consider a point moving in system S' with velocity $\boldsymbol{u}' = \mathrm{d}\boldsymbol{r}'/\mathrm{d}t'$. Its velocity relative to system S is $\boldsymbol{u} = \mathrm{d}\boldsymbol{r}/\mathrm{d}t$. Let systems S and S' be related by the Lorentz transformation (1.3.5) and (1.3.6) without rotation, i.e. $T = E$. Describing the inverse Lorentz transformation in terms of differentials of $t, t', \boldsymbol{r}, \boldsymbol{r}'$ one obtains

$$\boldsymbol{u} = \frac{(1 - v^2/c^2)^{1/2}}{1 + c^{-2}(\boldsymbol{v}\boldsymbol{u}')}\left\{\boldsymbol{u}' + \left[\left(\frac{1}{(1 - v^2/c^2)^{1/2}} - 1\right)\frac{\boldsymbol{v}\boldsymbol{u}'}{v^2} + \frac{1}{(1 - v^2/c^2)^{1/2}}\right]\boldsymbol{v}\right\}. \tag{1.3.11}$$

One may regard a point at hand as being at rest in some inertial system S'' defined by coordinates t'', $\boldsymbol{r}''$. Conversion from S' to S'' is given by the Lorentz transformation without rotation induced by velocity $\boldsymbol{u}'$. Then $\boldsymbol{u}$ is the velocity of S'' relative to S. To derive the velocity $\boldsymbol{u}''$ of S relative to S'' it should be taken into account that S' moves relative to S'' with velocity $-\boldsymbol{u}'$ and S moves relative to S' with velocity $-\boldsymbol{v}$. Hence, the velocity $\boldsymbol{u}''$ is expressed by (1.3.11) replacing $\boldsymbol{v}$ and $\boldsymbol{u}'$ by $-\boldsymbol{u}'$ and $-\boldsymbol{v}$ respectively. Equation (1.3.7) rewritten for this case as $T\boldsymbol{u}'' = -\boldsymbol{u}$ relates $\boldsymbol{u}$ and $\boldsymbol{u}''$ by means of a matrix T of spatial rotation S relative to S'' . In the general case one has $T \neq E$, i.e. combination of two Lorentz transformations without rotation (from S to S' and from S' to S'') leads to extra rotation in the resulting transformation (from S to S''). In the particular case of collinear velocities $\boldsymbol{v}$ and $\boldsymbol{u}'$ the velocities $\boldsymbol{u}$ and $\boldsymbol{u}''$ differ only by sign, hence $T = E$.

The velocities $\boldsymbol{v}$ and $\boldsymbol{u}'$ enter into (1.3.11) non-symmetrically (unless they are parallelly directed). This is due to the non-commutability of the Lorentz transformations. But the direction of the velocity alone depends on the order of the addition of velocities. The absolute magnitude of $\boldsymbol{u}$ is commutative since as seen from (1.3.11)

$$\boldsymbol{u}^2 = \frac{(\boldsymbol{u}' + \boldsymbol{v})^2 - c^{-2}|\boldsymbol{v} \times \boldsymbol{u}'|^2}{(1 + c^{-2}\boldsymbol{v}\boldsymbol{u}')^2}. \tag{1.3.12}$$

Formula (1.3.11) enables one to solve a set of problems of motion in the special theory of relativity. For example, let us find the relative velocity of two particles having in some system S' velocities $\boldsymbol{v}_1$ and $\boldsymbol{v}_2$ respectively. Evidently, the velocity of the second particle relative to the first one represents the velocity of the second particle in a reference system S in which the first particle is at rest. Hence, S moves relative to S' with velocity $\boldsymbol{v}_1$. Putting in (1.3.11) $\boldsymbol{v} = -\boldsymbol{v}_1, \boldsymbol{u}' = \boldsymbol{v}_2$ one finds the needed velocity $\boldsymbol{u}$. If $\boldsymbol{u}'$ is infinitesimal then within the terms of the first order in $|\boldsymbol{u}'|$

$$\boldsymbol{u} = \boldsymbol{v} + \mathrm{d}\boldsymbol{v}$$

$$\mathrm{d}\boldsymbol{v} = (1 - v^2/c^2)^{1/2} \left[\boldsymbol{u}' + \left((1 - v^2/c^2)^{1/2} - 1 \right) \frac{(\boldsymbol{v}\boldsymbol{u}')}{v^2} \boldsymbol{v} \right] \quad (1.3.13)$$

$$\boldsymbol{u}'' = -\left[\boldsymbol{v} + \boldsymbol{u}' - c^{-2}(\boldsymbol{v}\boldsymbol{u}')\boldsymbol{v}\right]. \quad (1.3.14)$$

Writing the Lorentz transformation from S to S' explicitly (no rotation, translatory velocity $\boldsymbol{v}$), from S' to S'' (no rotation, infinitesimal translatory velocity $\boldsymbol{u}'$) and from S to S'' (rotation T, translatory velocity $\boldsymbol{u}$), it may be seen that rotation T of the resulting transformation represents the infinitesimal rotation around vector Ω by angle $|\Omega|$

$$T\boldsymbol{r} = \boldsymbol{r} + (\Omega \times \boldsymbol{r}) \quad (1.3.15)$$

$$\Omega = -(1/v^2)[(1 - v^2/c^2)^{-1/2} - 1](\boldsymbol{v} \times \mathrm{d}\boldsymbol{v}). \quad (1.3.16)$$

The velocities $\boldsymbol{u}$ and $\boldsymbol{u}''$ thereby satisfy relation (1.3.7) or $T\boldsymbol{u}'' = -\boldsymbol{u}$.

Let a particle moving in system S be a gyroscope. If the velocity of the particle at moment t is $\boldsymbol{v}$ then for moment $t + \mathrm{d}t$ this velocity will be $\boldsymbol{u} = \boldsymbol{v} + \mathrm{d}\boldsymbol{v}$ with $\mathrm{d}\boldsymbol{v} = \dot{\boldsymbol{v}}dt$. S' and S'' are in this case the inertial rest systems of the particle at moments t and $t + \mathrm{d}t$ respectively. The direction of rotation of the gyroscope at moment t in system S' coincides with its direction at moment $t + \mathrm{d}t$ in system S'' provided that there is no external angular momentum. According to (1.3.15) the axis of gyroscope in system S undergoes precession with angular velocity

$$\omega = -(1/v^2)[(1 - v^2/c^2)^{-1/2} - 1)](\boldsymbol{v} \times \dot{\boldsymbol{v}}). \quad (1.3.17)$$

This effect is called the Thomas precession.

5. Aberration of light. Directions towards one and the same star as measured in two different reference frames differ from one another by aberration. Returning to (1.3.11) let us assume again that the velocity $\boldsymbol{v}$ of S' relative to S is directed along the x-axis, i.e. $v_x = v, v_y = v_z = 0$. If vector $\boldsymbol{u}'$ lies in the $x'y'$ plane at angle θ' with the x'-axis

$$u'_x = u' \cos\theta' \quad u'_y = u' \sin\theta' \quad u'_z = 0$$

then vector $\boldsymbol{u}$ lies in the xy plane at angle θ with the x axis

$$u_x = u\cos\theta \quad u_y = u\sin\theta \quad u_z = 0$$

and in virtue of (1.3.11) there results

$$\tan\theta = \frac{(1 - v^2/c^2)^{1/2} u' \sin\theta'}{u'\cos\theta' + v}. \quad (1.3.18)$$

For the light propagation $u = u' = c$ and the relativistic formula of aberration is

$$\tan\theta = \frac{(1 - v^2/c^2)^{1/2}\sin\theta'}{\cos\theta' + vc^{-1}}. \tag{1.3.19}$$

Inverting this formula one obtains

$$\tan\theta' = \frac{(1 - v^2/c^2)^{1/2}\sin\theta}{\cos\theta - vc^{-1}} \tag{1.3.20}$$

or after expanding in powers of v/c

$$\theta' - \theta = c^{-1}v\sin\theta + \tfrac{1}{2}c^{-2}v^2\sin\theta\cos\theta + \dots. \tag{1.3.21}$$

6. Doppler effect. Consider in system S' some periodic process related with the emission of light signals. These signals are registered in system S at moment t^*. Taking into account the signal delay due to the finite value of the light velocity the time of emission is t with

$$t^* = t + \frac{r(t)}{c}$$

$r(t)$ being the distance between the emitter and the receiver. This process is perceived in S as periodic with the frequency

$$\Delta t^* = \left(1 + \frac{v_r}{c}\right)\Delta t$$

$v_r = dr/dt$ being the radial velocity of the emitter. It may be assumed that the emitter is located at the origin of S', i.e. $\boldsymbol{r}' = 0$. Then from the inverse Lorentz transformation there results

$$t = \frac{t'}{(1 - v^2/c^2)^{1/2}}.$$

Therefore, the frequency Δt^* in S is related to the proper frequency $\Delta t'$ in system S' by the equation

$$\Delta t^* = \frac{1 + c^{-1}v_r}{(1 - v^2/c^2)^{1/2}}\Delta t'. \tag{1.3.22}$$

Here

$$v_r = -v\cos\theta \tag{1.3.23}$$

θ being the angle between the relative velocity of S' and the light direction.

These are basic kinematical effects associated with the Lorentz transformations. Their more detailed physical meaning is exposed in numerous textbooks on special relativity theory, in particular, in the monograph by Møller (1972). Our exposition is based on this monograph but here we confine ourselves to the most simple relations of special relativity theory needed for relativistic celestial mechanics.

1.3.3 Dynamics of a particle

A curve representing the process of motion of a particle in four-dimensional space-time is called the world line. Since the velocity of a material point cannot exceed the light velocity then for the world line of any material point

$$ds^2 = dx^{0^2} - dx^{1^2} - dx^{2^2} - dx^{3^2} > 0. \tag{1.3.24}$$

The world line of light is the isotropic straight line, i.e.

$$ds^2 = 0. \tag{1.3.25}$$

Taking s as the parameter of the world line of a material particle one may present its equations in a form

$$x^\alpha = x^\alpha(s). \tag{1.3.26}$$

Here and everywhere below the greek indices run from 0 to 3 and the latin indices run from 1 to 3. The unit tangent vector

$$u^\alpha = \frac{dx^\alpha}{ds} \tag{1.3.27}$$

defines the 4-velocity of the particle. From (1.3.24) it follows that

$$ds = c d\tau = c(1 - v^2/c^2)^{1/2} dt \tag{1.3.28}$$

τ the being proper time of the moving particle and $\boldsymbol{v}$ being an ordinary 3-vector of velocity with components $v^i = dx^i/dt$. Therefore,

$$u^0 = \frac{1}{(1 - v^2/c^2)^{1/2}} \qquad u^i = \frac{v^i}{c(1 - v^2/c^2)^{1/2}}. \tag{1.3.29}$$

A 4-vector $\boldsymbol{A}$ with components

$$A^\alpha = \frac{du^\alpha}{ds} = \frac{1}{c(1 - v^2/c^2)^{1/2}} \frac{du^\alpha}{dt} \tag{1.3.30}$$

represents the 4-acceleration of the moving particle. Because $\boldsymbol{u}^2 = 1$ vectors $\boldsymbol{u}$ and $\boldsymbol{A}$ are orthogonal

$$\boldsymbol{u}\boldsymbol{A} = 0. \tag{1.3.31}$$

By the second law of mechanics the equations of motion of the material particle in some inertial system may be written in the form

$$\frac{\mathrm{d}}{\mathrm{d}s}(m_0 u^\alpha) = c^{-2}\chi(v)F^\alpha. \tag{1.3.32}$$

m_0 is the constant quantity characterizing the inertial mass of the point and is called the rest mass. $\chi(v)$ is the factor determined below and is dependent only on the absolute magnitude of the 3-velocity $\boldsymbol{v}$. F^α are components of the external force. The 3-vector $\boldsymbol{F}$ with components F^i represents an ordinary vector of force. The condition $m_0 =$ constant implies

$$F^0 = c^{-1}\boldsymbol{F}\boldsymbol{v}.$$

The factor $\chi(v)$ is expressed as

$$\chi(v) = (1 - v^2/c^2)^{-1/2}.$$

Finally, equations (1.3.32) take the form

$$\frac{\mathrm{d}E}{\mathrm{d}t} = \boldsymbol{F}\boldsymbol{v} \qquad \frac{\mathrm{d}\boldsymbol{p}}{\mathrm{d}t} = \boldsymbol{F} \tag{1.3.33}$$

with kinetic energy $E = mc^2$, impulse $\boldsymbol{p} = m\boldsymbol{v}$ and mass m of the moving point satisfying the relation

$$m = \frac{m_0}{(1 - v^2/c^2)^{1/2}}. \tag{1.3.34}$$

The dynamical characteristic of the moving point is the energy–momentum vector $E_0\boldsymbol{u}$ with $E_0 = m_0c^2$ being the rest energy. The time component of this vector coincides with the energy of the point $E_0u^0 = mc^2$. The space components are components of the impulse multiplied by c, i.e. $E_0u^i = mv^ic$.

1.3.4 Energy–momentum tensor

Consider in some system S a stream of dust-like non-interacting particles. Denote by $\mathrm{d}\omega$ an elementary volume occupied by an elementary mass $\mathrm{d}m$. Let S_0 be a system co-moving with this elementary volume and having velocity $\boldsymbol{v}(v^1, v^2, v^3)$ relative to S. The elementary volume and mass at hand take in S_0 the values $\mathrm{d}\omega_0$ and $\mathrm{d}m_0$ characterizing elementary rest volume and rest mass respectively. From the preceding relations

$$\mathrm{d}\omega = (1 - v^2/c^2)^{1/2}\mathrm{d}\omega_0 \qquad \mathrm{d}m = \frac{\mathrm{d}m_0}{(1 - v^2/c^2)^{1/2}}. \tag{1.3.35}$$

Mass density may be characterized by any one of the quantities

$$\rho^* = \frac{dm_0}{d\omega_0} \qquad \rho = \frac{dm_0}{d\omega} \qquad \tilde{\rho} = \frac{dm}{d\omega}. \tag{1.3.36}$$

It is evident that ρ^* represents the rest mass density in a co-moving system S_0 (invariant density), ρ is the rest mass density in the system S and $\tilde{\rho}$ is the mass density in the system S (this is the total density including the mass corresponding to the kinetic energy of particles). The tensor of mass of a stream of particles is determined by

$$T^{\alpha\beta} = \rho^* u^\alpha u^\beta \tag{1.3.37}$$

with u^α being the 4-vector of velocity. Tensor $c^2 T^{\alpha\beta}$ is called the energy–momentum tensor of a stream of particles. Its components take the values

$$c^2 T^{00} = c^2 \tilde{\rho} \qquad c^2 T^{0i} = c\tilde{\rho} v^i \qquad c^2 T^{ij} = \tilde{\rho} v^i v^j. \tag{1.3.38}$$

A stream of dust particles represents the most simple case of moving material substance. In the general case the energy–momentum tensor is a symmetric covariant tensor of rank 2, the components of which have physical meaning as results from (1.3.38). The most important property of the total energy–momentum tensor taking into account all types of matter substantial for a specific physical problem is the conservation law. Analytically this law may be formulated in the form

$$\nabla_\beta T^{\alpha\beta} = 0. \tag{1.3.39}$$

Introducing of the covariant derivatives makes this relation valid in any curvilinear coordinates. For the tensor (1.3.37) this may be easily verified. In fact, one has

$$\nabla_\beta T^{\alpha\beta} = u^\alpha \frac{\partial(\rho^* u^\beta)}{\partial x^\beta} + \rho^* u^\beta \frac{\partial u^\alpha}{\partial x^\beta}.$$

The first term vanishes in virtue of the equation of continuity implying the conservation of the rest mass

$$\frac{\partial(\rho^* u^\beta)}{\partial x^\beta} = c^{-1}\left(\frac{\partial \rho}{\partial t} + \frac{\partial(\rho v^i)}{\partial x^i}\right) = 0. \tag{1.3.40}$$

Besides,

$$u^\beta \frac{\partial u^\alpha}{\partial x^\beta} = \frac{1}{c(1 - v^2/c^2)^{1/2}}\left(\frac{\partial u^\alpha}{\partial t} + v^i \frac{\partial u^\alpha}{\partial x^i}\right) = \frac{1}{c(1 - v^2/c^2)^{1/2}} \frac{du^\alpha}{dt} = A^\alpha$$

4-acceleration A^α being zero due to the absence of external forces.

Inside macroscopic bodies one may assume the validity of the equation of the continuity (1.3.40) and the equations of motion of a continuous mechanical medium

$$\rho \frac{\mathrm{d}v^i}{\mathrm{d}t} = \rho F^i + \frac{\partial p^{ij}}{\partial x^j} \tag{1.3.41}$$

F^i being the external force acting on the elementary volume and p^{ij} being the stress tensor: For an ideal fluid

$$p^{ij} = -p\delta^{ij} . \tag{1.3.42}$$

δ^{ij} as well as δ_{ij} and δ_i^j are determined by (1.2.5). An expression for the energy–momentum tensor of macroscopic bodies is obtained by generalizing (1.3.37) taking into account pressure p:

$$c^2 T^{\alpha\beta} = (c^2\mu^* + p)u^\alpha u^\beta - p\eta^{\alpha\beta} \tag{1.3.43}$$

with

$$\eta^{00} = 1 \qquad \eta^{0i} = 0 \qquad \eta^{ij} = -\delta^{ij} . \tag{1.3.44}$$

Density μ^* and pressure p are to be functionally related. For a perfect, non-viscous fluid the rest mass contains the mass proportional to the potential compressional energy Π. μ^* is the density of this total rest mass, to be distinguished from the density ρ^* of the invariable rest mass. Equations (1.3.39) applied to the tensor (1.3.43) yield

$$u^\alpha \frac{\partial}{\partial x^\beta}[(c^2\mu^* + p)u^\beta] + (c^2\mu^* + p)A^\alpha - \eta^{\alpha\beta}\frac{\partial p}{\partial x^\beta} = 0 \tag{1.3.45}$$

reducing in absence of the external forces to the equations of motion of continuous matter. Multiplication by u_α results in

$$(c^2\mu^* + p)\frac{\partial u^\beta}{\partial x^\beta} + c^2 u^\beta \frac{\partial \mu^*}{\partial x^\beta} = 0. \tag{1.3.46}$$

Defining ρ^* by the equation

$$\frac{\mathrm{d}\rho^*}{\rho^*} = \frac{\mathrm{d}\mu^*}{\mu^* + c^{-2}p} \tag{1.3.47}$$

with condition $\rho^* = \mu^*$ for $p = 0$, equation (1.3.46) is reduced to the equation of continuity

$$\rho^* \frac{\partial u^\beta}{\partial x^\beta} + u^\beta \frac{\mathrm{d}\rho^*}{\mathrm{d}\mu^*}\frac{\partial \mu^*}{\partial x^\beta} \equiv \frac{\partial(\rho^* u^\beta)}{\partial x^\beta} = 0.$$

Putting

$$\mu^* = \rho^*(1 + c^{-2}\Pi) \tag{1.3.48}$$

and substituting this expression into (1.3.47) one obtains the equation for the potential compressional energy Π

$$\mathrm{d}\Pi = (c^2/\rho^{*2})(\rho^*\mathrm{d}\mu^* - \mu^*\mathrm{d}\rho^*) = (p/\rho^{*2})\mathrm{d}\rho^*. \tag{1.3.49}$$

From this it follows that

$$\Pi = -\frac{p}{\rho^*} + \int_0^p \frac{\mathrm{d}p}{\rho^*}. \tag{1.3.50}$$

Considering that

$$u^\beta \frac{\partial p}{\partial x^\beta} = \frac{1}{c(1 - v^2/c^2)^{1/2}} \frac{\mathrm{d}p}{\mathrm{d}t}$$

and combining (1.3.45) and (1.3.46) one transforms the equations of motion to the form

$$(c^2\rho^* + \rho^*\Pi + p)A^\alpha = \eta^{\alpha\beta}\frac{\partial p}{\partial x^\beta} - \frac{u^\alpha}{c(1 - v^2/c^2)^{1/2}}\frac{\mathrm{d}p}{\mathrm{d}t}. \tag{1.3.51}$$

The energy–momentum tensor (1.3.43) takes the final form

$$c^2T^{\alpha\beta} = (c^2\rho^* + \rho^*\Pi + p)u^\alpha u^\beta - p\eta^{\alpha\beta}. \tag{1.3.52}$$

The representation of the energy–momentum tensor of macroscopic bodies used here was first founded by Fock (1955).

2

GRT Field Equations

2.1 BASIC PRINCIPLES OF GRT

2.1.1 Basic ideas

Currently, it might be possible to develop the main idea of the GRT from experimental results. Yet Einstein derived the basic statements of GRT by logical considerations proceeding from the special theory of relativity and the fundamental law of equality of gravitational and inertial mass. Even though the derivation by Einstein is regarded now by some physicists as not entirely devoid of deficiencies it should be remembered that just this approach has led to the greatest triumph of scientific logic.

Having completed the special theory of relativity Einstein successively put forward the principle of equivalence and the principle of general covariance. According to the principle of equivalence all physical processes follow the same pattern both in an inertial system under the action of the homogeneous gravitational field and in a non-inertial uniformly accelerated system in the absence of gravitation. The principle of equivalence is strictly local in contrast to the law of identity of gravitational and inertial mass underlying it. The principle of general covariance being of purely mathematical character implies that equations of physics should have the same form in all reference frames. Combination of these two principles enabled Einstein to formulate the principle of general relativity.

Proceeding from this, Einstein came to conclusion that the space-time of GRT is the pseudo-Riemannian space with the metric

$$\mathrm{d}s^2 = g_{\alpha\beta}\mathrm{d}x^\alpha \mathrm{d}x^\beta \tag{2.1.1}$$

(let us remember again that greek indices can have values 0, 1, 2, 3, while latin indices take the values 1, 2, 3). In the special theory of relativity it is possible, if one wishes, to put all equations in the covariant form and to use any curvilinear coordinates. But the space of events in the special theory of relativity is flat (affine), i.e. its tensor of curvature is zero. In this space there are preferable coordinate systems (affine) defined up to affine transformations (Lorentz transformations). In such affine systems the space-time metric of special relativity takes the form (2.1.1) with $g_{\alpha\beta} = \eta_{\alpha\beta}$ and

$$\eta_{00} = 1 \qquad \eta_{0i} = 0 \qquad \eta_{ij} = -\delta_{ij}. \tag{2.1.2}$$

It should be noted that no less often the signature $-, +, +, +$ is used. It is a pity that there is no agreement on the use of any one of these signatures. Conversion from one signature to another involves no difficulties but nevertheless this fact should be kept in mind.

The coordinates providing values (2.1.2) for the metric tensor are called Galilean. In GRT one cannot introduce the global Galilean coordinates (valid for the whole space-time). Instead, one may use quasi-Galilean coordinates such that

$$g_{\alpha\beta} = \eta_{\alpha\beta} + h_{\alpha\beta} \tag{2.1.3}$$

$h_{\alpha\beta}$ being functions of x^0, x^1, x^2, x^3 with $|h_{\alpha\beta}| \ll 1$. Representation (2.1.3) makes it evident that the GRT pseudo-Riemannian metric differs little from the pseudo-Galilean metric of special relativity. Quasi-Galilean coordinates x^α admit not only linear transformation but any non-linear transformation of the type

$$\tilde{x}^\alpha = x^\alpha + \xi^\alpha(x^\beta) \qquad \left|\frac{\partial \xi^\alpha}{\partial x^\beta}\right| \ll 1 \tag{2.1.4}$$

since metric (2.1.1) expressed in new coordinates $\tilde{x}^\alpha$ is again little distinct from the Galilean metric. But it should be underlined that this distinction looks different for each reference system.

At every point M of the GRT space-time one may introduce in accordance with (1.2.37) the locally geodesic coordinates ensuring $g_{\alpha\beta} = \eta_{\alpha\beta}$ and $g_{\alpha\beta,\mu} = 0$ at M (here and everywhere below the partial derivative with respect to some coordinate is denoted by a comma accompanied by the appropriate index). Moreover, the locally geodesic coordinates may be introduced in some vicinity of a given world line. Thus, in these coordinates in displacing from point M to an infinitesimally close point functions $h_{\alpha\beta}$ are at least of second order with respect to differences $x^\mu - x^\mu(M)$. Neglecting these infinitesimal terms of the second order one has in the infinitesimal vicinity of point M the space-time of special relativity. All relevant relations are valid in this infinitesimal vicinity. In particular, coordinates x^0, x^1, x^2, x^3 may be again interpreted as ct, x, y, z. Hence, the locally geodesic coordinates determine an inertial reference system where there is no gravitational field. Such a possibility of introducing the locally geodesic coordinates is due to the principle of equivalence valid only locally.

In a locally inertial system the tensor of mass $T^{\alpha\beta}$ has the same meaning as for special relativity. In the quasi-Galilean coordinates of GRT $T^{\alpha\beta}$

has approximately the same meaning. This tensor describes the total distribution and change of energy and momentum excluding the energy and momentum of non-gravitational origin. In special relativity the equations of conservation of energy and momentum have the form

$$\frac{\partial T^{\alpha\beta}}{\partial x^{\beta}} = 0 \tag{2.1.5}$$

for affine (Galilean) coordinates and

$$\nabla_{\beta} T^{\alpha\beta} = 0 \tag{2.1.6}$$

for arbitrary coordinates. In GRT $T^{\alpha\beta}$ also satisfies equations (2.1.6). However, as we cannot rewrite this as (2.1.5) in any coordinates, this equation does not result in the law of conservation. This is natural since the tensor $T^{\alpha\beta}$ does not include the energy and momentum of gravitational origin. To derive the law of conservation it is necessary to add to $T^{\alpha\beta}$ some non-tensorial quantity. This quantity is testimony to the energy and momentum of the gravitational field but its value depends on the choice of reference system.

2.1.2 Field equations

According to the basic idea of GRT the properties of space and time, i.e. the space-time metric, are determined by motion and distribution of masses and, conversely, motion and distribution of masses are governed by the field metric. This interrelation is revealed in the field equations for determining the tensor $g_{\alpha\beta}$. According to the number of the significantly different components of this tensor there should be ten field equations. Einstein set up their form using the following considerations.

(a) According to the Galileo law the gravitational mass is equal to the inertial mass which, in turn, is proportional to energy considered as one of the components of the tensor of mass. It may be suggested that the right-hand members of the field equations contain only the tensor of mass.

(b) Equations of the field should be formulated in the covariant form.

(c) In analogy with the Poisson equation for the Newtonian gravitational potential the field equations are expected to be of second order.

(d) If any solution of the covariant field equations is known then converting to any other coordinates one can construct infinitely many physically equivalent solutions. Therefore, the general solution of the field equations should admit four arbitrary functions and hence satisfy four identities. As the right-hand members of the field equations have to comply with the identities (2.1.6) the same should be true for their left-hand members.

These considerations lead to the following form of the field equations:

$$G^{\alpha\beta} + \Lambda g^{\alpha\beta} = -\kappa T^{\alpha\beta}. \tag{2.1.7}$$

The right-hand side of (2.1.7) contains the Einstein tensor

$$G^{\alpha\beta} = R^{\alpha\beta} - \tfrac{1}{2} R g^{\alpha\beta} \tag{2.1.8}$$

expressed in terms of the Ricci tensor $R^{\alpha\beta}$, scalar curvature R and metric tensor $g^{\alpha\beta}$. From the Bianchi identities (1.2.65) there results

$$\nabla_\beta R = 2 g_{\beta\mu} \nabla_\nu R^{\mu\nu}. \tag{2.1.9}$$

Therefore, the Einstein tensor satisfies the Bianchi identities

$$\nabla_\beta G^{\alpha\beta} = 0. \tag{2.1.10}$$

Hence, the whole left-hand side satisfies these identities and relations (2.1.6) are fulfilled for the right-hand side of equations (2.1.7).

Equations (2.1.7) contain a constant κ determined below in passing to the limiting case of the Newtonian field and the cosmological constant Λ which is of importance only in considering problems of cosmology. In all other applications of GRT it is sufficient to consider the field equations without a cosmological term

$$G^{\alpha\beta} = -\kappa T^{\alpha\beta}. \tag{2.1.11}$$

Contracting with $g_{\alpha\beta}$ there results

$$R = \kappa T \tag{2.1.12}$$

with

$$T = g_{\alpha\beta} T^{\alpha\beta}. \tag{2.1.13}$$

Therefore, equations (2.1.11) may be rewritten in the form

$$R^{\alpha\beta} = -\kappa (T^{\alpha\beta} - \tfrac{1}{2} T g^{\alpha\beta}). \tag{2.1.14}$$

In the domain outside the gravitating masses the tensor $T^{\alpha\beta}$ vanishes and the field equations in vacuum may be reduced simply to the vanishing Ricci tensor

$$R_{\alpha\beta} = 0. \tag{2.1.15}$$

The field equations represent ten non-linear second-order partial equations of the hyperbolic type. They determine ten unknown functions: six components of the metric tensor $g_{\alpha\beta}$ (four components remain arbitrary due to general covariance of the field equations), three components of velocity v^i of the matter, and mass density ρ (an equation of state relating the density and the pressure should be given separately).

2.1.3 Coordinate conditions

To derive the solution in some definite coordinate system it is necessary to add four non-tensorial equations called coordinate conditions. They are responsible for the choice of specific coordinates. For any problem at hand there may exist coordinate conditions which are preferable mathematically for solving this problem (providing, for example, a simple form of solution). In problems of relativistic celestial mechanics one often uses harmonic conditions determined by the equations

$$\frac{\partial(\sqrt{-g}g^{\alpha\beta})}{\partial x^\beta} = 0. \tag{2.1.16}$$

As seen below the field equations may be simplified under these conditions. But the explicit formulation of coordinate conditions is rather an exclusion and it is not always possible to establish coordinate conditions for a particular solution of the field equations.

The problem of coordinate conditions, particularly in relation with the role of harmonic conditions, has for a long while been the subject of rather emotional discussions. Relapses of these discussions occur even at present. According to Fock (1955), for the problem of motion of an isolated system of bodies the harmonic coordinates are preferable generalizing the inertial coordinates of special relativity and enabling them to be directly used in astronomical practice. For other authors, first of all for Infeld (Infeld and Plebanski 1960) the harmonic coordinates are mathematically convenient but the prescription to consider the field equations together with the harmonic conditions is contradictory to the essence of GRT. At the same time Infeld tried to find as opposite to the harmonic coordinates some other coordinates more 'objective' in the sense of their approximation to the Galilean coordinates of special relativity. At present all the discussion of Fock and Infeld seems meaningless. Each specific celestial mechanics problem may be solved in any coordinates (taking into account the possibility of mathematical simplification under some definite coordinate conditions) but in the final comparison with observations one should obtain coordinate-independent results. The problem of comparison between calculated and measurable quantities in astronomy will be considered below.

2.1.4 Equations of motion

For application in celestial mechanics the most important problem of GRT is the problem of motion of material bodies. Initially it was acceptable to postulate equations of the motion separately from the field equations as takes place in Newtonian gravitation. In fact, the Newtonian theory of gravitation includes separately the field theory (linear equations of Poisson and Laplace for Newtonian potential) and the equations of the motion (the laws of Newtonian mechanics). But Newtonian theory of the gravitational field is linear. In the linear theory the motion of the field sources (gravitating masses) does not depend on the field equations. In 1927 it was first demonstrated that the equations of the motion of material bodies in GRT are closely related to the field equations (Einstein and Grommer 1927). Starting in 1938, the group of Einstein and Infeld from one side and the group of Fock from another side succeeded in elaborating practical methods to derive the GRT equations of motion of material bodies.

2.1.5 Geodesic principle

For a test particle, i.e. for a particle with infinitesimal rest mass producing no influence on the surrounding field, the equations of motion are determined by the geodesic principle implying that the motion of such a particle is performed on the geodesic line in a given field. This law results from the field equations. In fact, for a stream of dust-like non-interacting particles the mass tensor is defined by (1.3.37),

$$T^{\alpha\beta} = \rho^* u^\alpha u^\beta$$

$u^\alpha = \mathrm{d}x^\alpha/\mathrm{d}s$ being the 4-velocity of a particle and

$$g_{\alpha\beta} u^\alpha u^\beta = 1. \tag{2.1.17}$$

The law of the conservation of rest mass expressed in covariant form is

$$\nabla_\beta(\rho^* u^\beta) = 0. \tag{2.1.18}$$

Therefore, from (2.1.6) it follows that

$$u^\beta \nabla_\beta u^\alpha = 0$$

or

$$\frac{\mathrm{D}u^\alpha}{\mathrm{d}s} \equiv \frac{\mathrm{d}^2 x^\alpha}{\mathrm{d}s^2} + \Gamma^\alpha_{\mu\nu} \frac{\mathrm{d}x^\mu}{\mathrm{d}s}\frac{\mathrm{d}x^\nu}{\mathrm{d}s} = 0. \tag{2.1.19}$$

Hence, the trajectory $x^\alpha = x^\alpha(s)$ of a test material particle is a non-isotropic geodesic. This result may be derived without the aid of the mass tensor and relation (2.1.6). As shown in Infeld and Schild (1949) the geodesic motion results from the vacuum field equations (2.1.15) regarding the particle as a field singularity. Due to the principle of equivalence the propagation of light is performed on isotropic geodesics.

Thus, there is no gravitational force in the Newtonian sense in GRT. The motion of test particles in the gravitational field is presented as the free inertial motion along the geodesic lines of the pseudo-Riemannian space with metric determined by the gravitating masses. Components of the metric tensor act as gravitation potentials, similar to the Newtonian potential for Newtonian gravitation. In this respect the field equations (2.1.11) and (2.1.15) may be regarded as generalizing Poisson and Laplace equations for the Newtonian potential. Irrespective of the smallness of the relativity effects in specific celestial mechanics problems the explanation of gravitation first achieved only by GRT is of paramount scientific importance. Celestial mechanics is primarily devoted to the motion of celestial bodies under gravitation and so it is inevitably relativistic, i.e. based on GRT.

2.1.6 Variational principle for the field equations

The remarkable property of equations (2.1.11) is that they may be derived from the condition of stationarity of some scalar invariant. This variational principle is of importance both from the theoretical point of view and for applications. Moreover, the operations needed for its derivation are of interest by themselves.

Consider the four-dimensional invariant integral

$$S_g = \int \sqrt{-g} R \, \mathrm{d}\Omega \tag{2.1.20}$$

taken over some 4-domain. Let us find the variation of this integral with changing $g_{\mu\nu}$ provided that the variations of $g_{\mu\nu}$ and their first derivatives vanish on the boundary of the domain at hand. From the definition (1.2.70) for the Ricci tensor

$$R_{\mu\nu} = \frac{\partial \Gamma^{\alpha}_{\alpha\nu}}{\partial x^{\mu}} - \frac{\partial \Gamma^{\alpha}_{\mu\nu}}{\partial x^{\alpha}} + \Gamma^{\alpha}_{\mu\beta}\Gamma^{\beta}_{\alpha\nu} - \Gamma^{\alpha}_{\alpha\beta}\Gamma^{\beta}_{\mu\nu} \tag{2.1.21}$$

it follows that

$$\delta R_{\mu\nu} = \nabla_{\mu}\delta\Gamma^{\alpha}_{\alpha\nu} - \nabla_{\alpha}\delta\Gamma^{\alpha}_{\mu\nu}. \tag{2.1.22}$$

Considering that $g^{\alpha\beta}$ are elements of the inverse matrix of $g_{\alpha\beta}$ one has for the differential of the determinant g

$$\mathrm{d}g = g g^{\alpha\beta} \, \mathrm{d}g_{\alpha\beta}. \tag{2.1.23}$$

From the definition (1.2.31) for the Christoffel symbols one has

$$\Gamma^{\alpha}_{\alpha\mu} = \frac{1}{2} g^{\alpha\beta} \frac{\partial g_{\alpha\beta}}{\partial x^{\mu}} = \frac{1}{2g} \frac{\partial g}{\partial x^{\mu}} = \frac{\partial \ln \sqrt{-g}}{\partial x^{\mu}}. \tag{2.1.24}$$

Therefore for any tensor a^{μ}

$$\nabla_{\mu} a^{\mu} = \frac{1}{\sqrt{-g}} \frac{\partial (\sqrt{-g} a^{\mu})}{\partial x^{\mu}}. \tag{2.1.25}$$

Applying (2.1.25) to (2.1.22) one obtains

$$g^{\mu\nu} \delta R_{\mu\nu} = \frac{1}{\sqrt{-g}} \frac{\partial}{\partial x^{\mu}} [\sqrt{-g} (g^{\mu\nu} \partial \Gamma^{\alpha}_{\alpha\nu} - g^{\alpha\nu} \delta \Gamma^{\mu}_{\alpha\nu})].$$

Therefore in the variation of the integral (2.1.20)

$$\delta S_g = \int \sqrt{-g} g^{\mu\nu} \delta R_{\mu\nu} \, \mathrm{d}\Omega + \int R_{\mu\nu} \delta(\sqrt{-g} g^{\mu\nu}) \, \mathrm{d}\Omega.$$

The first term makes no contribution since $\delta g_{\mu\nu}$ together with their first derivatives vanish on the domain boundary. From

$$g_{\mu\nu} g^{\beta\nu} = \delta^{\beta}_{\mu}$$

one has

$$g_{\mu\nu} \mathrm{d} g^{\beta\nu} = -g^{\beta\nu} \mathrm{d} g_{\mu\nu}. \tag{2.1.26}$$

Hence, alongside with (2.1.23) the differential of g may be presented in the form

$$\mathrm{d}g = -g g_{\alpha\beta} \mathrm{d} g^{\alpha\beta}. \tag{2.1.27}$$

Contracting (2.1.26) with $g^{\alpha\beta}$ one finds

$$\mathrm{d} g^{\alpha\beta} = -g^{\alpha\mu} g^{\beta\nu} \mathrm{d} g_{\mu\nu}. \tag{2.1.28}$$

For any tensor $a_{\alpha\beta}$ one then obtains

$$a_{\alpha\beta} \mathrm{d} g^{\alpha\beta} = -a^{\alpha\beta} \mathrm{d} g_{\alpha\beta}. \tag{2.1.29}$$

From (2.1.24)

$$\delta(\sqrt{-g} g^{\mu\nu}) = \sqrt{-g} (\delta g^{\mu\nu} - \tfrac{1}{2} g^{\mu\nu} g_{\alpha\beta} \delta g^{\alpha\beta}) \tag{2.1.30}$$

so that

$$R_{\mu\nu} \delta(\sqrt{-g} g^{\mu\nu}) = \sqrt{-g} (R_{\mu\nu} - \tfrac{1}{2} R g_{\mu\nu}) \delta g^{\mu\nu}.$$

Therefore, taking into account (2.1.29) it is possible to present the variation of (2.1.20) in the form

$$\delta S_g = -\int (R^{\mu\nu} - \tfrac{1}{2} R g^{\mu\nu}) \sqrt{-g}\, \delta g_{\mu\nu} \, \mathrm{d}\Omega. \tag{2.1.31}$$

Thus, the condition of stationarity of integral (2.1.20) yields the left-hand side of (2.1.11), enabling us to consider this integral as the action for the gravitational field. The right-hand side of (2.1.11), i.e. the mass tensor, also may be obtained as a variation of the appropriate action integral

$$S_m = \int (c^2 \rho^* + \rho^* \Pi) \sqrt{-g} \, \mathrm{d}\Omega \tag{2.1.32}$$

with ρ^* being an invariant density satisfying the equation of the continuity (2.1.18) and Π being the potential compressional energy (1.3.50). Using (2.1.25) the equation of continuity may be rewritten as

$$\frac{\partial}{\partial x^\alpha} (\sqrt{-g} \rho^* u^\alpha) = 0. \tag{2.1.33}$$

Defining the function $\rho = \rho(x^\alpha)$ by means of the relation

$$\sqrt{-g} \rho^* = \rho \frac{\mathrm{d}s}{\mathrm{d}x^0} \tag{2.1.34}$$

equation (2.1.33) reduces to the equation of continuity in the standard form (1.3.40). Hence, ρ is again the rest mass density and equation (2.1.34) generalizes the relation between ρ and ρ^* given by (1.3.35) and (1.3.36). Taking the variation of (2.1.34) with respect to $g_{\mu\nu}$ yields

$$\delta \rho^* = \tfrac{1}{2} \rho^* (u^\alpha u^\beta - g^{\alpha\beta}) \delta g_{\alpha\beta}. \tag{2.1.35}$$

From (1.3.49) one has

$$\delta S_m = \int \left[\left(c^2 + \Pi + \frac{p}{\rho^*} \right) \delta \rho^* + \frac{1}{2g} (c^2 \rho^* + \rho^* \Pi) \delta g \right] \sqrt{-g} \, \mathrm{d}\Omega$$

or using (2.1.23) and (2.1.35)

$$\delta S_m = \tfrac{1}{2} c^2 \int T^{\mu\nu} \sqrt{-g} \, \delta g_{\mu\nu} \, \mathrm{d}\Omega \tag{2.1.36}$$

with

$$c^2 T^{\mu\nu} = (c^2 \rho^* + \rho^* \Pi + p) u^\mu u^\nu - p g^{\mu\nu}. \tag{2.1.37}$$

In the absence of the gravitational field this expression coincides with (1.3.52). Combination of (2.1.31) and (2.1.36) enables one to conclude that the field equations (2.1.11) follow from the variational principle

$$\delta(2c^{-2}\kappa S_m - S_g) = 0 \tag{2.1.38}$$

where $g_{\mu\nu}$ are to be considered as arbitrary functions of the coordinates.

2.1.7 Gravitational energy and conservation laws

The gravitational radiation of celestial bodies plays a minor role and in most problems of relativistic celestial mechanics it may be ignored. Nevertheless, the phenomenon of gravitational radiation by itself is one of the principal features of relativistic celestial mechanics. According to the basic statements of GRT the gravitational energy does not enter into the energy–momentum tensor and is taken indirectly by means of the gravitational potentials. One may add the gravitational energy as the extra terms to the energy–momentum tensor only in an artificial manner by specifying a coordinate system. These extra terms are not tensorial quantities. At one and the same point, depending on the choice of coordinates, they may take any value. For an isolated system of bodies like the Solar System with large distances between the masses the effect of the gravitational field becomes negligible and in the limit there appears the homogeneous and isotropic Galilean space. Disregarding the gravitational radiation one has in this case 10 classic integrals of motion. If the condition of the Euclidean background at infinity is not fulfilled the system of bodies cannot be regarded as isolated and the integrals, at least some of them, do not exist. Just this case is realized, for example, in considering the Solar System at the cosmological background. But such a problem in spite of its theoretical interest has not so far been studied in detail.

Evidently, to speak rigorously, any actual physical system cannot be isolated due to the loss of energy in form of electromagnetic or gravitational radiation. But for the Solar System the loss of the electromagnetic energy is only a small amount of the total energy of the system and the loss of gravitational energy accounts for 10^{-23}–10^{-24} of the electromagnetic radiation

The GRT energy–momentum tensor satisfies relations (2.1.6) or

$$(\sqrt{-g}T^{\mu\nu})_{,\mu} + \sqrt{-g}\Gamma^{\nu}_{\alpha\beta}T^{\alpha\beta} = 0. \tag{2.1.39}$$

These relations do not lead to the conservation laws since their left-hand sides cannot be presented as divergencies of some symmetrical quantities.

But if one separates in the Einstein tensor the terms with the second derivatives then after some transformations this tensor may be represented in the form (Fock 1955)

$$-gG^{\mu\nu} = \tfrac{1}{2}(\tilde{g}^{\alpha\mu}\tilde{g}^{\beta\nu} - \tilde{g}^{\alpha\beta}\tilde{g}^{\mu\nu})_{,\alpha\beta} + L^{\mu\nu} \tag{2.1.40}$$

with

$$\tilde{g}^{\alpha\beta} = \sqrt{-g}g^{\alpha\beta}.$$

The quantities $L^{\mu\nu}$ are non-linear in the first derivatives and may be calculated by the formulae

$$\begin{aligned} L^{\mu\nu} =& \tfrac{1}{2}(\tilde{g}^{\alpha\beta}_{,\beta}\tilde{g}^{\mu\nu}_{,\alpha} - \tilde{g}^{\alpha\mu}_{,\beta}\tilde{g}^{\beta\nu}_{,\alpha} - \tilde{g}^{\alpha\mu}_{,\alpha}\tilde{g}^{\beta\nu}_{,\beta}) \\ &- g(\Pi^{\mu\alpha\beta}\Pi^{\nu}_{\alpha\beta} - \tfrac{1}{2}g^{\alpha\mu}g^{\beta\nu}\Gamma^{\lambda}_{\alpha\lambda}\Gamma^{\delta}_{\beta\delta} + \tfrac{1}{2}g^{\mu\nu}H) \end{aligned} \tag{2.1.41}$$

with

$$\begin{aligned} \Pi^{\mu\alpha\beta} &= \frac{1}{2g}(\tilde{g}^{\alpha\nu}\tilde{g}^{\mu\beta}_{,\nu} + \tilde{g}^{\beta\nu}\tilde{g}^{\mu\alpha}_{,\nu} - \tilde{g}^{\mu\nu}\tilde{g}^{\alpha\beta}_{,\nu}) \\ \Pi^{\nu}_{\alpha\beta} &= g_{\alpha\lambda}g_{\beta\mu}\Pi^{\nu\lambda\mu} \end{aligned}$$

and

$$H = g^{\mu\nu}(\Gamma^{\alpha}_{\beta\nu}\Gamma^{\beta}_{\alpha\mu} - \Gamma^{\alpha}_{\mu\nu}\Gamma^{\beta}_{\alpha\beta}). \tag{2.1.42}$$

Substituting (2.1.11) into (2.1.40) and denoting

$$U^{\mu\nu} = -c^{-2}gT^{\mu\nu} + \frac{1}{8\pi G}L^{\mu\nu} \tag{2.1.43}$$

one obtains

$$(\tilde{g}^{\alpha\beta}\tilde{g}^{\mu\nu} - \tilde{g}^{\alpha\mu}\tilde{g}^{\beta\nu})_{,\alpha\beta} = 16\pi GU^{\mu\nu}. \tag{2.1.44}$$

It is evident that

$$U^{\mu\nu}_{,\nu} = 0. \tag{2.1.45}$$

This relation, actually equivalent to (2.1.39), is presented as the divergence of a symmetrical, non-tensorial quantity. It is easily verified that $L^{\mu\nu}$ are expressed in terms of the Christoffel symbols. Therefore, under linear transformations these quantities and hence $U^{\mu\nu}$ act as tensors. In a specific coordinate system the first term in $U^{\mu\nu}$ may be interpreted as the tensor due to the energy of the matter and all fields except for the gravitational field whereas the second term represents the pseudo-tensor of the gravitational field (the Landau–Lifshitz pseudo-tensor). Describing (2.1.45) separately for $\nu = 0$ and $\nu = i$ one obtains after integration over the domain occupied by the masses

$$\frac{\mathrm{d}}{\mathrm{d}t}\int U^{00}\,\mathrm{d}^3x = -c\int n_kU^{0k}\,\mathrm{d}S \tag{2.1.46}$$

$$\frac{\mathrm{d}}{\mathrm{d}t}\int U^{0i}\,\mathrm{d}^3x = -c\int n_kU^{ik}\,\mathrm{d}S \tag{2.1.47}$$

with $\mathrm{d}S$ being the element of the space surface and n_k being components of the unit normal vector to the surface. Besides, from (2.1.45)

$$(x^\mu U^{\alpha\nu} - x^\nu U^{\alpha\mu})_{,\alpha} = 0. \tag{2.1.48}$$

Putting here $\mu = i$ and integrating one finds for $\nu = 0$ and $\nu = j$

$$\frac{\mathrm{d}}{\mathrm{d}t}\int (x^i U^{00} - ctU^{0i})\,\mathrm{d}^3x = -c\int n_k(x^i U^{0k} - ctU^{ik})\,\mathrm{d}S \tag{2.1.49}$$

$$\frac{\mathrm{d}}{\mathrm{d}t}\int (x^i U^{0j} - x^j U^{0i})\,\mathrm{d}^3x = -c\int n_k(x^i U^{jk} - x^j U^{ik})\mathrm{d}S. \tag{2.1.50}$$

Relations (2.1.46), (2.1.47), (2.1.49) and (2.1.50) generalize the conservation laws of classical mechanics and demonstrate that the change of the total amount of the quantity at hand inside some volume is due only to the flow of this quantity through the boundary surface. If this flow may be ignored there result the common sense conservation laws implying the constancy of the quantities at hand.

In accordance with the left-hand sides of relations (2.1.46), (2.1.47), (2.1.49) and (2.1.50) the mass $\tilde{M}$ of the material system, its momentum $\tilde{P}^i$, the coordinates of its centre of mass $\tilde{X}^i$and its angular momentum $\tilde{M}^{ij}$ can be determined by means of

$$\tilde{M} = c^2\int U^{00}\,\mathrm{d}^3x \tag{2.1.51}$$

$$\tilde{P}^i = c^2\int U^{0i}\,\mathrm{d}^3x \tag{2.1.52}$$

$$\tilde{M}\tilde{X}^i = c^2\int x^i U^{00}\,\mathrm{d}^3x \tag{2.1.53}$$

$$\tilde{M}^{ij} = c^2\int (x^i U^{0j} - x^j U^{0i})\,\mathrm{d}^3x. \tag{2.1.54}$$

The tilde in these designations indicates that the relevant quantities include contributions due to the gravitational field.

2.2 WEAK GRAVITATIONAL FIELD

2.2.1 Linearized theory

The simplest approximate solution of the field equations is obtained in the linearized theory. This theory physically corresponds to the case when corrections $h_{\alpha\beta}$ to the Galilean values $\eta_{\alpha\beta}$ are so small that their squares and products may be neglected. Separating linear and non-linear parts in the Ricci tensor one has from (1.2.67) and (1.2.71)

$$R_{\mu\nu} = \tfrac{1}{2}\eta^{\alpha\beta}(h_{\alpha\beta,\mu\nu} + h_{\mu\nu,\alpha\beta} - h_{\alpha\mu,\beta\nu} - h_{\beta\nu,\alpha\mu}) + L'_{\mu\nu} \tag{2.2.1}$$

$$L'_{\mu\nu} = \tfrac{1}{2}h^{\alpha\beta}(h_{\alpha\beta,\mu\nu} + h_{\mu\nu,\alpha\beta} - h_{\alpha\mu,\beta\nu} - h_{\beta\nu,\alpha\mu}) + g^{\alpha\beta}g^{\lambda\delta}(\Gamma_{\lambda\mu\nu}\Gamma_{\delta\alpha\beta} - \Gamma_{\lambda\alpha\mu}\Gamma_{\delta\beta\nu}). \tag{2.2.2}$$

Based on (2.1.14) the field equations may be rewritten in the form

$$R_{\alpha\beta} = -\kappa T^*_{\alpha\beta} \tag{2.2.3}$$

with

$$T^*_{\alpha\beta} = T_{\alpha\beta} - \tfrac{1}{2}Tg_{\alpha\beta}. \tag{2.2.4}$$

Substituting the Galilean values of the metric tensor into the energy–momentum tensor (2.1.37) , ignoring the terms dependent on the internal structure ($p = \Pi = 0$) and retaining only the first-order terms in v^i/c one has

$$T^{00} = \rho \qquad T^{0i} = c^{-1}\rho v^i \qquad T^{ij} = 0 \qquad T = \rho$$

$$T_{00} = \rho \qquad T_{0i} = -c^{-1}\rho v^i \qquad T_{ij} = 0$$

so that right-hand members of (2.2.3) become

$$T^*_{00} = \tfrac{1}{2}\rho \qquad T^*_{0i} = -c^{-1}\rho v^i \qquad T^*_{ij} = \tfrac{1}{2}\rho\delta_{ij}. \tag{2.2.5}$$

Introducing the functions

$$L_{\mu\nu} = L'_{\mu\nu} + \kappa T^*_{\mu\nu} \tag{2.2.6}$$

equations (2.2.3) may be rewritten as follows:

$$h_{00,ss} - 2h_{0s,0s} + h_{ss,00} = 2L_{00} \tag{2.2.7}$$

$$h_{0m,ss} - h_{0s,ms} + h_{ss,0m} - h_{ms,0s} = 2L_{0m} \tag{2.2.8}$$

$$h_{mn,ss} - h_{mn,00} + h_{ss,mn} - h_{ms,ns} - h_{ns,ms} - h_{00,mn} + h_{0m,0n} + h_{0n,0m} = 2L_{mn}. \tag{2.2.9}$$

As earlier, each index occurring twice implies summation, for example, $h_{00,ss}$ stands for the Laplacian of h_{00}. In the linearized theory $L'_{\mu\nu}$ are ignored and the right-hand members of (2.2.7)–(2.2.9) become the known functions determined by (2.2.5) and (2.2.6). Equations (2.2.7)–(2.2.9) themselves represent in this case second-order linear partial differential equations. These equations are significantly simplified under harmonic conditions (2.1.16). Defining $h^{\mu\nu}$ by

$$g^{\mu\nu} = \eta^{\mu\nu} + h^{\mu\nu} \tag{2.2.10}$$

and using the relation between contravariant and covariant components of the metric tensor one finds

$$\begin{aligned} h^{00} &= -h_{00} + (h_{00})^2 + \dots \\ h^{0m} &= h_{0m} - h_{00}h_{0m} + h_{0s}h_{ms} + \dots \\ h^{mn} &= -h_{mn} - h_{ms}h_{ns} + \dots . \end{aligned} \tag{2.2.11}$$

It will be clear below that in arbitrary non-rotating quasi-Galilean coordinates the components h_{00} and h_{mn} are of second order in v/c and h_{0m} are of third order (v being the characteristic velocity of bodies). Formulae (2.2.11) yield the relation between $h_{\mu\nu}$ and $h^{\mu\nu}$ up to the terms of fifth order inclusively. With the same accuracy the determinant g is

$$-g = 1 + h_{00} - h_{ss} - h_{00}h_{ss} + \tfrac{1}{2}(h_{ss})^2 - \tfrac{1}{2}h_{rs}h_{rs} + \dots \tag{2.2.12}$$

and hence

$$\begin{aligned} \sqrt{-g} = 1 &+ \tfrac{1}{2}(h_{00} - h_{ss}) - \tfrac{1}{8}(h_{00})^2 - \tfrac{1}{4}h_{00}h_{ss} \\ &+ \tfrac{1}{8}(h_{ss})^2 - \tfrac{1}{4}h_{rs}h_{rs} + \dots . \end{aligned} \tag{2.2.13}$$

Therefore,

$$\begin{aligned} \sqrt{-g}g^{00} = 1 &- \tfrac{1}{2}(h_{00} + h_{ss}) + \tfrac{3}{8}(h_{00})^2 + \tfrac{1}{4}h_{00}h_{ss} \\ &+ \tfrac{1}{8}(h_{ss})^2 - \tfrac{1}{4}h_{rs}h_{rs} + \dots . \end{aligned} \tag{2.2.14}$$

$$\sqrt{-g}g^{0m} = h_{0m} - \tfrac{1}{2}(h_{00} + h_{ss})h_{0m} + h_{0s}h_{ms} + \dots . \tag{2.2.15}$$

$$\begin{aligned} \sqrt{-g}g^{mn} = &-\delta_{mn} - h_{mn} - \tfrac{1}{2}(h_{00} - h_{ss})\delta_{mn} \\ &+ [\tfrac{1}{8}(h_{00})^2 + \tfrac{1}{4}h_{00}h_{ss} - \tfrac{1}{8}(h_{ss})^2 + \tfrac{1}{4}h_{rs}h_{rs}]\delta_{mn} \\ &- \tfrac{1}{2}(h_{00} - h_{ss})h_{mn} - h_{ms}h_{ns} + \dots . \end{aligned} \tag{2.2.16}$$

Differentiation of these expressions enables us to deduce explicitly the harmonic conditions up to the terms of fifth order inclusively. For the linearized theory the harmonic conditions are

$$h_{00,0} + h_{ss,0} - 2h_{0s,s} = 0 \tag{2.2.17}$$

$$h_{00,m} - h_{ss,m} + 2h_{ms,s} - 2h_{0m,0} = 0. \tag{2.2.18}$$

Under these conditions equations (2.2.7)–(2.2.9) take the form

$$h_{\mu\nu,ss} - h_{\mu\nu,00} = 2L_{\mu\nu}. \tag{2.2.19}$$

Thus, the field equations for the linearized theory in the harmonic coordinates become the wave equations. As is known the solution of the wave equation

$$\psi_{,ss} - \psi_{,00} = -4\pi\sigma(t, \boldsymbol{r}) \tag{2.2.20}$$

is expressed by the volume integral

$$\psi = \int \frac{\sigma(t', \boldsymbol{r}')}{|\boldsymbol{r} - \boldsymbol{r}'|} \mathrm{d}^3x' \tag{2.2.21}$$

with d^3x' being an elementary 3-volume and the density being evaluated at the current point $\boldsymbol{r}'$ at the retarded moment

$$t' = t - c^{-1}|\boldsymbol{r} - \boldsymbol{r}'|. \tag{2.2.22}$$

If the distance $|\boldsymbol{r} - \boldsymbol{r}'|$ between the external point $\boldsymbol{r}$ and the current point $\boldsymbol{r}'$ is small as compared with distance ct then the exact solution (2.2.21) may be expanded in series

$$\begin{aligned} \psi = & \int \frac{\sigma(t, \boldsymbol{r}')}{|\boldsymbol{r} - \boldsymbol{r}'|} \mathrm{d}^3x - c^{-1}\frac{\mathrm{d}}{\mathrm{d}t}\int \sigma(t, \boldsymbol{r}')\,\mathrm{d}^3x' \\ & + \frac{1}{2}c^{-2}\frac{\mathrm{d}^2}{\mathrm{d}t^2}\int \sigma(t, \boldsymbol{r}')|\boldsymbol{r} - \boldsymbol{r}'|\; d^3x' + \ldots . \end{aligned} \tag{2.2.23}$$

In the linearized theory it is sufficient to take only the first term. Then the solution of (2.2.19) takes the form

$$h_{\mu\nu} = -\frac{\kappa}{2\pi}\int \frac{(T^*_{\mu\nu})'}{|\boldsymbol{r} - \boldsymbol{r}'|} \mathrm{d}^3x' \tag{2.2.24}$$

with primes indicating that the appropriate value is to be taken for the moment t at the current point $\boldsymbol{r}'$. In the next section the comparison with the Newtonian equations of motion of a test particle results in a determining constant κ

$$\kappa = 8\pi G c^{-2}. \tag{2.2.25}$$

With the aid of (2.2.5) the solution (2.2.24) of the linearized field equations in harmonic coordinates is presented in the form

$$h_{00} = -c^{-2}2U \qquad h_{0i} = c^{-3}4U^i \qquad h_{ij} = -c^{-2}2U\delta_{ij} \tag{2.2.26}$$

with Newtonian potential U and vector potential U^i

$$U = G\int \frac{\rho'\mathrm{d}^3x'}{|\boldsymbol{r} - \boldsymbol{r}'|} \qquad U^i = G\int \frac{\rho' v'^i}{|\boldsymbol{r} - \boldsymbol{r}'|}\mathrm{d}^3x. \tag{2.2.27}$$

It may be directly verified that, within the adopted accuracy, relation (2.2.18) is fulfilled identically and relation (2.2.17) is satisfied in virtue of the equation of continuity (1.3.40).

In all celestial mechanics applications of GRT the velocities of bodies are much smaller than the light velocity. Also, the square of the characteristic velocity v^2 and Newtonian potential U are of the same order. Therefore, the small parameters in problems of relativistic celestial mechanics are v^2/c^2 and U/c^2. From (2.2.26) it follows that h_{00} and h_{ij} are of second order and h_{0i} are of third order of smallness. This is true in harmonic coordinates and in all practically employed quasi-Galilean coordinates. In fact, returning to the linearized equations (2.2.7)–(2.2.9) in arbitrary quasi-Galilean coordinates and denoting by $h^*_{00}, h^*_{0i}, h^*_{ij}$ the solution (2.2.26) for the harmonic coordinates one obtains the solution of (2.2.7)–(2.2.9) in arbitrary coordinates in the form

$$h_{00} = h^*_{00} \qquad h_{0i} = h^*_{0i} + a_{0,i} + a_{i,0}$$

$$h_{ij} = h^*_{ij} + a_{i,j} + a_{j,i} \tag{2.2.28}$$

a_μ being four arbitrary functions of x^o, x^1, x^2, x^3. In dealing with (2.2.28) a_0 is to be regarded as a third-order function and a_i as second-order functions. Substitution of (2.2.28) into (2.2.7)–(2.2.9), taking into account that $h^*_{\mu\nu}$ satisfy equations (2.2.19) and relations (2.2.17) and (2.2.18), demonstrates the validity of (2.2.28). In doing this it is to be considered that differentiating with respect to x^0 increases the order of smallness of the corresponding quantity. Therefore, all terms on the left-hand sides (2.2.7)–(2.2.9) with first- and second-order derivatives with respect to x^0 should be rejected as having higher orders than required.

Expansions (2.2.28) may be obtained from the coordinate transformation from harmonic coordinates $x^{*\mu}$ to arbitrary quasi-Galilean coordinates x^μ. This transformation is of the form

$$x^{*0} = x^0 + a_0 \tag{2.2.29}$$

$$x^{*i} = x^i - a_i. \tag{2.2.30}$$

In fact, the usual tensor formula

$$g_{\mu\nu}(x^\delta) = \frac{\partial x^{*\alpha}}{\partial x^\mu}\frac{\partial x^{*\beta}}{\partial x^\nu} g^*_{\alpha\beta}(x^{*\delta}) \tag{2.2.31}$$

yields (2.2.28) within the adopted accuracy. Finally, the linearized metric of the weak gravitational field in quasi-Galilean coordinates is given by

$$\begin{aligned} ds^2 = {} & (1 - c^{-2}2U)c^2\,dt^2 + 2(c^{-3}4U^i + a_{0,i} + a_{i,0})c\,dt dx^i \\ & + [-(1 + c^{-2}2U)\delta_{ij} + a_{i,j} + a_{j,i}]\,dx^i dx^j. \end{aligned} \tag{2.2.32}$$

Remember once again that arbitrary functions a_i, a_0 due to specific coordinate conditions are small functions of the second and third order respectively. With $h_{\mu\nu} = 0$ the corresponding metric becomes Galilean (or Minkowski) determining the space-time of special relativity. The Galilean metric yields the Newtonian equations of light propagation (motion in a straight line) but cannot give the equations of motion of material bodies. Retaining in (2.2.32) only the term $h_{00} = -c^{-2}2U$ and rejecting all h_{0i}, h_{ij} the metric then obtained gives the Newtonian equations of motion of bodies and an incorrect approximation for the light propagation equations. The correct post-Newtonian equations of light propagation are obtained with the full metric (2.2.32). However, this metric is insufficient to derive the post-Newtonian equations of motion of material bodies. To do this it is necessary to take into account in h_{00} the fourth-order term.

2.2.2 Post-Newtonian metric

As shown, for example, by Fock (1955) the fourth-order correction term in the expansion $h_{00} = c^{-2}h_{00}^{(2)} + c^{-4}h_{00}^{(4)} + \dots$ in harmonic coordinates takes the form

$$h_{00}^{(4)} = 2\left(U^2 - \frac{\partial^2 \chi}{\partial t^2} - \tilde{U}\right). \tag{2.2.33}$$

The auxiliary functions χ and $\tilde{U}$ satisfy the Poisson equations

$$\chi_{,ss} = U \tag{2.2.34}$$

$$\tilde{U}_{,ss} = -4\pi G(\tfrac{3}{2}\rho v^2 - \rho U + \rho\Pi + 3p). \tag{2.2.35}$$

In converting by means of (2.2.29) and (2.2.30) from harmonic coordinates to arbitrary quasi-Galilean coordinates it is necessary also to transform in (2.2.31) $g^*_{\alpha\beta}(x^{*\delta})$ to $g^*_{\alpha\beta}(x^\delta)$. For the general case of the gravitational field created by N bodies with coordinates x^s_A and velocity components v^s_A $(A = 1, 2, \dots, N)$ this transformation implies

$$\begin{aligned} g^*_{\alpha\beta}(x^{*\delta}, x^{*s}_A, v^{*s}_A) &= g^*_{\alpha\beta}(x^\delta, x^s_A, v^s_A) + g^*_{\alpha\beta,\delta}(x^{*\delta} - x^\delta) \\ &+ \sum_A \left(\frac{\partial g^*_{\alpha\beta}}{\partial x^s_A}(x^{*s}_A - x^s_A) + \frac{\partial g^*_{\alpha\beta}}{\partial v^s_A}(v^{*s}_A - v^s_A)\right) + \dots . \end{aligned} \tag{2.2.36}$$

Considering that the Newtonian potential U depends both on the coordinates x^α of the current point of the field and on the coordinates x^s_A of the gravitating bodies and denoting (2.2.33) by $h_{00}^{*(4)}$ one finds the correction term $h_{00}^{(4)}$ in arbitrary coordinates:

$$c^{-4}h_{00}^{(4)} = c^{-4}h_{00}^{*(4)} + 2a_{0,0} + c^{-2}2U_{,s}a_s + c^{-2}2\sum_A \frac{\partial U}{\partial x^s_A}(\tilde{a}_s)_A. \tag{2.2.37}$$

$(\tilde{a}_s)_A$ denotes the regular part of the function a_s in substituting $x^s = x^s_A$. This is equivalent to using the generalized δ-function of Infeld and Plebanski (1960). Adding (2.2.37) to the component g_{00} of the linearized metric (2.2.32) one obtains the post-Newtonian metric determining the motion of the bodies for arbitrary quasi-Galilean coordinates of GRT in the post-Newtonian approximation.

It is of interest to compare the solution (2.2.32) and (2.2.37) with the post-Newtonian solution of Will (1985). Following Will let $\Phi_1, \Phi_2, \Phi_3, \Phi_4$ be functions determined by the equations

$$\Phi_{1,ss} = -4\pi G\rho v^2 \qquad \Phi_{2,ss} = -4\pi G\rho U$$

$$\Phi_{3,ss} = -4\pi G\rho\Pi \qquad \Phi_{4,ss} = -4\pi Gp.$$

Then the post Newtonian metric (2.2.32) and (2.2.37) will be

$$\begin{aligned} g_{00} =& 1 - c^{-2}2U + c^{-4}\left(-3\Phi_1 + 2\Phi_2 - 2\Phi_3 - 6\Phi_4 - 2\frac{\partial^2\chi}{\partial t^2} + 2U^2\right) \\ &+ 2a_{0,0} + c^{-2}2U_{,s}a_s + c^{-2}2\sum_A \frac{\partial U}{\partial x^s_A}(\tilde{a}_s)_A \end{aligned} \tag{2.2.38}$$

$$g_{0i} = c^{-3}4U^i + a_{0,i} + a_{i,0} \tag{2.2.39}$$

$$g_{ij} = -(1 + c^{-2}2U)\delta_{ij} + a_{i,j} + a_{j,i} \tag{2.2.40}$$

with

$$a_0 = c^{-3}\frac{\partial\chi}{\partial t} \qquad a_i = 0 \tag{2.2.41}$$

for the metric by Will. In addition, the definitions given by Will involve the density which is denoted here by ρ^*. Using (2.1.34) with (2.2.41) one has

$$\rho = \rho^*[1 + c^{-2}(\tfrac{1}{2}v^2 + 3U) + \ldots]. \tag{2.2.42}$$

From this it follows that

$$U = U^* + c^{-2}(\tfrac{1}{2}\Phi_1 + 3\Phi_2) + \ldots. \tag{2.2.43}$$

Here U and U^* stand for the Newtonian potential defined with ρ and ρ^* respectively. Substitution of (2.2.43) into (2.2.38) results in the expressions of Will (taking into account the change in signature of the metric).

2.2.3 Post-Newtonian equations of motion of a test particle

The motion of a test particle in the given gravitational field (2.1.1)–(2.1.3) is described by the geodesic motion (2.1.19). The Christoffel symbols are determined by the expressions

$$
\begin{aligned}
\Gamma^0_{00} &= \tfrac{1}{2}h_{00,0} + \dots \qquad \Gamma^0_{0k} = \tfrac{1}{2}h_{00,k} + \dots \\
\Gamma^0_{km} &= \tfrac{1}{2}(h_{0k,m} + h_{0m,k} - h_{km,0}) + \dots \\
\Gamma^i_{00} &= \tfrac{1}{2}h_{00,i} + \tfrac{1}{2}h_{ij}h_{00,j} - h_{0i,0} + \dots \\
\Gamma^i_{0k} &= \tfrac{1}{2}(h_{0k,i} - h_{0i,k} - h_{ik,0}) + \dots \\
\Gamma^i_{km} &= \tfrac{1}{2}(h_{km,i} - h_{ik,m} - h_{im,k}) + \dots .
\end{aligned} \tag{2.2.44}
$$

These functions are given here in the linear approximation except for Γ^i_{00} retaining one non-linear term of the same order (four) as the following linear term. The quantity Γ^i_{00} enters into the equations of motion with the factor c^2 requiring more accurate determination than the other components.

Let us start with equations (2.1.19) with s as the independent argument. These equations are related by (2.1.17) which may be rewritten as a quadratic equation relative to $\mathrm{d}x^0/\mathrm{d}s$

$$
(1+h_{00})\left(\frac{\mathrm{d}x^0}{\mathrm{d}s}\right)^2 + 2h_{0k}\frac{\mathrm{d}x^k}{\mathrm{d}s}\frac{\mathrm{d}x^0}{\mathrm{d}s} + (-\delta_{km}+h_{km})\frac{\mathrm{d}x^k}{\mathrm{d}s}\frac{\mathrm{d}x^m}{\mathrm{d}s} - 1 = 0. \tag{2.2.45}
$$

The positive root of this equation in the post-Newtonian approximation is

$$
\begin{aligned}
\frac{\mathrm{d}x^0}{\mathrm{d}s} = {} & 1 - \frac{1}{2}h_{00} + \frac{1}{2}\frac{\mathrm{d}x^k}{\mathrm{d}s}\frac{\mathrm{d}x^k}{\mathrm{d}s} + \frac{3}{8}(h_{00})^2 - \frac{1}{4}h_{00}\frac{\mathrm{d}x^k}{\mathrm{d}s}\frac{\mathrm{d}x^k}{\mathrm{d}s} \\
& - \frac{1}{2}h_{km}\frac{\mathrm{d}x^k}{\mathrm{d}s}\frac{\mathrm{d}x^m}{\mathrm{d}s} - \frac{1}{8}\left(\frac{\mathrm{d}x^k}{\mathrm{d}s}\frac{\mathrm{d}x^k}{\mathrm{d}s}\right)^2 - h_{0k}\frac{\mathrm{d}x^k}{\mathrm{d}s} + \dots .
\end{aligned} \tag{2.2.46}
$$

Considering now the geodesic equations for the space components and excluding $\mathrm{d}x^0/\mathrm{d}s$ with the aid of (2.2.46) one obtains

$$
\begin{aligned}
& \frac{\mathrm{d}^2x^i}{\mathrm{d}s^2} + \frac{1}{2}h_{00,i} - \frac{1}{2}h_{00}h_{00,i} + \frac{1}{2}h_{ik}h_{00,k} - h_{0i,0} + (h_{0k,i} - h_{0i,k} - h_{ik,0})\frac{\mathrm{d}x^k}{\mathrm{d}s} \\
& + \frac{1}{2}h_{00,i}\frac{\mathrm{d}x^k}{\mathrm{d}s}\frac{\mathrm{d}x^k}{\mathrm{d}s} + \left(\frac{1}{2}h_{km,i} - h_{ik,m}\right)\frac{\mathrm{d}x^k}{\mathrm{d}s}\frac{\mathrm{d}x^m}{\mathrm{d}s} = 0.
\end{aligned} \tag{2.2.47}
$$

Let us derive the equations of motion of a test particle referred to coordinate time t. Using the following relations:

$$
\frac{\mathrm{d}x^i}{\mathrm{d}x^0} = \frac{\mathrm{d}x^i}{\mathrm{d}s}\frac{\mathrm{d}s}{\mathrm{d}x_0} \qquad \frac{\mathrm{d}^2x^i}{\mathrm{d}x^{0^2}} = \frac{\mathrm{d}^2x^i}{\mathrm{d}s^2}\left(\frac{\mathrm{d}s}{\mathrm{d}x^0}\right)^2 + \frac{\mathrm{d}x^i}{\mathrm{d}s}\frac{\mathrm{d}^2s}{\mathrm{d}x^{0^2}}
$$

$$
\frac{\mathrm{d}s}{\mathrm{d}x^0} = \left(\frac{\mathrm{d}x^0}{\mathrm{d}s}\right)^{-1} \qquad \frac{\mathrm{d}^2s}{\mathrm{d}x^{0^2}} = -\frac{\mathrm{d}^2x^0}{\mathrm{d}s^2}\left(\frac{\mathrm{d}x^0}{\mathrm{d}s}\right)^{-3}
$$

the geodesic equations referred to x^0 take the form

$$\frac{d^2x^i}{dx^{0^2}} = -\Gamma^i_{\alpha\beta}\frac{dx^\alpha}{dx^0}\frac{dx^\beta}{dx^0} + \Gamma^0_{\alpha\beta}\frac{dx^\alpha}{dx^0}\frac{dx^\beta}{dx^0}\frac{dx^i}{dx^0}. \tag{2.2.48}$$

Using (2.2.44) one obtains

$$\begin{aligned}\ddot{x}^i =& -\tfrac{1}{2}c^2 h_{00,i} - \tfrac{1}{2}c^2 h_{ik}h_{00,k} + c^2 h_{0i,0} + \tfrac{1}{2}c h_{00,0}\dot{x}^i \\ &+ c h_{ik,0}\dot{x}^k + c(h_{0i,k} - h_{0k,i})\dot{x}^k + h_{00,k}\dot{x}^k\dot{x}^i \\ &+ (h_{ik,m} - \tfrac{1}{2}h_{km,i})\dot{x}^k\dot{x}^m.\end{aligned} \tag{2.2.49}$$

Here a dot denotes differentiation with respect to t. Remember again that the zero index after a comma means partial differentiation with respect to $x^0 = ct$. The terms linear in $h_{\mu\nu}$ and caused by the term $c^{-1}\Gamma^0_{km}\dot{x}^i\dot{x}^k\dot{x}^m$ are omitted because they are beyond the order of the post-Newtonian approximation. Equations (2.2.49) may also be derived from (2.2.41) by transforming to argument t.

The leading term on the right-hand side of (2.2.49) is the first one. This term is to be compared with the Newtonian equations

$$\ddot{x}^i = U_{,i}. \tag{2.2.50}$$

Equations (2.2.49) reduce to equations (2.2.50) in the Newtonian approximation provided that the constant κ is determined by (2.2.25). h_{00} in the first term of the right-hand side of (2.2.49) should be given up to the fourth-order terms inclusive. For all other terms of (2.2.49) one may use the values from (2.2.32). With the aid of the geodesic variational principle in the form (1.2.50) equations (2.2.49) may be put into the Lagrange form

$$\frac{d}{dt}\frac{\partial L}{\partial \dot{x}^i} - \frac{\partial L}{\partial x^i} = 0. \tag{2.2.51}$$

Considering that

$$\left(\frac{ds}{dx^0}\right)^2 = 1 + h_{00} + c^{-1}2h_{0k}\dot{x}^k - c^{-2}\dot{x}^k\dot{x}^k + c^{-2}h_{km}\dot{x}^k\dot{x}^m + \dots \tag{2.2.52}$$

and raising this expression to power 1/2, omitting the constant term and multiplying by c^{-2} one obtains

$$\begin{aligned}L =& \tfrac{1}{2}\dot{x}^k\dot{x}^k - \tfrac{1}{2}c^2 h_{00} + \tfrac{1}{8}c^{-2}(\dot{x}^k\dot{x}^k)^2 - \tfrac{1}{4}h_{00}\dot{x}^k\dot{x}^k \\ &+ \tfrac{1}{8}c^2(h_{00})^2 - \tfrac{1}{2}h_{km}\dot{x}^k\dot{x}^m - c h_{0k}\dot{x}^k + \dots.\end{aligned} \tag{2.2.53}$$

This Lagrangian is correct up to second-order terms inclusive and is completely equivalent to (2.2.49). Evidently, h_{00} in the second term of (2.2.53) should be calculated within fourth-order terms inclusive. In deriving equations (2.2.49) from (2.2.53) it should be kept in mind that in the second-order terms the second derivatives $\ddot{x}^i$ are to be excluded with the aid of the Newtonian equations of motion. This results in equations (2.2.49) with explicit form relative to $\ddot{x}^i$. Let us add that relation (2.2.46) expressed in terms of coordinate time t takes the form

$$\frac{\mathrm{d}x^0}{\mathrm{d}s} = 1 - \tfrac{1}{2}h_{00} + \tfrac{1}{2}c^{-2}\dot{x}^k\dot{x}^k + \tfrac{3}{8}(h_{00})^2 - \tfrac{3}{4}c^{-2}h_{00}\dot{x}^k\dot{x}^k - \tfrac{1}{2}c^{-2}h_{km}\dot{x}^k\dot{x}^m + \tfrac{3}{8}c^{-4}(\dot{x}^k\dot{x}^k)^2 - c^{-1}h_{0k}\dot{x}^k + \ldots. \quad (2.2.54)$$

Combination of the second and the last terms in the Lagrangian (2.2.53) enables one to conclude that an arbitrary function a_0 enters into the Lagrangian only as the total derivative in time. Therefore, a_0 has no influence on the post-Newtonian equations of motion of a test particle.

The simplicity of equations (2.2.49) makes them very useful in solving practical problems. Solution of (2.2.49) determines the motion of a test particle in some specific coordinate system given by components $g_{\alpha\beta}$. In the general case, for comparison with observations it is necessary then to describe the technique of observations in the same coordinate system. This aim is achieved with the use of the equations of light propagation considered below. Only afterwards is it possible to exclude the physically meaningless coordinates x^i and to deal with the directly measurable quantities in the proper time of an observer.

2.2.4 Post-Newtonian equations of light propagation

Let us derive now the equations of light propagation in the post-Newtonian approximation. The propagation of light referred to the coordinate time t is described by the equations of isotropic geodesics (1.2.47) and (1.2.48), i.e.

$$\ddot{x}^i + \Gamma^i_{\alpha\beta}\dot{x}^\alpha\dot{x}^\beta = -\frac{\mathrm{d}^2t}{\mathrm{d}\lambda^2}\left(\frac{\mathrm{d}t}{\mathrm{d}\lambda}\right)^{-2}\dot{x}^i \quad (2.2.55)$$

$$g_{\alpha\beta}\dot{x}^\alpha\dot{x}^\beta = 0 \quad (2.2.56)$$

with λ being the canonical parameter. In more detail, relation (2.2.56) yields

$$c^2(1+h_{00}) + 2ch_{0k}\dot{x}^k + (-\delta_{km} + h_{km})\dot{x}^k\dot{x}^m = 0. \quad (2.2.57)$$

Differentiating with respect to time and considering that $\dot{x}^i = O(c)$, $\ddot{x}^i = O(1)$ one obtains

$$c^3h_{00,0} + c^2h_{00,k}\dot{x}^k + c(2h_{0k,m} + h_{km,0})\dot{x}^k\dot{x}^m + h_{ik,m}\dot{x}^i\dot{x}^k\dot{x}^m = 2\dot{x}^k\ddot{x}^k \quad (2.2.58)$$

retaining only terms of order $O(c)$ and $O(1)$. Substituting into (2.2.55) the Christoffel symbols one finds within the same accuracy

$$\ddot{x}^i + \tfrac{1}{2}c^2 h_{00,i} + c(h_{0k,i} - h_{0i,k} - h_{ik,0})\dot{x}^k + (\tfrac{1}{2}h_{km,i} - h_{ik,m})\dot{x}^k\dot{x}^m = -\frac{d^2t}{d\lambda^2}\left(\frac{dt}{d\lambda}\right)^{-2}\dot{x}^i. \tag{2.2.59}$$

As seen from (2.2.55) and (2.2.57), $\dot{x}^k\dot{x}^k = c^2$ and $d^2t/d\lambda^2 = 0$ in the Newtonian approximation. Multiplying equations (2.2.59) by $\dot{x}^i$, adding them up and excluding $\dot{x}^i\ddot{x}^i$ with the aid of (2.2.58) one has

$$-\frac{d^2t}{d\lambda^2}\left(\frac{dt}{d\lambda}\right)^{-2} = \tfrac{1}{2}ch_{00,0} + h_{00,k}\dot{x}^k + c^{-1}(h_{0k,m} - \tfrac{1}{2}h_{km,0})\dot{x}^k\dot{x}^m. \tag{2.2.60}$$

Substituting this expression into (2.2.59) one obtains the post-Newtonian equations of light propagation

$$\begin{aligned}\ddot{x}^i = &-\tfrac{1}{2}c^2 h_{00,i} + h_{00,k}\dot{x}^k\dot{x}^i + (h_{ik,m} - \tfrac{1}{2}h_{km,i})\dot{x}^k\dot{x}^m + \tfrac{1}{2}ch_{00,0}\dot{x}^i \\ &+ c(h_{0i,k} - h_{0k,i} + h_{ik,0})\dot{x}^k + c^{-1}(h_{0k,m} - \tfrac{1}{2}h_{km,0})\dot{x}^k\dot{x}^m\dot{x}^i.\end{aligned} \tag{2.2.61}$$

In spite of apparent differences the equations of light propagation (2.2.61) and the equations of motion of a test particle (2.2.49) turn out to be identical. The difference appears only in the order of smallness of the separate terms on the right-hand sides of these equations. This is due to the fact that velocities $\dot{x}^i$ in equations (2.2.49) are small compared with the light velocity c but they have order $O(c)$ in equations (2.2.61). Hence,there are terms negligible in equations (2.2.49) and non-negligible in (2.2.61) and vice versa. Besides, all $h_{\mu\nu}$ occurring in (2.2.61), including h_{00}, may be taken from (2.2.32). The first four terms on the right-hand side of (2.2.61) are of order $O(1)$ whereas the other six terms are of order $O(c^{-1})$.

In the case of the constant field with $h_{\mu\nu}$ independent of x^0 equations (2.2.61) may be put into the Lagrange form (2.2.51) with Lagrangian

$$L = \tfrac{1}{2}\dot{x}^k\dot{x}^k - \tfrac{1}{2}h_{00}\dot{x}^k\dot{x}^k - \tfrac{1}{2}h_{km}\dot{x}^k\dot{x}^m - \tfrac{1}{2}ch_{0k}\dot{x}^k - \tfrac{1}{2}c^{-1}h_{0k}\dot{x}^k\dot{x}^m\dot{x}^m. \tag{2.2.62}$$

Performing differentiation of L it should be remembered that the terms with $\ddot{x}^k$ on the right-hand side are at least of second order of smallness and have to be rejected whereas the quantity $\dot{x}^k\dot{x}^k$ may be replaced by the Newtonian value c^2. The equations obtained in this manner coincide with equations (2.2.61) (under the condition $h_{\mu\nu,0} = 0$).

2.3 PROBLEM OF MEASURABLE QUANTITIES IN GRT

2.3.1 Measurement of infinitesimal time intervals and distances

As stated above the curvilinearity of the metric of the GRT space-time manifests itself differently in different coordinate systems. The problem is how to determine coordinate-independent distances and time intervals from the coordinates x^0, x^1, x^2, x^3 of some given system. For infinitesimal time intervals and distances the problem is comparatively simple. A set of relations useful in practice is considered below.

(1) The interval ds^2 for two infinitely close events at the same space point is $c^2 d\tau^2$, $d\tau$ being a proper time interval of a system. Putting $dx^1 = dx^2 = dx^3 = 0$ in the expression of ds^2 one finds

$$d\tau = c^{-1}\sqrt{g_{00}}dx^0. \tag{2.3.1}$$

This formula determines the proper time of the rest particle (a particle at rest with respect to a given system).

(2) Similarly, the time of a clock moving in a given system is determined by

$$d\tau = c^{-1}\left(g_{00} + 2g_{0i}\frac{dx^i}{dx^0} + g_{ik}\frac{dx^i}{dx^0}\frac{dx^k}{dx^0}\right)^{1/2} dx^0. \tag{2.3.2}$$

This formula yields the proper time of the moving particle and represents a generalization of (1.3.10).

(3) In special relativity the distance between points is determined as the interval between events occurring at the same moment x^0. In GRT one cannot put $dx^0 = 0$ in the expression of interval since in accordance with (2.3.1) the proper time at different points depends differently on x^0. The distance between two infinitely close points $A(x^i + dx^i)$ and $B(x^i)$ may be determined, for example, by the following procedure (Landau and Lifshitz 1962). Let a light signal be emitted from point A at moment $x^0 + dx^0_{(1)}$ of the coordinate time. This signal reaches point B at moment x^0 and after reflection from B returns to A at moment $x^0 + dx^0_{(2)}$. The distance between points A and B is defined as the interval of the proper time of a system elapsed between emission and back reception of the signal in A multiplied by $c/2$. Considering the condition $ds^2 = 0$ of the light propagation as an equation with respect to dx^0 one obtains two roots

$$dx^0_{(1,2)} = -\frac{g_{0i}}{g_{00}}dx^i \mp \frac{1}{g_{00}}\left[(g_{0i}g_{0k} - g_{00}g_{ik})dx^i dx^k\right]^{1/2}.$$

The interval of the coordinate time between emission and return of the signal to A is $dx^0_{(2)} - dx^0_{(1)}$. Multiplying this difference first by $c^{-1}\sqrt{g_{00}}$ to transform to the proper time of the system and then by $c/2$ one finds the expression for the square of the infinitesimal space distance in GRT:

$$d\ell^2 = \gamma_{ik}dx^i dx^k \tag{2.3.3}$$

with

$$\gamma_{ik} = \frac{1}{g_{00}} g_{0i} g_{0k} - g_{ik}. \tag{2.3.4}$$

The quadratic form (2.3.3) generally changes in time. Therefore, the integral of $d\ell$ depends on the world line between given space points. Only in a constant field with $g_{\alpha\beta}$ independent of x^0 does the integral of $d\ell$ have quite definite sense and determines the space distance between points. By its meaning the proper time $d\tau$ defined by (2.3.1) should be real and the quadratic form $d\ell^2$ should be positively definite. From this it follows that

$$g_{00} > 0 \qquad \begin{vmatrix} g_{00} & g_{01} \\ g_{10} & g_{11} \end{vmatrix} < 0 \qquad \begin{vmatrix} g_{00} & g_{01} & g_{02} \\ g_{10} & g_{11} & g_{12} \\ g_{20} & g_{21} & g_{22} \end{vmatrix} > 0 \qquad g < 0. \tag{2.3.5}$$

These conditions are necessary in order that the appropriate reference system might be physically realized.

The quantities γ_{ik} may be considered as components of the three-dimensional metric tensor determining the metric (2.3.3). Along with this they may be used to facilitate operations in transforming from covariant components $g_{\alpha\beta}$ to contravariant components $g^{\alpha\beta}$ (the inversion of the 4×4 matrix is replaced thereby by inversion of the 3×3 matrix). Indeed, if contravariant components γ^{ik} are known then

$$g^{0i} = \frac{1}{g_{00}} g_{0k} \gamma^{ik} \qquad g^{ik} = -\gamma^{ik} \qquad g = -g_{00}\gamma \tag{2.3.6}$$

with $\gamma = \det \|\gamma_{ik}\|$.

(4) In the special theory of relativity time is different for moving clocks. In GRT it is different even for clocks at rest in different space points of one and the same reference system. A comparison of clock readings at infinitely close points $A(x^i + dx^i)$ and $B(x^i)$, i.e. their synchronization, may be again performed by the previously considered light signals (Einstein synchronization). Let the moment $x^0 + \Delta x^0$ at point A be the middle reading of clock A between emission and return of the light signal. It is this moment which is adopted as being simultaneous with moment x^0 of clock B. Therefore,

$$\Delta x^0 = \tfrac{1}{2}(dx^0_{(1)} + dx^0_{(2)}) = -\frac{g_{0i} dx^i}{g_{00}}. \tag{2.3.7}$$

For a closed infinitesimal loop this synchronization is generally impossible since after completing transfer of A over the whole loop Δx^0 takes a value different from the initial one. In a reference system with $g_{0i} \neq 0$ Einstein synchronization in the finite domain is impossible. Only under $g_{0i} = 0$ is Einstein synchronization possible in the whole space.

(5) The relation (2.3.1) defines the proper time for events occurring at the same point of space. Consider now the events occurring at different space points. Let x^0 and $x^0 + \mathrm{d}x^0$ be the moments of departure and arrival of the same particle from B to A respectively (A and B being again infinitely close points). To obtain a coordinate time interval between these events one should take the difference between the actual reading $x^0+\mathrm{d}x^0$ at A and the moment $x^0 + \Delta x^0$ of A which is simultaneous with the moment x^0 of B. Thus,

$$x^0 + \mathrm{d}x^0 - (x^0 + \Delta x^0) = \frac{1}{g_{00}} g_{0\alpha} \mathrm{d}x^\alpha .$$

Multiplying this expression by $c^{-1}\sqrt{g_{00}}$ one obtains the interval of the proper time of A between the events occurring at different space points:

$$\mathrm{d}\tau = c^{-1}(g_{00})^{-1/2} g_{0\alpha} \mathrm{d}x^\alpha . \qquad (2.3.8)$$

2.3.2 Chronometric invariants

The equations of motion of a test particle in the a gravitational field are determined by equations (2.2.49) resulting from the geodesic principle. But these equations are different in different reference systems and their separate terms do not admit an invariant interpretation. Physical interpretation may be achieved in using quantities having a definite meaning not only for one specific reference system x^α but within a set of reference systems related by the transformation

$$\tilde{x}^0 = \tilde{x}^0(x^0, x^1, x^2, x^3) \qquad (2.3.9)$$

$$\tilde{x}^i = \tilde{x}^i(x^1, x^2, x^3) \qquad \frac{\partial \tilde{x}^i}{\partial x^0} = 0. \qquad (2.3.10)$$

This means that the systems are fixed relative to each other (using physical terminology such reference systems are said to belong to one and the same reference frame). Quantities invariant under these transformations are called chronometric invariants and their theory, elements of which are expounded here, has been elaborated by Zel'manov (1956). The theory of chronometric invariants has a direct bearing on the problem of measurement of time and distances. It is easy to verify that the proper time element $\mathrm{d}\tau$ (2.3.8) for the infinitely close events at the infinitely close points and the elementary distance $\mathrm{d}\ell$ (2.3.3) are chronometric invariants. The field metric is described with the aid of these quantities in the form

$$\mathrm{d}s^2 = c^2 \mathrm{d}\tau^2 - \mathrm{d}\ell^2 . \qquad (2.3.11)$$

This representation of the metric is called 3 + 1 splitting on the proper space and proper time. All operations with three-dimensional vectors and tensors are performed in the space with metric (2.3.3). Introducing the synchronized 3-velocity

$$q^i = \mathrm{d}x^i/\mathrm{d}\tau \tag{2.3.12}$$

its scalar square will be

$$q^2 = \gamma_{ik}q^i q^k = \mathrm{d}\ell^2/\mathrm{d}\tau^2 \tag{2.3.13}$$

and, hence,

$$\mathrm{d}s^2 = c^2\mathrm{d}\tau^2(1 - q^2/c^2). \tag{2.3.14}$$

Therefore, the 4-velocity $u^\alpha = \mathrm{d}x^\alpha/\mathrm{d}s$ of a moving particle is expressed in the form

$$u^\alpha = c^{-1}\left(1 - \frac{q^2}{c^2}\right)^{-1/2}\frac{\mathrm{d}x^\alpha}{\mathrm{d}\tau} \tag{2.3.15}$$

or, excluding $\mathrm{d}x^0/\mathrm{d}\tau$ from (2.3.8),

$$u^0 = \left(1 - \frac{q^2}{c^2}\right)^{-1/2}\left(\frac{1}{\sqrt{g_{00}}} - \frac{g_{0i}}{g_{00}}\frac{q^i}{c}\right) \tag{2.3.16}$$

$$u^i = c^{-1}\left(1 - \frac{q^2}{c^2}\right)^{-1/2} q^i. \tag{2.3.17}$$

The geodesic equations (2.1.19) imply for the space components

$$\frac{\mathrm{d}}{\mathrm{d}\tau}\frac{q^i}{(1 - q^2/c^2)^{1/2}} = \left(1 - \frac{q^2}{c^2}\right)^{-1/2}[F^i - 2(D^i_k + A^{\cdot i}_{k\cdot})q^k - \Delta^i_{km}q^k q^m] \tag{2.3.18}$$

with

$$F^i = -\frac{c^2}{g_{00}}\Gamma^i_{00} \tag{2.3.19}$$

$$D^i_k + A^{\cdot i}_{k\cdot} = \frac{c}{\sqrt{g_{00}}}\left(\Gamma^i_{0k} - \frac{g_{0k}}{g_{00}}\Gamma^i_{00}\right) \tag{2.3.20}$$

$$\Delta^i_{km} = \Gamma^i_{km} - \frac{g_{0m}}{g_{00}}\Gamma^i_{0k} - \frac{g_{0k}}{g_{00}}\Gamma^i_{0m} + \frac{1}{g_{00}^2}g_{0k}g_{0m}\Gamma^i_{00}. \tag{2.3.21}$$

Multiplying equations (2.3.18) by the rest mass m_0 and defining the mass of a particle as

$$m = (1 - q^2/c^2)^{-1/2}m_0$$

one obtains the generalizations of the equations of special relativity. The vector (2.3.19) represents acceleration of a point at rest at a given moment.

The condition $F^i = 0$ turns out to be necessary and sufficient for converting by transformation of x^0 alone to the coordinates providing

$$g_{00} = 1 \qquad \partial g_{0i}/\partial x^0 = 0.$$

The relation (2.3.20) determines simultaneously the symmetric tensor D^{ik} and the antisymmetric tensor A^{ik}. Indeed, from

$$D^{ki} + A^{ki} = \gamma^{km}(D^i_m + A^{\cdot i}_{m\cdot}) = -\frac{c}{\sqrt{g_{00}}} g^{\alpha k}\Gamma^i_{0\alpha}$$

one has

$$D^{ik} = -\frac{c}{2\sqrt{g_{00}}}(g^{\alpha k}\Gamma^i_{0\alpha} + g^{\alpha i}\Gamma^k_{0\alpha}) = \frac{c}{2\sqrt{g_{00}}}\frac{\partial g^{ik}}{\partial x^0} \qquad (2.3.22)$$

$$A^{ik} = \frac{c}{2\sqrt{g_{00}}}(g^{\alpha k}\Gamma^i_{0\alpha} - g^{\alpha i}\Gamma^k_{0\alpha}) = \frac{c}{2\sqrt{g_{00}}} g^{\alpha i} g^{\beta k}\left(\frac{\partial g_{0\alpha}}{\partial x^\beta} - \frac{\partial g_{0\beta}}{\partial x^\alpha}\right). \qquad (2.3.23)$$

The tensor D^{ik} is called the tensor of deformation of the reference system. The invariant

$$D = \gamma_{ik} D^{ik} = \frac{c}{\sqrt{g_{00}}}\frac{\partial \ln\sqrt{\gamma}}{\partial x^0} \qquad (2.3.24)$$

characterizes the speed of volume expansion of the space element. The tensor A^{ik} represents the tensor of angular velocity of rotation of the system. The equation $A^{ik} = 0$ gives a necessary and sufficient condition for eliminating all components g_{0i} by a suitable transformation of x^0. Finally, the term (2.3.21) admits an elegant formula

$$\gamma^{km}\gamma^{jn}\Delta^i_{kj} = g^{\alpha m} g^{\beta n}\Gamma^i_{\alpha\beta}.$$

In spite of theoretical interest equations (2.3.18) are seldom used in practical problems due to purely technical difficulties.

Let us note in addition that the condition of the light propagation $\mathrm{d}\tau = c^{-1}\mathrm{d}\ell$ resulting from (2.3.11) leads, using (2.3.8), to the relation

$$c\mathrm{d}t = \frac{1}{\sqrt{g_{00}}}\left(1 + c^{-1}\frac{g_{0i}}{g_{00}}\frac{\mathrm{d}x^i}{\mathrm{d}t}\right)^{-1} \mathrm{d}\ell \qquad (2.3.25)$$

involving the coordinate time interval $\mathrm{d}t$, coordinate velocity of light $\mathrm{d}x^i/\mathrm{d}t$ and invariant (proper) distance $\mathrm{d}\ell$.

Let us note once again that F^i, A^{ik}, D^{ik} do not depend on the choice of time coordinate, i.e. they are not changed under transformation (2.3.9). Under transformations (2.3.10) of the space coordinates they act as three-dimensional vectors and tensors.

It is useful to indicate particular types of gravitational fields. If a given field enables one to choose such a coordinate system that all $g_{\alpha\beta}$ do not depend on x^0 then such a field, as mentioned above, is called constant and x^0 is called the world time. A constant field may be caused only by one body. If there are at least two bodies they should be in relative motion and hence their field cannot be constant. If the field is caused by one body and this body is at rest then both directions of time are equivalent and one can choose such coordinates that all $g_{0i} = 0$. Such a field is called static and for this field Einstein clock synchronization is possible for the whole space. If the field is constant but at least one of the components $g_{0i} \neq 0$ then the field is called stationary. An example of the stationary field is the field due to an axial symmetrical body in rotation around the axis of symmetry.

The theory of chronometric invariants is one of the new branches of GRT designed to elaborate the mathematical formalism for solving the problem of measurement in GRT. This problem is discussed in many references. In particular, in the textbooks by Synge (1960), Møller (1972) and Vladimirov (1982) GRT is consistently treated under this point of view. Only the elements of tetrad formalism closely related to this branch are considered below.

2.3.3 Tetrads

At every point of the GRT space-time one may introduce locally, i.e. in the infinitesimal vicinity of the point, a pseudo-Cartesian reference system. This system may be presented by means of a tetrad composed of four orthonormal vectors $\lambda^{\nu}_{(\mu)}$. The lower index taken in parentheses means the number of the vector ($\mu = 0, 1, 2, 3$). The upper index means, as usual, the contravariant component. The vector $\lambda^{\nu}_{(0)}$ is timelike whereas $\lambda^{\nu}_{(i)}$ form a space triad, consisting of three orthonormal spacelike vectors. All relations of special relativity are locally valid in such a reference system and all tensor operations with local tetrad indices (indices in parentheses) are performed with the aid of the Minkowski tensor $\eta_{\mu\nu}$. Orthonormality of the tetrad vectors implies

$$g_{\mu\nu}\lambda^{\mu}_{(\alpha)}\lambda^{\nu}_{(\beta)} = \delta_{\alpha\beta}. \qquad (2.3.26)$$

Raising local lower indices and lowering global upper indices gives

$$\lambda^{(\beta)}_{\alpha} = \eta^{\beta\mu} g_{\alpha\nu}\lambda^{\nu}_{(\mu)}. \qquad (2.3.27)$$

From this it follows that

$$g_{\mu\nu} = \lambda^{(\alpha)}_{\mu}\lambda^{(\beta)}_{\nu}\eta_{\alpha\beta} \qquad (2.3.28)$$

and

$$ds^2 = \eta_{\mu\nu}dx^{(\mu)}dx^{(\nu)} \qquad (2.3.29)$$

with

$$dx^{(\mu)} = \lambda_\nu^{(\mu)} dx^\nu \qquad dx^\mu = \lambda_{(\nu)}^\mu dx^{(\nu)}. \tag{2.3.30}$$

The main merit of the tetrad formalism is the local Galilean form (2.3.29) for the field metric. All quantities being projected onto the tetrad have physical meaning. In particular, $dx^{(0)}$ and $dx^{(i)}$ are local measurable intervals of time and distance respectively. For a point moving in a given gravitational field the simplest example is the co-moving tetrad. Its time-like vector coincides with the 4-velocity of the point

$$\lambda_{(0)}^\alpha = u^\alpha = dx^\alpha/ds \tag{2.3.31}$$

and the triad vectors correspond to coordinate directions and satisfy the orthonormality conditions (2.3.26). If $v^i = dx^i/dt$ are components of the coordinate 3-velocity of a moving point then

$$\begin{aligned}
\lambda_{(0)}^0 &= 1 - \frac{1}{2}h_{00} + \frac{1}{2}\frac{v^2}{c^2} + \frac{3}{8}(h_{00})^2 + \frac{3}{8}\frac{v^4}{c^4} \\
&\quad - h_{0k}\frac{v^k}{c} - \frac{3}{4}h_{00}\frac{v^2}{c^2} - \frac{1}{2}h_{ik}\frac{v^i v^k}{c^2} + \dots \\
\lambda_{(0)}^i &= \frac{v^i}{c}\left(1 - \frac{1}{2}h_{00} + \frac{1}{2}\frac{v^2}{c^2} + \dots\right) \\
\lambda_{(i)}^0 &= \frac{v^i}{c}\left(1 - h_{00} + \frac{1}{2}\frac{v^2}{c^2}\right) - \frac{1}{2}h_{ik}\frac{v^k}{c} - h_{0i} + \dots \\
\lambda_{(k)}^i &= \delta_{ik} + \frac{1}{2}h_{ik} + \frac{1}{2c^2}v^i v^k - \frac{1}{2c^2}h_{00}v^i v^k + \frac{3}{8}h_{im}h_{km} \\
&\quad + \frac{3}{8c^4}v^i v^k v^2 - \frac{1}{8c^2}h_{im}v^k v^m - \frac{1}{8c^2}h_{km}v^i v^m + \dots .
\end{aligned} \tag{2.3.32}$$

Using (2.3.27) one obtains

$$\begin{aligned}
\lambda_0^{(0)} &= 1 + \frac{1}{2}h_{00} + \frac{1}{2}\frac{v^2}{c^2} - \frac{1}{8}(h_{00})^2 + \frac{3}{8}\frac{v^4}{c^4} \\
&\quad - \frac{1}{4}h_{00}\frac{v^2}{c^2} - \frac{1}{2}h_{ik}\frac{v^i v^k}{c^2} + \dots \\
\lambda_i^{(0)} &= -\frac{v^i}{c}\left(1 - \frac{1}{2}h_{00} + \frac{1}{2}\frac{v^2}{c^2} + \dots\right) + h_{ik}\frac{v^k}{c} + h_{0i} + \dots \\
\lambda_0^{(i)} &= -\frac{v^i}{c}\left(1 + \frac{1}{2}\frac{v^2}{c^2}\right) + \frac{1}{2}h_{ik}\frac{v^k}{c} + \dots \\
\lambda_k^{(i)} &= \delta_{ik} - \frac{1}{2}h_{ik} + \frac{1}{2c^2}v^i v^k - \frac{1}{2c^2}h_{00}v^i v^k - \frac{1}{8}h_{im}h_{km} \\
&\quad - h_{0k}\frac{v^i}{c} + \frac{3}{8c^4}v^i v^k v^2 - \frac{1}{8c^2}h_{im}v^k v^m - \frac{5}{8c^2}h_{km}v^i v^m + \dots .
\end{aligned} \tag{2.3.33}$$

Evidently, the tetrad components of the 4-velocity in a co-moving system are

$$\begin{aligned} u^{(0)} &= \lambda^{(0)}_{\alpha} u^{\alpha} = \eta^{0\mu} g_{\alpha\beta} \lambda^{\beta}_{(\mu)} u^{\alpha} = g_{\alpha\beta} u^{\alpha} u^{\beta} = 1 \\ u^{(i)} &= \lambda^{(i)}_{\alpha} u^{\alpha} = \eta^{i\mu} g_{\alpha\beta} \lambda^{\beta}_{(\mu)} u^{\alpha} = -g_{\alpha\beta} \lambda^{\beta}_{(i)} \lambda^{\alpha}_{(0)} = 0. \end{aligned} \qquad (2.3.34)$$

2.3.4 Measurable quantities in the tetrad formalism

$3+1$ splitting (2.3.11) for the weak field has the form

$$c\mathrm{d}\tau = (1 + \tfrac{1}{2}h_{00} - \tfrac{1}{8}h^2_{00})c\mathrm{d}t + (h_{0i} - \underline{\tfrac{1}{2}h_{00}h_{0i}})\mathrm{d}x^i \qquad (2.3.35)$$

$$\gamma_{ik} = \delta_{ik} - h_{ik} + \underline{h_{0i}h_{0k}}. \qquad (2.3.36)$$

h_{00} and h_{ik} are of the second order of smallness. As for h_{0i}, they are of the third order in a quasi-inertial coordinate system. In accelerated moving (for example, rotating) systems they are of first order. Therefore, in (2.3.35) and (2.3.36) and in the formulae given below the underlined terms are to be rejected if h_{0i} are of third order. The tetrad associated with the splitting (2.3.11) is

$$\lambda^{(0)}_0 = 1 \qquad \lambda^{(0)}_i = 0 \qquad \lambda^{(k)}_0 = 0 \qquad \lambda^{(k)}_i = \delta_{ik} - \tfrac{1}{2}h_{ik} + \underline{\tfrac{1}{2}h_{0i}h_{0k}}. \qquad (2.3.37)$$

Using it one may compose the space triad

$$\mathrm{d}x^{(i)} = \lambda^{(i)}_k \mathrm{d}x^k = \mathrm{d}x^i + \tfrac{1}{2}(-h_{ik} + \underline{h_{0i}h_{0k}})\mathrm{d}x^k \qquad (2.3.38)$$

enabling us to present the element of space distance in the form

$$\mathrm{d}\ell^2 = \delta_{ik}\mathrm{d}x^{(i)}\mathrm{d}x^{(k)}. \qquad (2.3.39)$$

$\mathrm{d}\tau$ and $\mathrm{d}x^{(i)}$ may be regarded as physically measurable infinitesimal intervals of time and distance respectively.

It is of interest to see that the coordinate-dependent Galilean transformations with no physical meaning induce physically meaningful transformations of infinitesimal intervals of time and distance.

Let t, x^i be the coordinates of the reference system (b) related to the Solar System barycentre. Elements of time $\mathrm{d}\tau_b$ and distance $\mathrm{d}\ell_b$ for this system are calculated using (2.3.35), (2.3.38) and (2.3.39), rejecting h_{0i} which are of third order for this case. Consider now a formally Newtonian transformation

$$x^i = R^i(t) + \xi^i \qquad \mathrm{d}x^i = \dot{R}^i\mathrm{d}t + \mathrm{d}\xi^i \qquad (2.3.40)$$

$R^i(t)$ being a function describing the motion of an observer in system (b). If $R^i(t)$ represents the motion of the geocentre then t, ξ^i are the coordinates of the formally geocentric reference system (g) with the metric

$$\begin{aligned}
ds^2 =& (1 + h_{00} - c^{-2}\dot{\boldsymbol{R}}^2 + 2c^{-1}h_{0k}\dot{R}^k + c^{-2}h_{km}\dot{R}^k\dot{R}^m)c^2dt^2 \\
&+ 2(-c^{-1}\dot{R}^i + h_{0i} + c^{-1}h_{ik}\dot{R}^k)c\,dt d\xi^i + (-\delta_{ik} + h_{ik})d\xi^i d\xi^k.
\end{aligned} \tag{2.3.41}$$

$3 + 1$ splitting for this metric determines the elementary intervals of time $d\tau_g$ and length $d\ell_g$. With the aid of (2.3.35) and (2.3.36) applied to (2.3.41) one obtains

$$\begin{aligned}
c d\tau_g =& (1 + \tfrac{1}{2}h_{00} - \tfrac{1}{2}c^{-2}\dot{\boldsymbol{R}}^2 + c^{-1}h_{0k}\dot{R}^k + \tfrac{1}{2}c^{-2}h_{km}\dot{R}^k\dot{R}^m \\
&- \tfrac{1}{8}h_{00}^2 + \tfrac{1}{4}c^{-2}h_{00}\dot{\boldsymbol{R}}^2 - \tfrac{1}{8}c^{-4}\dot{\boldsymbol{R}}^4)c\,dt \\
&+ (-c^{-1}\dot{R}^i + h_{0i} + \tfrac{1}{2}c^{-1}h_{00}\dot{R}^i + c^{-1}h_{ik}\dot{R}^k - \tfrac{1}{2}c^{-3}\dot{R}^2\dot{R}^i)d\xi^i
\end{aligned} \tag{2.3.42}$$

$$d\ell_g^2 = (\delta_{ik} - h_{ik} + c^{-2}\dot{R}^i\dot{R}^k)d\xi^i d\xi^k. \tag{2.3.43}$$

In (2.3.41)–(2.3.43) the quantities $h_{\mu\nu}$ have values corresponding to system (b). $d\tau_g$ may be expressed in terms of $d\tau_b$ and dx^i. To do this it is sufficient to substitute into (2.3.42) the expression of $d\xi^i$ from (2.3.40) and then the expression of cdt from (2.3.35). One obtains

$$\begin{aligned}
d\tau_g =& (1 + \tfrac{1}{2}c^{-2}\dot{\boldsymbol{R}}^2 + \tfrac{3}{8}c^{-4}\dot{\boldsymbol{R}}^4 - \tfrac{1}{2}c^{-2}h_{00}\dot{\boldsymbol{R}}^2 - \tfrac{1}{2}c^{-2}h_{km}\dot{R}^k\dot{R}^m)d\tau_b \\
&+ c^{-2}(-\dot{R}^i - \tfrac{1}{2}c^{-2}\dot{\boldsymbol{R}}^2\dot{R}^i + \tfrac{1}{2}h_{00}\dot{R}^i + h_{ik}\dot{R}^k)dx^i.
\end{aligned} \tag{2.3.44}$$

Now let the observer be in motion in system (g) according to some functional law $\rho^i = \rho^i(t)$. One may perform the formally Newtonian transformation to a reference system (s) moving with the observer

$$\xi^i = \rho^i(t) + \eta^i \qquad d\xi^i = \dot{\rho}^i dt + d\eta^i. \tag{2.3.45}$$

System (s) with coordinates t, η^i means transformation to a ground observer (topocentric system) or an observer on a satellite (satellite system). In terms of t, η^i the field metric becomes

$$\begin{aligned}
ds^2 =& [1 + h_{00} - c^{-2}(\dot{\boldsymbol{R}} + \dot{\rho})^2 + 2c^{-1}h_{0k}(\dot{R}^k + \dot{\rho}^k) \\
&+ c^{-2}h_{km}(\dot{R}^k + \dot{\rho}^k)(\dot{R}^m + \dot{\rho}^m)]c^2dt^2 \\
&+ 2[-c^{-1}(\dot{R}^i + \dot{\rho}^i) + h_{0i} + c^{-1}h_{ik}(\dot{R}^k + \dot{\rho}^k)]c\,dt d\eta^i \\
&+ (-\delta_{ik} + h_{ik})d\eta^i d\eta^k.
\end{aligned} \tag{2.3.46}$$

3 + 1 splitting for this metric determines the elementary intervals of time $d\tau_s$ and length $d\ell_s$. With the aid of (2.3.35) and (2.3.36) applied to (2.3.46) one obtains

$$\begin{aligned} c d\tau_s = & [1 + \tfrac{1}{2}h_{00} - \tfrac{1}{2}c^{-2}(\dot{\boldsymbol{R}} + \dot{\rho})^2 + c^{-1}h_{0k}(\dot{R}^k + \dot{\rho}^k) \\ & + \tfrac{1}{2}c^{-2}h_{km}(\dot{R}^k + \dot{\rho}^k)(\dot{R}^m + \dot{\rho}^m) - \tfrac{1}{8}h_{00}^2 + \tfrac{1}{4}c^{-2}h_{00}(\dot{\boldsymbol{R}} + \dot{\rho})^2 \\ & - \tfrac{1}{8}c^{-4}(\dot{\boldsymbol{R}} + \dot{\rho})^4] c dt + [-c^{-1}(\dot{R}^i + \dot{\rho}^i) + h_{0i} + c^{-1}h_{ik}(\dot{R}^k + \dot{\rho}^k) \\ & + \tfrac{1}{2}c^{-1}h_{00}(\dot{R}^i + \dot{\rho}^i) - \tfrac{1}{2}c^{-3}(\dot{\boldsymbol{R}} + \dot{\rho})^2(\dot{R}^i + \dot{\rho}^i)] d\eta^i \end{aligned} \qquad (2.3.47)$$

$$d\ell_s^2 = [\delta_{ik} - h_{ik} + c^{-2}(\dot{R}^i + \dot{\rho}^i)(\dot{R}^k + \dot{\rho}^k)] d\eta^i d\eta^k. \qquad (2.3.48)$$

As above, (2.3.47) may be reduced to the form

$$\begin{aligned} d\tau_s = & [1 + \tfrac{1}{2}c^{-2}\dot{\rho}^2 - \tfrac{1}{2}c^{-2}h_{00}\dot{\rho}^2 - \tfrac{1}{2}c^{-2}h_{km}\dot{\rho}^k\dot{\rho}^m + \tfrac{1}{2}c^{-4}\dot{\boldsymbol{R}}^2\dot{\rho}^2 \\ & + \tfrac{1}{2}c^{-4}(\dot{\boldsymbol{R}}\dot{\rho})^2 + c^{-4}\dot{\rho}^2(\dot{\boldsymbol{R}}\dot{\rho}) + \tfrac{3}{8}c^{-4}\dot{\rho}^4] d\tau_g \\ & + c^{-2}[-\dot{\rho}^i + \tfrac{1}{2}h_{00}\dot{\rho}^i + h_{ik}\dot{\rho}^k - c^{-2}(\dot{\boldsymbol{R}}\dot{\rho})\dot{R}^i \\ & - \tfrac{1}{2}c^{-2}\dot{\boldsymbol{R}}^2\dot{\rho}^i - c^{-2}(\dot{\boldsymbol{R}}\dot{\rho})\dot{\rho}^i - \tfrac{1}{2}c^{-2}\dot{\rho}^2\dot{\rho}^i] d\xi^i. \end{aligned} \qquad (2.3.49)$$

As $d\tau_b$, $d\tau_g$, $d\tau_s$ are infinitesimal physically measurable time intervals they should be related by the Lorentz transformations. Indeed, relations (2.3.44) and (2.3.49) may be presented as the Lorentz transformations if the coordinate distances entered into them are expressed in terms of the chronometric invariant quantities. If $\dot{R}^i$ is the coordinate velocity of system (g) with respect to system (b) then the corresponding chronometric invariant velocity is $dR^{(i)}/d\tau_b$ with $dR^{(i)}$ to be calculated by (2.3.38) and $d\tau_b$ to be evaluated at the point $\xi^i = 0$. Hence,

$$dR^{(i)} = dR^i - \tfrac{1}{2}h_{ik}dR^k \qquad (d\tau_b)_{\xi^i=0} = (1 + \tfrac{1}{2}h_{00} + \dots)\, dt \qquad (2.3.50)$$

$$\frac{dR^{(i)}}{d\tau_b} = \dot{R}^i - \tfrac{1}{2}h_{00}\dot{R}^i - \tfrac{1}{2}h_{ik}\dot{R}^k. \qquad (2.3.51)$$

Then expression (2.3.44) may be put in the form

$$d\tau_g = \left(1 - c^{-2}\frac{dR^{(k)}}{d\tau_b}\frac{dR^{(k)}}{d\tau_b}\right)^{-1/2} \left(d\tau_b - c^{-2}\frac{dR^{(k)}}{d\tau_b}\, dx^{(k)}\right) \qquad (2.3.52)$$

which is the Lorentz transformation.

Consider now the transformation (2.3.49). The coordinate velocity of system (s) with respect to (g) is given by the function $\dot{\rho}^i$. The tetrad in (g) analogous to (2.3.37) is formed by

$$\mu_0^{[0]} = 1 \qquad \mu_i^{[0]} = 0 \qquad \mu_0^{[k]} = 0$$

$$\mu_i^{[k]} = \delta_{ik} - \tfrac{1}{2}h_{ik} + \tfrac{1}{2}c^{-2}\dot{R}^i\dot{R}^k. \tag{2.3.53}$$

Elementary physical distances in (g) are then represented by the expressions

$$\mathrm{d}\xi^{[i]} = \mu_k^{[i]}\mathrm{d}\xi^k = \mathrm{d}\xi^i - \tfrac{1}{2}h_{ik}\mathrm{d}\xi^k + \tfrac{1}{2}c^{-2}(\dot{R}^k\mathrm{d}\xi^k)\dot{R}^i. \tag{2.3.54}$$

The chronometric invariant velocity of (s) relative to (g) is given by the derivative $\mathrm{d}\rho^{[i]}/\mathrm{d}\tau_g$ with $\mathrm{d}\rho^{[i]}$ calculated from (2.3.54) and $\mathrm{d}\tau_g$ evaluated at the point $\eta^i = 0$. Therefore,

$$\mathrm{d}\rho^{[i]} = \mathrm{d}\rho^i - \tfrac{1}{2}h_{ik}\mathrm{d}\rho^k + \tfrac{1}{2}c^{-2}(\dot{R}^k\mathrm{d}\rho^k)\dot{R}^i \tag{2.3.55}$$

$$(\mathrm{d}\tau_g)_{\eta^i=0} = (1 + \tfrac{1}{2}h_{00} - \tfrac{1}{2}c^{-2}\dot{\boldsymbol{R}}^2 - c^{-2}\dot{\boldsymbol{R}}\dot{\rho} + \ldots)\,\mathrm{d}t \tag{2.3.56}$$

$$\begin{aligned}\frac{\mathrm{d}\rho^{[i]}}{\mathrm{d}\tau_g} &= \dot{\rho}^i - \tfrac{1}{2}h_{00}\dot{\rho}^i - \tfrac{1}{2}h_{ik}\dot{\rho}^k + \tfrac{1}{2}c^{-2}(\dot{\boldsymbol{R}}\dot{\rho})\dot{R}^i \\ &\quad + \tfrac{1}{2}c^{-2}\dot{\boldsymbol{R}}^2\dot{\rho}^i + c^{-2}(\dot{\boldsymbol{R}}\dot{\rho})\dot{\rho}^i.\end{aligned} \tag{2.3.57}$$

On the basis of (2.3.54) and (2.3.57) expression (2.3.49) takes the form of the Lorentz transformation

$$\mathrm{d}\tau_s = \left(1 - c^{-2}\frac{\mathrm{d}\rho^{[k]}}{\mathrm{d}\tau_g}\frac{\mathrm{d}\rho^{[k]}}{\mathrm{d}\tau_g}\right)^{-1/2}\left(\mathrm{d}\tau_g - c^{-2}\frac{\mathrm{d}\rho^{[k]}}{\mathrm{d}\tau_g}\mathrm{d}\xi^{[k]}\right). \tag{2.3.58}$$

In changing from (b) to (g) transformation (2.3.40) of the coordinate space intervals $\mathrm{d}x^i$ and $\mathrm{d}\xi^i$ may be presented in terms of measurable quantities by the Lorentz transformation without rotation (boost)

$$\begin{aligned}\mathrm{d}\xi^{[i]} &= \mathrm{d}x^{(i)} + \frac{1}{2}c^{-2}\left(\frac{\mathrm{d}R^{(k)}}{\mathrm{d}\tau_b}\mathrm{d}x^{(k)}\right)\frac{\mathrm{d}R^{(i)}}{\mathrm{d}\tau_b} \\ &\quad - \left(1 + \frac{1}{2}c^{-2}\frac{\mathrm{d}R^{(k)}}{\mathrm{d}\tau_b}\frac{\mathrm{d}R^{(k)}}{\mathrm{d}\tau_b}\right)\frac{\mathrm{d}R^{(i)}}{\mathrm{d}\tau_b}\mathrm{d}\tau_b.\end{aligned} \tag{2.3.59}$$

Introducing in system (s) the tetrad

$$\begin{aligned}&\nu_0^{\{0\}} = 1 \qquad \nu_i^{\{0\}} = 0 \qquad \nu_0^{\{k\}} = 0 \\ &\nu_i^{\{k\}} = \delta_{ik} - \tfrac{1}{2}h_{ik} + \tfrac{1}{2}c^{-2}(\dot{R}^i + \dot{\rho}^i)(\dot{R}^k + \dot{\rho}^k)\end{aligned} \tag{2.3.60}$$

elementary physical distances in (s) are expressed by

$$\mathrm{d}\eta^{\{i\}} = \nu_k^{\{i\}}\mathrm{d}\eta^k = \mathrm{d}\eta^i - \tfrac{1}{2}h_{ik}\mathrm{d}\eta^k + \tfrac{1}{2}c^{-2}[(\dot{\boldsymbol{R}} + \dot{\rho})\mathrm{d}\eta](\dot{R}^i + \dot{\rho}^i). \tag{2.3.61}$$

The coordinate conversion (2.3.45) from $\mathrm{d}\xi^i$ to $\mathrm{d}\eta^i$ is written in terms of $\mathrm{d}\xi^{[i]}$ and $\mathrm{d}\eta^{\{i\}}$ as the Lorentz transformation with rotation

$$\mathrm{d}\eta^i = \mathrm{d}\xi^{[i]} + \frac{1}{2}c^{-2}\left(\frac{\mathrm{d}\rho^{[k]}}{\mathrm{d}\tau_g}\mathrm{d}\xi^{[k]}\right)\frac{\mathrm{d}\rho^{[i]}}{\mathrm{d}\tau_g} - \left(1+\frac{1}{2}c^{-2}\frac{\mathrm{d}\rho^{[k]}}{\mathrm{d}\tau_g}\frac{\mathrm{d}\rho^{[k]}}{\mathrm{d}\tau_g}\right)\frac{\mathrm{d}\rho^{[i]}}{\mathrm{d}\tau_g}\mathrm{d}\tau_g + \frac{1}{2}c^{-2}[\mathrm{d}\boldsymbol{\eta}\times(\dot{\boldsymbol{R}}\times\dot{\boldsymbol{\rho}})]^{[i]}. \tag{2.3.62}$$

It is to be noted that direct transformation from (b) to (s) is expressed by the Lorentz boost

$$\mathrm{d}\eta^{\{i\}} = \mathrm{d}x^{(i)} + \frac{1}{2}c^{-2}\left[\left(\frac{\mathrm{d}R^{(k)}}{\mathrm{d}\tau_b}+\frac{\mathrm{d}\rho^{(k)}}{\mathrm{d}\tau_b}\right)\mathrm{d}x^{(k)}\right]\left(\frac{\mathrm{d}R^{(i)}}{\mathrm{d}\tau_b}+\frac{\mathrm{d}\rho^{(i)}}{\mathrm{d}\tau_b}\right) - \left[1+\frac{1}{2}c^{-2}\left(\frac{\mathrm{d}R^{(k)}}{\mathrm{d}\tau_b}+\frac{\mathrm{d}\rho^{(k)}}{\mathrm{d}\tau_b}\right)^2\right]\left(\frac{\mathrm{d}R^{(i)}}{\mathrm{d}\tau_b}+\frac{\mathrm{d}\rho^{(i)}}{\mathrm{d}\tau_b}\right)\mathrm{d}\tau_b. \tag{2.3.63}$$

Thus, the purely Newtonian transformations (2.3.40) and (2.3.45) induce in terms of measurable quantities Lorentz transformations (2.3.52), (2.3.58), (2.3.59), (2.3.62) and (2.3.63). These transformations are adequate in solving astronomical problems related to infinitesimal intervals of time and distance.

2.3.5 Rotation in a weak field

To illustrate the use of tetrad quantities let us consider the problem of rotation in a weak field (Lightman *et al* 1975). A test particle characterized only by its mass moves in a given gravitational field on the geodesic. If the test particle represents a gyroscope then its axes are subjected to the Fermi–Walker transport described by equations (1.2.56). Mathematically, a gyroscope represents a particle with spin. The spin vector S^α is space-like and orthogonal to the 4-velocity vector of a particle. Therefore, the transport of S^α is described by the equations

$$\mathrm{D}S^\alpha/\mathrm{d}s = -A^\beta S_\beta u^\alpha. \tag{2.3.64}$$

The 4-acceleration of a particle $A^\beta = \mathrm{D}u^\beta/\mathrm{d}s$ is zero for the geodesic motion when there are no forces of non-gravitational origin. Describing equations (2.3.64) in a co-moving locally Lorentzian system determined by the tetrad (2.3.32) one has

$$\mathrm{d}S^{(\alpha)}/\mathrm{d}t = -\Gamma^{(\alpha)}_{(\mu)(\nu)}S^{(\mu)}u^{(\nu)} - A_{(\mu)}S^{(\mu)}u^{(\alpha)}. \tag{2.3.65}$$

Since in the co-moving tetrad $u^{(0)} = 1$, $u^{(i)} = 0$, $A_{(0)} = u^\mu A_\mu = 0$ then

$$dS^{(\alpha)}/dt = -\Gamma^{(\alpha)}_{(\mu)(0)} S^{(\mu)} - A_{(k)} S^{(k)} u^{(\alpha)}.$$

In a co-moving system $S_{(0)} = u^\mu S_\mu = 0$ so that finally the equations of spin transport take the form

$$dS^{(i)}/dt = -\Gamma^{(i)}_{(0)(j)} S^{(j)}. \tag{2.3.66}$$

Christoffel symbols in coordinates x^α are expressed by (2.2.44). In the co-moving system taking into account (1.2.26) one obtains

$$\begin{aligned}\Gamma^{(\alpha)}_{(\beta)(\gamma)} &= \frac{\partial^2 x^\mu}{\partial x^{(\beta)} \partial x^{(\gamma)}} \frac{\partial x^{(\alpha)}}{\partial x^\mu} + \frac{\partial x^\mu}{\partial x^{(\beta)}} \frac{\partial x^\nu}{\partial x^{(\gamma)}} \frac{\partial x^{(\alpha)}}{\partial x^\sigma} \Gamma^\sigma_{\mu\nu} \\ &= \lambda^{(\alpha)}_\mu \lambda^\mu_{(\gamma),(\beta)} + \lambda^\mu_{(\beta)} \lambda^\nu_{(\gamma)} \lambda^{(\alpha)}_\sigma \Gamma^\sigma_{\mu\nu}.\end{aligned} \tag{2.3.67}$$

Substitution of (2.2.44) for the Christoffel symbols and (2.3.32), (2.3.33) for the tetrad components results in

$$\begin{aligned}\Gamma^{(i)}_{(0)(j)} &= \tfrac{1}{2} c^{-2} (v^j v^i_{,(0)} - v^i v^j_{,(0)}) + \tfrac{1}{2} c^{-1} (v^j h_{00,i} - v^i h_{00,j}) \\ &+ \tfrac{1}{2} (h_{0j,i} - h_{0i,j}) + \tfrac{1}{2} c^{-1} v^k (h_{jk,i} - h_{ik,j})\end{aligned} \tag{2.3.68}$$

within third-order accuracy with respect to v/c. Here

$$v^i_{,(0)} = v^i_{,0} + c^{-1} v^k v^i_{,k} = c^{-1} \dot{v}^i.$$

In virtue of the Newtonian equations of motion

$$v^i_{,0} = -\tfrac{1}{2} c h_{00,i} + c^{-1} A^i \tag{2.3.69}$$

A^i being a 3-acceleration caused by the non-gravitational forces. Therefore,

$$\begin{aligned}\Gamma^{(i)}_{(0)(j)} &= \tfrac{1}{4} c^{-1} (v^j h_{00,i} - v^i h_{00,j}) + \tfrac{1}{2} c^{-1} v^k (h_{jk,i} - h_{ik,j}) \\ &+ \tfrac{1}{2} (h_{0j,i} - h_{0i,j}) + \tfrac{1}{2} c^{-3} (v^j A^i - v^i A^j).\end{aligned}$$

Substitution of the values (2.2.32) for a weak field gives finally

$$\begin{aligned}\Gamma^{(i)}_{(0)(j)} &= \tfrac{1}{2} c^{-3} (v^j A^i - v^i A^j) - \tfrac{3}{2} c^{-3} (v^j U_{,i} - v^i U_{,j}) \\ &+ \tfrac{1}{2} c^{-1} v^k (a_{j,ik} - a_{i,jk}) + 2 c^{-3} (U^j_{,i} - U^i_{,j}) \\ &+ \tfrac{1}{2} (a_{j,i0} - a_{i,j0}).\end{aligned} \tag{2.3.70}$$

In three-dimensional vector notation $\boldsymbol{S} = (S^{(1)}, S^{(2)}, S^{(3)})$ and

$$\boldsymbol{\Omega} = (\Gamma^{(2)}_{(0)(3)}, \Gamma^{(3)}_{(0)(1)}, \Gamma^{(1)}_{(0)(2)}). \tag{2.3.71}$$

Equation (2.3.66) may be put into the form

$$\frac{\mathrm{d}\boldsymbol{S}}{\mathrm{d}t} = \boldsymbol{\Omega} \times \boldsymbol{S}. \tag{2.3.72}$$

Substituting (2.3.70) into (2.3.71) one finds

$$\begin{aligned}\boldsymbol{\Omega} = & -\tfrac{1}{2}c^{-3}(\boldsymbol{v} \times \boldsymbol{A}) + \tfrac{3}{2}c^{-3}(\boldsymbol{v} \times \nabla U) + 2c^{-3}(\nabla \times \boldsymbol{U}) \\ & + \tfrac{1}{2}c^{-1}\boldsymbol{v}\nabla(\nabla \times \boldsymbol{a}) + \tfrac{1}{2}c^{-1}\frac{\partial}{\partial t}(\nabla \times \boldsymbol{a}).\end{aligned} \tag{2.3.73}$$

Here $\boldsymbol{A} = (A^i)$ is the non-gravitational acceleration, U is the Newtonian potential, $\boldsymbol{U} = (U^i)$ is the vector potential, $\nabla = (\partial/\partial x^i)$ is the vector gradient, and $\boldsymbol{a} = (a_1, a_2, a_3)$ is a triplet of arbitrary coordinate functions. Equation (2.3.72) describes the precession of spin relative to the co-moving system whose axes are assumed to be directed towards fixed distant celestial objects. If the reference system associated with the gyroscope is considered as an analogue of the inertial dynamical reference system and the co-moving system is treated as an analogue of the inertial kinematic system then the space rotation of one system with respect to the other is determined by the angular velocity (2.3.73). The first term in (2.3.73) corresponds to the Thomas precession (1.3.17) of special relativity. The second term in (2.3.73) due to the velocity of the particle at hand is called geodesic precession or de Sitter–Fokker precession. If the vector potential $\boldsymbol{U}$ is caused by rotation of the central body determining the motion of a particle then the third term in (2.3.73) is called Lense–Thirring precession. The fourth and the fifth terms in (2.3.73) give a contribution from arbitrary coordinate functions entering into the weak field metric (2.2.32).

Relativistic effects in rotational motion of celestial bodies are far less than the effects in translatory motion. Only translatory motion will be treated further but it may be noted that the relativistic effects in the Earth's rotation have been investigated by Voinov (1988).

3

One-body Problem

3.1 SCHWARZSCHILD PROBLEM

3.1.1 Schwarzschild metric

Among many exact solutions of the Einstein field equations known at present only a few solutions are used in astronomy, mainly in relativistic cosmology and relativistic astrophysics. In relativistic celestial mechanics it is possible to use only three rigorous solutions related to the case of one gravitating body. These are the Schwarzschild solution for a fixed spherical body, the Kerr solution for a rotating spherical body and the Weyl–Levi-Civita solution for a fixed spheroid. By its application the Schwarzschild solution is the most important one.

A fixed body of spherical structure produces a spherically symmetric gravitational field with a metric of the form

$$ds^2 = p(r)c^2dt^2 + 2b(r)cdtdr - q(r)dr^2 - a^2(r)(d\theta^2 + \sin^2\theta d\varphi^2). \quad (3.1.1)$$

t is the coordinate time, r, φ, θ are spherical coordinates, p, q, a, b are functions of r to be determined from the field equations. As 'time' t and 'radial distance' r may be chosen arbitrarily not violating the field spherical symmetry, two of these functions, a and b for example, may remain arbitrary. Then functions p and q are expressed by the field equations in terms of a and b. One may consider a more general case with p, q, a, and b dependent on t as well (Brumberg 1972) but this case does not occur in applications to celestial mechanics. In terms of rectangular coordinates

$$x^1 = r\sin\theta\cos\varphi \qquad x^2 = r\sin\theta\sin\varphi \qquad x^3 = r\cos\theta$$

the metric (3.1.1) takes the form

$$\begin{aligned} ds^2 =& p(r)c^2dt^2 + 2b(r)\frac{x^i}{r}cdtdx^i \\ &- \frac{1}{r^2}\left[a^2(r)\delta_{ik} + \left(q(r) - \frac{a^2(r)}{r^2}\right)x^ix^k\right]dx^idx^k. \end{aligned} \quad (3.1.2)$$

To simplify (3.1.1) or (3.1.2) two transformations are suitable. The first one is aimed at excluding the mixed terms, while the second one reduces the spatial part of the metric to the isotropic form. The first transformation is performed by

$$dt^* = dt + \frac{1}{c}\frac{b(r)}{p(r)}dr. \tag{3.1.3}$$

Then metric (3.1.1) becomes

$$ds^2 = p(r)c^2 dt^{*^2} - \left(q(r) + \frac{b^2(r)}{p(r)}\right) dr^2 - a^2(r)(d\theta^2 + \sin^2\theta d\varphi^2). \tag{3.1.4}$$

The second transformation involves only the radial coordinate

$$r^* = f(r) \tag{3.1.5}$$

and reduces (3.1.4) to the form

$$ds^2 = A(r^*)c^2 dt^{*^2} - B(r^*)(dr^{*^2} + r^{*^2}d\theta^2 + r^{*^2}\sin^2\theta d\varphi^2) \tag{3.1.6}$$

with

$$A(r^*) = p(r) \tag{3.1.7}$$

and

$$B(r^*) = \frac{a^2(r)}{f^2(r)} = \frac{1}{f'^2(r)}\left(q(r) + \frac{b^2(r)}{p(r)}\right). \tag{3.1.8}$$

A prime denotes differentiation with respect to r. The solution of (3.1.8) relative to $f(r)$ is

$$f(r) = \exp\int \frac{1}{a(r)}\left(q(r) + \frac{b^2(r)}{p(r)}\right)^{1/2} dr. \tag{3.1.9}$$

In rectangular coordinates the metric (3.1.6) implies

$$g_{00} = A \qquad g_{0i} = 0 \qquad g_{ik} = -B\delta_{ik}. \tag{3.1.10}$$

For the moment A and B are regarded as functions of all four coordinates. For such a metric calculation of quantities characterizing gravitational fields presents no difficulties. Using tensor algebra formalism one finds the following.

Contravariant metric tensor:

$$g^{00} = \frac{1}{A} \qquad g^{0i} = 0 \qquad g^{ik} = -\frac{1}{B}\delta_{ik}. \tag{3.1.11}$$

Christoffel symbols of the first kind:

$$\Gamma_{000} = \tfrac{1}{2}A_{,0} \qquad \Gamma_{00i} = \tfrac{1}{2}A_{,i} \qquad \Gamma_{0ik} = \tfrac{1}{2}\delta_{ik}B_{,0} \qquad \Gamma_{k00} = -\tfrac{1}{2}A_{,k}$$
$$\Gamma_{k0i} = -\tfrac{1}{2}\delta_{ik}B_{,0} \qquad \Gamma_{kij} = \tfrac{1}{2}(\delta_{ij}B_{,k} - \delta_{ik}B_{,j} - \delta_{jk}B_{,i}) \tag{3.1.12}$$

Christoffel symbols of the second kind:

$$\Gamma^0_{00} = \frac{1}{2A}A_{,0} \qquad \Gamma^i_{00} = \frac{1}{2B}A_{,i} \qquad \Gamma^0_{0i} = \frac{1}{2A}A_{,i}$$
$$\Gamma^k_{0i} = \frac{1}{2B}B_{,0}\delta_{ik} \qquad \Gamma^0_{ik} = \frac{1}{2A}B_{,0}\delta_{ik}$$
$$\Gamma^k_{ij} = \frac{1}{2B}(\delta_{ik}B_{,j} + \delta_{jk}B_{,i} - \delta_{ij}B_{,k}) \tag{3.1.13}$$

Ricci tensor:

$$R_{00} = -\frac{1}{2B}\left(A_{,ss} - \frac{1}{2A}A_{,s}A_{,s} + \frac{1}{2B}A_{,s}B_{,s} - 3B_{,00}\right.$$
$$\left. + \frac{3}{2A}A_{,0}B_{,0} + \frac{3}{2B}B_{,0}B_{,0}\right)$$
$$R_{0i} = -\frac{1}{2B}\left(-2B_{,0i} + \frac{1}{A}A_{,i}B_{,0} + \frac{2}{B}B_{,i}B_{,0}\right)$$
$$R_{ik} = -\frac{1}{2B}\left[-B_{,ik} - \delta_{ik}B_{,ss} + \frac{1}{2A}(A_{,i}B_{,k} + A_{,k}B_{,i}\right.$$
$$- \delta_{ik}A_{,s}B_{,s} + \delta_{ik}B_{,0}B_{,0}) + \frac{1}{2B}(3B_{,i}B_{,k} + \delta_{ik}B_{,s}B_{,s})$$
$$\left. - \frac{B}{A}\left(A_{,ik} - \delta_{ik}B_{,00} + \frac{1}{2A}\delta_{ik}A_{,0}B_{,0} - \frac{1}{2A}A_{,i}A_{,k}\right)\right]. \tag{3.1.14}$$

Returning to the Schwarzschild problem one has $A_{,0} = B_{,0} = 0$. The field equations (2.1.15) for the empty space, i.e. for the external relative to the gravitating body space, yield

$$A_{,ss} - \frac{1}{2A}A_{,s}A_{,s} + \frac{1}{2B}A_{,s}B_{,s} = 0 \tag{3.1.15}$$

$$B_{,ik} + \delta_{ik}B_{,ss} + \frac{1}{2A}(\delta_{ik}A_{,s}B_{,s} - A_{,i}B_{,k} - A_{,k}B_{,i})$$
$$- \frac{1}{2B}(3B_{,i}B_{,k} + \delta_{ik}B_{,s}B_{,s}) + \frac{B}{A}\left(A_{,ik} - \frac{1}{2A}A_{,i}A_{,k}\right) = 0. \tag{3.1.16}$$

Equations (3.1.15) and (3.1.16) may be satisfied by putting

$$A = \left(\frac{1-\psi/2}{1+\psi/2}\right)^2 \qquad B = (1+\tfrac{1}{2}\psi)^4 \tag{3.1.17}$$

where the function ψ of the spatial coordinates is to satisfy the relation

$$\psi\psi_{,ik} - 3\psi_{,i}\psi_{,k} + \delta_{ik}\psi_{,s}\psi_{,s} = 0 \tag{3.1.18}$$

including, following from (3.1.18) for contraction $i = k$, the Laplace equation

$$\psi_{,ss} = 0. \tag{3.1.19}$$

For the function ψ one may take the Newtonian potential

$$\psi = m/r^* \qquad m = GM/c^2. \tag{3.1.20}$$

G is the gravitational constant, M is the mass of the gravitating body, the constant m is chosen to provide the coincidence with the Newtonian limit for h_{00}. From (3.1.5), (3.1.7)–(3.1.9) there results

$$a(r) = r^*\left(1+\frac{m}{2r^*}\right)^2 \qquad r^* = \tfrac{1}{2}[(a^2-2ma)^{1/2} + a - m] \tag{3.1.21}$$

$$p(r) = 1 - \frac{2m}{a(r)} \qquad q(r) = \frac{a'^2(r) - b^2(r)}{p(r)}. \tag{3.1.22}$$

Thus the static Schwarzschild metric is determined by expressions (3.1.1) or (3.1.2) with two arbitrary functions $a(r), b(r)$. Two other functions $p(r)$ and $q(r)$ are determined by (3.1.22). Depending on the choice of $a(r)$ and $b(r)$ one may obtain different coordinate forms of this metric. The forms most generally employed are those associated with the following six sets of values for $a(r)$ and $b(r)$:

	(I)	(II)	(III)	(IV)	(V)	(VI)
$a(r) =$	r	$r+m$	$r(1+m/2r)^2$	$a^3 = r^2(a-2m)$	r	r
$b(r) =$	0	0	0	0	$2m/r$	$(2m/r)^{1/2}$

(3.1.23)

The values (I) correspond to the so-called standard coordinates of the Schwarzschild problem. This form is of great use in investigating the Schwarzschild problem. The values (II) correspond to the harmonic coordinates defined by (2.1.16). If harmonic coordinates are denoted by a tilde then the transformation from harmonic to arbitrary coordinates is represented by the relations

$$d\tilde{t} = dt + c^{-1}\frac{b(r)}{p(r)}dr \tag{3.1.24}$$

$$\tilde{r} = a(r) - m \qquad \tilde{\theta} = \theta \qquad \tilde{\varphi} = \varphi \tag{3.1.25}$$

or in rectangular coordinates

$$\tilde{x}^i = \frac{a(r) - m}{r} x^i. \tag{3.1.26}$$

The values (III) in (3.1.23) are associated with the isotropic coordinates marked here by an asterisk. In these coordinates the space part of the interval is distinguished from the Euclidean metric only by a factor dependent on a point of the field.

Form (IV) is obtained by taking $a(r)$ to be a root of the indicated cubic equation. In these coordinates introduced by Painlevé the gravitational field is equivalent by its action to a central force since the coefficients of metric (3.1.1) in $c^2 \mathrm{d}t^2$ and $r^2(\mathrm{d}\theta^2 + \sin^2\theta \mathrm{d}\varphi^2)$ are equal in magnitude and differ only by sign. This form has been investigated by Ganea (1973).

Forms (V) and (VI) introduced respectively by Eddington and Painlevé are examples of stationary metrics for the Schwarzschild problem. In practical problems the relevant reference systems are not used since the presence of the mixed terms makes the solution more complicated. These mixed terms are of artificial, mathematical origin and are not due to physical reasons.

In what follows only the static case $b(r) = 0$ of the Schwarzschild problem is considered. Then there remains in (3.1.1) or (3.1.2) only one arbitrary function $a(r)$ satisfying the condition of the Galilean metric at infinity: $a(r)/r \to 1$, $a'(r) \to 1$ with $r \to \infty$.

Historical remarks concerning all systems (3.1.23) may be found, for example, in the treatise by Chazy (1928, 1930).

Solution (3.1.1) or (3.1.2) with (3.1.22) relates to the external Schwarzschild problem, i.e. the determination of the gravitational field outside a fixed spherical body. For cosmology and relativistic astrophysics the internal Schwarzschild problem, i.e. the determination of the gravitational field inside a body, is of no less importance. But for practical purposes of relativistic celestial mechanics this solution is of no interest and is not considered here.

The external Schwarzschild solution is valid as long as the component g_{00} is positive. The value of r which vanishes g_{00} is called the gravitational radius (radius of the the Schwarzschild sphere inside which the external solution is not valid). This value is determined by the equation $p(r) = 0$. Therefore, the gravitational radius is equal to $2m$ in standard coordinates, m in harmonic coordinates and $m/2$ in isotropic coordinates.

3.1.2 Motion of a test particle and measurable quantities

The Schwarzschild solution is the exact solution of the field equations. Therefore it is possible with the aid of the variational geodesic principle to derive the exact equations of motion of a test particle and light propagation. These equations may be rigorously solved in elliptic functions. Indeed, taking the plane $\theta = \pi/2$ as plane of motion and applying the principle (1.2.55) to the metric (3.1.1) one may deduce the Lagrange function in the form

$$L = -p(r)c^2 \left(\frac{dt}{ds}\right)^2 + q(r) \left(\frac{dr}{ds}\right)^2 + a^2(r) \left(\frac{d\varphi}{ds}\right)^2. \tag{3.1.27}$$

L being explicitly independent of ct, φ and s there exist three first integrals

$$p(r)c\frac{dt}{ds} = E \qquad a^2(r)\frac{d\varphi}{ds} = K \qquad L = I \tag{3.1.28}$$

with $I = 1$ for the material particle and $I = 0$ for the light particle. Thus, the motion is described by the system with one degree of freedom with the Lagrangian

$$L = q(r) \left(\frac{dr}{ds}\right)^2 + K^2 W(r) \tag{3.1.29}$$

where

$$W(r) = \frac{1}{a^2(r)} - \frac{A}{p(r)} \qquad A = (E/K)^2. \tag{3.1.30}$$

Consider, first of all, the circular solutions of the Schwarzschild problem. The circular solution $r =$ constant is evidently determined by the condition

$$W'(r) = 0. \tag{3.1.31}$$

For the stability of the circular solution one should have

$$W''(r) \geq 0. \tag{3.1.32}$$

Relation (3.1.31) results in

$$\frac{p^2(r)}{a(r)} = mA. \tag{3.1.33}$$

In addition, the condition $L = I$ involves the restriction on a radius of the circular motion as follows:

$$\frac{1}{a(r)} \left(1 - \frac{3m}{a(r)}\right) = mB \qquad B = \frac{I}{K^2}. \tag{3.1.34}$$

Finally, relation (3.1.32) leads to the inequality

$$1 - \frac{6m}{a(r)} \geq 0. \tag{3.1.35}$$

From (3.1.34) and (3.1.35) it follows that the radius of the nearest circular orbit to the gravitating body is determined by the equation $a(r) = 3m$ (with the motion with the light velocity for this orbit) and the radius of the nearest stable circular orbit is determined by the equation $a(r) = 6m$. These results are usually formulated in the standard coordinates. Introducing the function $a(r)$ one can rewrite them in terms of arbitrary quasi-Galilean coordinates.

For the circular motion

$$\varphi = nt + \text{constant} \tag{3.1.36}$$

with $n = \mathrm{d}\varphi/\mathrm{d}t$ being the mean motion of the particle. Relation (3.1.33) represents the generalized Kepler third law

$$n^2 a^3(r) = GM. \tag{3.1.37}$$

In accordance with (3.1.1) the proper time of a particle moving in a circular orbit is determined by

$$(\mathrm{d}\tau/\mathrm{d}t)^2 = p(r) - c^{-2} n^2 a^2(r)$$

or

$$\frac{\mathrm{d}\tau}{\mathrm{d}t} = \left(1 - \frac{3m}{a(r)}\right)^{1/2}. \tag{3.1.38}$$

The mean motion $n' = \mathrm{d}\varphi/\mathrm{d}\tau$ referred to the proper time is evidently the measurable quantity since the sidereal period of revolution, expressed in the proper time, $T' = 2\pi/n'$, is directly obtained from astronomical observations. Define now two auxiliary quantities r'_N and r_N by means of the Newtonian formulae

$$n'^2 r'^3_N = GM \qquad n^2 r^3_N = GM. \tag{3.1.39}$$

From (3.1.39) r'_N represents an indirectly measurable quantity. r_N is also an indirectly measurable quantity because from (3.1.37)–(3.1.39) it follows that

$$r'_N = r_N \left(1 - \frac{3m}{r_N}\right)^{1/3}. \tag{3.1.40}$$

The physical constants GM and m do not depend on the coordinate system and are to be considered as the measurable quantities. Therefore, the mean motion n expressed in terms of GM and r_N is the indirectly measurable quantity. The angular coordinate φ in the Schwarzschild problem may be regarded as a measurable quantity. Therefore, in virtue of the relation $n = \mathrm{d}\varphi/\mathrm{d}t$ the coordinate time t on the circular motion may also be treated as an indirectly measurable quantity. Only the radius r of the circular orbit determined by the equation

$$a(r) = r_N \tag{3.1.41}$$

is a coordinate-dependent, unmeasurable quantity. Solving equation (3.1.41) with respect to r one obtains an explicit expression of r depending on the indirectly measurable quantity r_N and on the coordinate conditions employed.

Here, we use the notions of 'measurable quantity', 'indirectly measurable quantity', 'coordinate-dependent (unmeasurable) quantity'. With each term being conventional let us explain the meaning attributed here to these notions. Measurable quantity is to be meant as a quantity directly obtainable from observation without involving any theoretical data (such as, for example, the theories of motion of celestial bodies or the laws of light propagation). Indirectly measurable quantity means a quantity resulting from calculations involving only measurable quantities. Finally, a coordinate-dependent (unmeasurable) quantity is referred to as a quantity resulting from calculations involving quantities due to application of the reference system formalism. In particular, such a quantity may depend on explicitly introduced arbitrary functions or parameters characterizing the choice of coordinate system (coordinate conditions).

Returning to the general case and drawing on (3.1.28) and (3.1.29) it is easy to obtain the differential equation of the trajectory in the form

$$\left(\frac{\mathrm{d}(1/a(r))}{\mathrm{d}\varphi}\right)^2 = A - B + \frac{2mB}{a(r)} - \frac{1}{a^2(r)} + \frac{2m}{a^3(r)} \tag{3.1.42}$$

with A and B determined again by (3.1.30) and (3.1.34) in terms of the initial constants E and K. The right-hand side of (3.1.42) is a polynomial of the third degree with respect to $1/a(r)$. Thus, this equation may be rigorously solved in elliptic functions. The detailed celestial mechanics analysis of such solution has been performed by Cherny (1949). Among numerous papers dealing with the equation specifically in GRT let us note, for example, the investigations by Bogorodsky (1962) and Bogdan and Plebanski (1962), as well as a more recent paper by Ashby (1986). Exact solution of (3.1.42) involves many highly interesting questions having no analogy in the Newtonian two-body problem, i.e. the capture of the particle, temporary capture (close encounter comprising several revolutions around a gravitating body) and so on. But all these phenomena occur at the distance not exceeding several gravitational radii from the primary and are applied in problems of relativistic astrophysics considered, for example, in the treatise by Zel'dovich and Novikov (1967).

For the real bodies of the Solar System the gravitational radii are extremely small as compared with their linear sizes and all these interesting cases cannot be realized in practice. For relativistic celestial mechanics it is sufficient to deal with an approximate solution. However, its distinction from the Newtonian solution should be investigated in detail. In particular, the study of influence of the coordinate conditions on a given relativistic relation is here of great importance. Therefore, instead of the exact equation (3.1.42) an approximate equation of the trajectory of the particle is studied below and this analysis extends to all reference systems (I)–(IV) of (3.1.23).

3.1.3 Post-Newtonian approximation

For most practically employed quasi-Galilean reference systems one may represent the function $a(r)$ by the expansion

$$a(r) = r\left(1 + (1-\alpha)\frac{m}{r} + \epsilon\frac{m^2}{r^2} + \dots\right) \tag{3.1.43}$$

$\alpha, \epsilon, \dots$ being the coordinate parameters defining specific coordinate conditions. The values $\alpha = 1, \epsilon = 0$ correspond to standard coordinates (I) of (3.1.23). Harmonic coordinates (II) are obtained with $\alpha = \epsilon = 0$. Isotropic coordinates (III) are associated with $\alpha = 0, \epsilon = 1/4$. Painlevé coordinates (IV) result from $\alpha = 2, \epsilon = -3/2$. Confining ourselves to post-Newtonian accuracy, i.e. retaining in g_{00} terms of second order and in g_{ik} terms of first order with respect to m/r, one may reject in (3.1.43) all terms of second and higher order in m/r. In the post-Newtonian approximation isotropic coordinates coincide with harmonic ones. In this approximation it is suitable to consider the metric (3.1.2) with (3.1.43) taking into account the main parameters β and γ of the PPN formalism (Will 1985). Then the generalized Schwarzschild metric in the post-Newtonian approximation takes the form

$$\begin{aligned} \mathrm{d}s^2 &= \left(1 - \frac{2m}{r} + 2(\beta-\alpha)\frac{m^2}{r^2} + \dots\right)c^2\mathrm{d}t^2 \\ &\quad - \left[\delta_{ij} + \frac{2m}{r}\left((\gamma-\alpha)\delta_{ij} + \alpha\frac{x^i x^j}{r^2}\right) + \dots\right]\mathrm{d}x^i\mathrm{d}x^j . \end{aligned} \tag{3.1.44}$$

For GRT one has $\beta = \gamma = 1$. At present, astronomical observation data (radio ranging of the internal planets, lunar laser ranging, trajectory measurements of space probes) confirm these theoretical values within a precision of 0.1 to 1.5% (Will 1986). Nevertheless, considering β and γ in the literal form enables one to obtain a better understanding of the contribution of separate terms of the metric to the relativistic perturbations. Under $\alpha = 0$ the form (3.1.44) reduces to the well-known Eddington–Robertson metric.

Allowing the coordinate time t to be an independent argument and applying (2.2.53) to (3.1.44) one may present the Lagrangian of the equations of the moving test particle in the form

$$L = \frac{1}{2}\dot{\boldsymbol{r}}^2 + \frac{GM}{r} + \frac{1}{8}c^{-2}(\dot{\boldsymbol{r}}^2)^2 + \frac{m}{r}\left[\left(\gamma + \frac{1}{2} - \alpha\right)\dot{\boldsymbol{r}}^2 + \left(-\beta + \frac{1}{2} + \alpha\right)\frac{GM}{r} + \alpha\frac{(\boldsymbol{r}\dot{\boldsymbol{r}})}{r^2}\right]. \tag{3.1.45}$$

$\boldsymbol{r}$ denotes the triplet of the space coordinates x^1, x^2, x^3 of the moving particle. It is possible to apply the usual vector operations to $\boldsymbol{r}$ but it is self-evident that this quantity is not a physically meaningful vector. The word 'vector' here and below means just the triplet of the corresponding components. Lagrangian (3.1.45) results in the equations of motion (2.2.49)

$$\ddot{\boldsymbol{r}} + \frac{GM}{r^3}\boldsymbol{r} = \frac{m}{r^3}\left[\left(2(\beta + \gamma - \alpha)\frac{GM}{r} - (\gamma + \alpha)\dot{\boldsymbol{r}}^2 + 3\alpha\frac{(\boldsymbol{r}\dot{\boldsymbol{r}})}{r^2}\right)\boldsymbol{r} + 2(\gamma + 1 - \alpha)(\boldsymbol{r}\dot{\boldsymbol{r}})\dot{\boldsymbol{r}}\right]. \tag{3.1.46}$$

These equations in rectangular coordinates are convenient for numerical integration. To solve them analytically introduce polar coordinates r and u in the plane of motion. This is achieved by the transformation

$$\boldsymbol{r} = X\boldsymbol{l} + Y\boldsymbol{m} \qquad \dot{\boldsymbol{r}} = \dot{X}\boldsymbol{l} + \dot{Y}\boldsymbol{m} \tag{3.1.47}$$

$\boldsymbol{l}$ and $\boldsymbol{m}$ being unit vectors determining the orientation of the plane of motion by (1.1.6). Putting then

$$X = r\cos u \qquad Y = r\sin u \tag{3.1.48}$$

one derives the equations of motion in the orbital plane

$$\ddot{r} - r\dot{u}^2 + \frac{GM}{r^2} = m\left(2(\beta + \gamma - \alpha)\frac{GM}{r^3} - (\gamma + \alpha)\dot{u}^2 + (\gamma + 2)\frac{\dot{r}^2}{r^2}\right)$$
$$\frac{\mathrm{d}}{\mathrm{d}t}(r^2\dot{u}) = 2m(\gamma + 1 - \alpha)\dot{r}\dot{u}. \tag{3.1.49}$$

These equations admit, first of all, a particular circular solution

$$r = a \qquad u = nt + \text{constant} \tag{3.1.50}$$

where the mean motion n is related to the radius a of a circular orbit by the formula

$$n = \left(\frac{GM}{a^3}\right)^{1/2}\left[1+\frac{m}{a}\left(-\frac{1}{2}\gamma-\beta+\frac{3}{2}\alpha\right)\right]. \tag{3.1.51}$$

Equations (3.1.49) possess two integrals, the area integral and the *vis viva* (energy) integral. It is very simple to derive them from the Lagrangian (3.1.45) by introducing spherical coordinates r, θ and φ, taking combinations $\partial L/\partial\dot{\varphi}$ and $\dot{r}(\partial L/\partial\dot{r})+\dot{\varphi}(\partial L/\partial\dot{\varphi})-L$ for $\theta=\pi/2$ and putting afterwards $\varphi=u$. The integral of area has the form

$$r^2\dot{u}\left(1+\frac{1}{2}c^{-2}(\dot{r}^2+r^2\dot{u}^2)+(2\gamma+1-2\alpha)\frac{m}{r}\right)=P. \tag{3.1.52}$$

The *vis viva* integral is

$$\frac{1}{2}(\dot{r}^2+r^2\dot{u}^2)-\frac{GM}{r}+\frac{3}{8}c^{-2}(\dot{r}^2+r^2\dot{u}^2)^2+\frac{m}{r}\left[\left(\beta-\frac{1}{2}-\alpha\right)\frac{GM}{r}\right.$$
$$\left.+\left(\gamma+\frac{1}{2}\right)\dot{r}^2+\left(\gamma+\frac{1}{2}-\alpha\right)r^2\dot{u}^2\right]=h \tag{3.1.53}$$

Substituting into the relativistic terms the Newtonian expressions one gets

$$r^2\dot{u}=P\left(1-2(\gamma+1-\alpha)\frac{m}{r}-\frac{h}{c^2}\right) \tag{3.1.54}$$

$$\dot{r}^2=-r^2\dot{u}^2+2\frac{GM}{r}+2h$$
$$+2m\left(-\frac{3}{2}\frac{h^2}{GM}-2(\gamma+2)\frac{h}{r}+(-2\gamma-\beta-2+\alpha)\frac{GM}{r^2}+\alpha\frac{P^2}{r^3}\right). \tag{3.1.55}$$

From this there results the differential equation of the trajectory in the form

$$\left(\frac{\mathrm{d}(1/r)}{\mathrm{d}u}\right)^2=\frac{2h}{P^2}+\frac{2GM}{P^2}\frac{1}{r}-\frac{1}{r^2}+m\left(\frac{h^2}{GMP^2}+\frac{4h}{P^2}(\gamma+1-2\alpha)\frac{1}{r}\right.$$
$$\left.+\frac{2GM}{P^2}(2\gamma+2-\beta-3\alpha)\frac{1}{r^2}+2\alpha\frac{1}{r^3}\right). \tag{3.1.56}$$

It is evident that this equation coincides with equation (3.1.42), deduced above, with $\beta=\gamma=1$, $\varphi=u$, using (3.1.43) and relating the constants of integration by putting

$$A-B=\frac{2h}{P^2}+\frac{h^2}{c^2P^2} \qquad B=\frac{c^2}{P^2}.$$

Instead of P and h one introduces into (3.1.56) new constants a and e such that $a(1-e)$ and $a(1+e)$ are the boundary limits for changing r in the motion of the particle. Thus, a and e act as the semi-major axis and the eccentricity of the orbit. Then,

$$\left(\frac{\mathrm{d}(1/r)}{\mathrm{d}u}\right)^2 = \left(\frac{1}{r} - \frac{1}{a(1+e)}\right)\left(\frac{1}{a(1-e)} - \frac{1}{r}\right) \times \left(1 - 2(2\gamma - \beta + 2 - \alpha)\frac{m}{a(1-e^2)} - 2\alpha\frac{m}{r}\right) \quad (3.1.57)$$

with

$$P^2 = GMa(1-e^2)\left[1 + \left(-\gamma - 1 + \alpha + 2\frac{2\gamma - \beta + 2 - \alpha}{1-e^2}\right)\frac{m}{a}\right] \quad (3.1.58)$$

$$h = -\frac{GM}{2a}\left(1 - \frac{1}{4}(4\gamma + 3 - 4\alpha)\frac{m}{a}\right). \quad (3.1.59)$$

Equation (3.1.57) may be solved in the same form as for Newtonian theory, i.e.

$$r = \frac{a(1-e^2)}{1 + e\cos f}. \quad (3.1.60)$$

Substitution of this expression into (3.1.57) yields

$$\left(\frac{\mathrm{d}f}{\mathrm{d}u}\right)^2 = 1 - 2(2\gamma - \beta + 2 - \alpha)\frac{m}{a(1-e^2)} - 2\alpha\frac{m}{r} \quad (3.1.61)$$

or within the adopted accuracy

$$\frac{\mathrm{d}f}{\mathrm{d}u} = \nu\left(1 - \alpha\frac{m}{a(1-e^2)}e\cos f\right) \quad (3.1.62)$$

with

$$\nu = 1 - (2\gamma - \beta + 2)\frac{m}{a(1-e^2)}. \quad (3.1.63)$$

The approximate solution of (3.1.62) is described in the form

$$f = \psi - \alpha\frac{m}{a(1-e^2)}e\sin\psi \quad (3.1.64)$$

and

$$\psi = \nu(u - \omega) \quad (3.1.65)$$

ω being an arbitrary constant. Thus, the equation of the trajectory of a test particle in the field with the metric (3.1.44) in the post-Newtonian approximation is given by (3.1.60) and (3.1.63)–(3.1.65).

For relativistic celestial mechanics only the case of the elliptic type motion is of practical interest. In this case between two consecutive passages through the pericentre, $f = 0$ and $f = 2\pi$, the argument of latitude u changes from $u = \omega$ to $u = \omega + 2\pi/\nu = \omega + 2\pi + \Delta\omega$ with

$$\Delta\omega = (2\gamma + 2 - \beta)\frac{2\pi m}{a(1-e^2)} \qquad (3.1.66)$$

reducing for $\beta = \gamma = 1$ to the famous formula of the Schwarzschild advancement of the pericentre for one revolution.

Having obtained the trajectory of the particle it is not difficult to find the dependence of the true anomaly f on time. Using (3.1.54), (3.1.64) and (3.1.65) one gets

$$n dt = \frac{r^2}{a^2(1-e^2)^{1/2}}\left[1 + \frac{m}{a}\left(-\beta + \alpha + (2\gamma + 2 - \alpha)\frac{a}{r}\right)\right] \mathrm{d}f \qquad (3.1.67)$$

with the mean motion n determined by (3.1.51). Introducing the eccentric anomaly E related to the true anomaly f by the usual formula (1.1.14) and integrating (3.1.67) one obtains the relativistic analogue of the Kepler equation

$$E - \left(1 + (-2\gamma - 2 + \alpha)\frac{m}{a}\right) e \sin E = l \qquad (3.1.68)$$

the mean anomaly l being the linear function of time t with the frequency

$$\frac{\mathrm{d}l}{\mathrm{d}t} = n\left(1 - (2\gamma + 2 - \beta)\frac{m}{a}\right). \qquad (3.1.69)$$

With the aid of (3.1.68) it is easy to find the expression for the period of revolution of the particle around the central body. In doing this one should clearly define what period is kept in mind. For the anomalistic period T_1 defined as the time interval of the increase of f or E by 2π one has

$$T_1 = \frac{2\pi}{n}\left(1 + (2\gamma + 2 - \beta)\frac{m}{a}\right). \qquad (3.1.70)$$

The sidereal period T_2' is defined as the time interval of the increase of u by 2π. ψ and f increase thereby by $2\pi\nu$ and the increase of E is

$$2\pi\left(1 + (\nu - 1)\frac{(1-e^2)^{1/2}}{1 + e\cos f_0}\right)$$

with f_0 being the initial value of the true anomaly. Therefore,

$$T_2' = \frac{2\pi}{n}\left[1 + (2\gamma + 2 - \beta)\frac{m}{a}\left(1 - \frac{(1-e^2)^{1/2}}{(1 + e\cos f_0)^2}\right)\right]. \qquad (3.1.71)$$

To the end of this period the position vector of the particle in the chosen reference system coincides by its direction with the initial position. T_2' depends on the initial position of the particle. Using

$$\frac{1}{2\pi}\int_0^{2\pi}\frac{\mathrm{d}f}{(1+e\cos f)^2}=\frac{1}{2\pi}\int_0^{2\pi}\frac{\mathrm{d}l}{(1-e^2)^{3/2}}=\frac{1}{(1-e^2)^{3/2}}$$

the mean value of the sidereal period in changing the initial position f_0 from 0 to 2π is determined by

$$T_2=\frac{2\pi}{n}\left(1-(2\gamma+2-\beta)\frac{m}{a}\frac{e^2}{1-e^2}\right). \tag{3.1.72}$$

Let us remember that in (3.1.69)–(3.1.71) the mean motion n is given by (3.1.51). In deriving (3.1.68) and subsequently from it formulae (3.1.70)–(3.1.72) the coordinate time t is used as the independent argument. For some problems it may be more suitable to use the proper time τ defined by (2.3.2). In application to (3.1.44) in the post-Newtonian approximation there results

$$\left(\frac{\mathrm{d}\tau}{\mathrm{d}t}\right)^2=1-\frac{2m}{r}-c^{-2}\dot{r}^2+\ldots$$

or with the use of the energy integral

$$\frac{\mathrm{d}\tau}{\mathrm{d}t}=1-\frac{2m}{r}+\frac{1}{2}\frac{m}{a}+\ldots. \tag{3.1.73}$$

Therefore, equation (3.1.67) referred to the proper time τ reduces to

$$n'\mathrm{d}\tau=\frac{r^2}{a^2(1-e^2)^{1/2}}\left[1+\frac{m}{a}\left(2-\beta+\alpha+(2\gamma-\alpha)\frac{a}{r}\right)\right]\mathrm{d}f \tag{3.1.74}$$

with the mean motion n' for the proper time of the particle, i.e.

$$n'=\left(\frac{GM}{a^3}\right)^{1/2}\left[1+\frac{m}{a}\left(-\frac{1}{2}\gamma-\beta+\frac{3}{2}+\frac{3}{2}\alpha\right)\right]. \tag{3.1.75}$$

By integrating one obtains the relativistic analogue of the Kepler equation referred to the proper time of the moving particle

$$E-\left(1+(-2\gamma+\alpha)\frac{m}{a}\right)e\sin E=l' \tag{3.1.76}$$

and l' is the linear function of τ with the frequency

$$\frac{\mathrm{d}l'}{\mathrm{d}\tau}=n'\left(1-(2\gamma+2-\beta)\frac{m}{a}\right). \tag{3.1.77}$$

Therefore, the anomalistic and sidereal periods similar to (3.1.70)–(3.1.72) but referred to the proper time are respectively

$$\mathcal{T}_1 = \frac{2\pi}{n'}\left(1 + (2\gamma + 2 - \beta)\frac{m}{a}\right) \tag{3.1.78}$$

$$\mathcal{T}_2' = \frac{2\pi}{n'}\left[1 + (2\gamma + 2 - \beta)\frac{m}{a}\left(1 - \frac{(1-e^2)^{1/2}}{(1 + e\cos f_0)^2}\right)\right] \tag{3.1.79}$$

$$\mathcal{T}_2 = \frac{2\pi}{n'}\left(1 - (2\gamma + 2 - \beta)\frac{m}{a}\frac{e^2}{1-e^2}\right). \tag{3.1.80}$$

Thus, $\mathcal{T}_1$, $\mathcal{T}_2'$, $\mathcal{T}_2$ are expressed by the same formulae as T_1, T_2', T_2 replacing n by n'.

3.1.4 Solution for small eccentricities

If the eccentricity is small it is convenient to use expansions in powers of the eccentricity enabling us to represent the coordinates of the particle as explicit functions of time t. Such expressions are deduced here up to the terms of the second degree inclusively. The relation between r and E gives, from (3.1.68),

$$\begin{aligned}\frac{r}{a} &= 1 + \frac{1}{2}\left(1 + (-2\gamma - 2 + \alpha)\frac{m}{a}\right)e^2 - e\cos l \\ &\quad - \frac{1}{2}\left(1 + (-2\gamma - 2 + \alpha)\frac{m}{a}\right)e^2\cos 2l + \dots\end{aligned} \tag{3.1.81}$$

Using the expression (3.1.60) one may integrate (3.1.67) as follows:

$$\begin{aligned}l &= f - \left(2 + (-2\gamma - 2 + \alpha)\frac{m}{a}\right)e\sin f \\ &\quad + \left[\frac{3}{4} + \left(-\gamma - 1 + \frac{1}{2}\alpha\right)\frac{m}{a}\right]e^2\sin 2f + \dots\end{aligned} \tag{3.1.82}$$

Inversion of this relation yields

$$\begin{aligned}f &= l + \left(2 + (-2\gamma - 2 + \alpha)\frac{m}{a}\right)e\sin l \\ &\quad + \left[\frac{5}{4} + \left(-3\gamma - 3 + \frac{3}{2}\alpha\right)\frac{m}{a}\right]e^2\sin 2l + \dots\end{aligned} \tag{3.1.83}$$

From this, using (3.1.64), one has

$$\begin{aligned}\psi &= l + 2\left(1 + (-\gamma - 1 + \alpha)\frac{m}{a}\right)e\sin l \\ &\quad + \left[\frac{5}{4} + \left(-3\gamma - 3 + \frac{5}{2}\alpha\right)\frac{m}{a}\right]e^2\sin 2l + \dots\end{aligned} \tag{3.1.84}$$

Finally, with the aid of (3.1.65), one obtains

$$u = \omega + \frac{l}{\nu} + 2\left(1 + (\gamma + 1 - \beta + \alpha)\frac{m}{a}\right) e \sin l + \left[\frac{5}{4} + \left(-\frac{1}{2}\gamma - \frac{1}{2} - \frac{5}{4}\beta + \frac{5}{4}\alpha\right)\frac{m}{a}\right] e^2 \sin 2l + \ldots. \quad (3.1.85)$$

Thus, the rectangular coordinates (3.1.48) of the moving particle contain two angular variables l and l/ν associated with the periods T_1 and T_2, respectively. Instead, one may introduce the mean longitude λ and the longitude of pericentre π which are linear functions of time t with frequencies

$$\dot{\lambda} = n^* = \left(1 + (2\gamma + 2 - \beta)\frac{m}{a}e^2\right) n \quad (3.1.86)$$

$$\dot{\pi} = (2\gamma + 2 - \beta)\frac{m}{a(1-e^2)} n. \quad (3.1.87)$$

Then, in (3.1.81) and (3.1.85) one may replace l as the argument of trigonometric functions by $\lambda - \pi$. The combination $\omega + l/\nu$ in (3.1.85) may be replaced by $\lambda - \Omega$ with the constant Ω representing the longitude of the ascending node of the orbit.

Instead of a and e one may introduce any other similar quantities a^* and e^*. For example, according to the traditional choice of mean elements e^* might be chosen as being one half the coefficient of $\sin(\lambda - \pi)$ in the expansion for u and a^* might be chosen as being related to n^* by the same functional dependence (3.1.51) between n and a. With such a choice one has

$$a^* = a\left(1 + \frac{2}{3}(-2\gamma - 2 + \beta)\frac{m}{a}e^2\right) \quad (3.1.88)$$

$$e^* = \left(1 + (\gamma + 1 - \beta + \alpha)\frac{m}{a}\right) e. \quad (3.1.89)$$

Expansions (3.1.81) and (3.1.85) become

$$\frac{r}{a^*} = 1 + \frac{1}{2}\left[1 + \left(-\frac{4}{3}\gamma - \frac{4}{3} + \frac{2}{3}\beta - \alpha\right)\frac{m}{a^*}\right] e^{*2} - \left(1 + (-\gamma - 1 + \beta - \alpha)\frac{m}{a^*}\right) e^* \cos(\lambda - \pi) - \frac{1}{2}\left(1 + (-4\gamma - 4 + 2\beta - \alpha)\frac{m}{a^*}\right) e^{*2} \cos(2\lambda - \pi) + \ldots \quad (3.1.90)$$

$$u = \lambda - \Omega + 2e^* \sin(\lambda - \pi) + \left[\frac{5}{4} + \left(-3\gamma - 3 + \frac{5}{4}\beta\right)\frac{m}{a^*}\right] e^{*2} \sin(2\lambda - \pi) + \ldots. \quad (3.1.91)$$

Let us remember once again the expression of angular variables in terms of the old and new constants:

$$l = \left(1 - (2\gamma + 2 - \beta)\frac{m}{a}\right) n(t - t_0) + l_0 = \nu n^*(t - t_0) + l_0 \qquad (3.1.92)$$

$$\lambda = \left(1 + (2\gamma + 2 - \beta)\frac{m}{a}e^2\right) n(t - t_0) + \lambda_0 = n^*(t - t_0) + \lambda_0 \qquad (3.1.93)$$

$$\pi = (2\gamma + 2 - \beta)\frac{m}{a(1 - e^2)}\lambda + \nu(\omega + \Omega) \qquad (3.1.94)$$

where l_0 is the mean anomaly at the epoch and $\lambda_0 = \omega + \Omega + l_0/\nu$ is the mean longitude at the epoch. From (3.1.92)–(3.1.94) we obtain the relations

$$l = \lambda - \pi \qquad \lambda = \omega + \Omega + l/\nu \qquad (3.1.95)$$

used above.

3.1.5 Orbital elements and measurable quantities

To illustrate the problem of measurable quantities let us examine the expression for spherical coordinates. The radius vector r is given by (3.1.81) or (3.1.90). On the basis of the expressions (3.1.85) or (3.1.91) for u, it is easy to derive the corresponding expressions for θ and φ. In fact,

$$\cos\theta = \sin i \sin u$$

and for small inclinations

$$\varphi = u + \Omega - \tfrac{1}{4}\sin^2 i \sin 2u + \ldots.$$

Substitution of (3.1.85) yields

$$\begin{aligned}\cos\theta = \sin i \Bigg[&\sin(\lambda - \Omega) \\ &+ \left(1 + (\gamma + 1 - \beta + \alpha)\frac{m}{a}\right) e[\sin(2\lambda - \pi - \Omega) - \sin(\pi - \Omega)] \\ &- \left(1 + 2(\gamma + 1 - \beta + \alpha)\frac{m}{a}\right) e^2 \sin(\lambda - \Omega) \\ &+ \frac{1}{8}\left(1 + (-10\gamma - 10 + 3\beta + 2\alpha)\frac{m}{a}\right) e^2 \sin(\lambda - 2\pi + \Omega) \\ &+ \frac{1}{8}\left(9 + (6\gamma + 6 - 13\beta + 18\alpha)\frac{m}{a}\right) e^2 \sin(3\lambda - 2\pi - \Omega) + \ldots \Bigg]\end{aligned} \qquad (3.1.96)$$

$$\varphi = \lambda + 2\left(1 + (\gamma + 1 - \beta + \alpha)\frac{m}{a}\right) e \sin(\lambda - \pi)$$
$$+ \left[\frac{5}{4} + \left(-\frac{1}{2}\gamma - \frac{1}{2} - \frac{5}{4}\beta + \frac{5}{2}\alpha\right)\frac{m}{a}\right] e^2 \sin 2(\lambda - \pi)$$
$$- \frac{1}{4}\sin^2 i \left[\sin 2(\lambda - \Omega)\right.$$
$$- 2\left(1 + (\gamma + 1 - \beta + \alpha)\frac{m}{a}\right) e \sin(\lambda + \pi - 2\Omega)$$
$$+ 2\left(1 + (\gamma + 1 - \beta + \alpha)\frac{m}{a}\right) e \sin(3\lambda - \pi - 2\Omega)$$
$$- 4\left(1 + 2(\gamma + 1 - \beta + \alpha)\frac{m}{a}\right) e^2 \sin 2(\lambda - \Omega)$$
$$+ \frac{1}{4}\left(3 + (18\gamma + 18 - 11\beta + 6\alpha)\frac{m}{a}\right) e^2 \sin 2(\pi - \Omega)$$
$$+ \frac{1}{4}\left(13 + (14\gamma + 14 - 21\beta + 26\alpha)\frac{m}{a}\right)$$
$$\left.\times e^2 \sin(4\lambda - 2\pi - 2\Omega) + \ldots\right] + \ldots. \tag{3.1.97}$$

Introducing now e^* instead of e, one obtains

$$\cos\theta = \sin i \left[\sin(\lambda - \Omega) + e^*[(\sin(2\lambda - \pi - \Omega) - \sin(\pi - \Omega)]\right.$$
$$- e^{*2}\sin(\lambda - \Omega) + \frac{1}{8}\left(1 + (-12\gamma - 12 + 5\beta)\frac{m}{a^*}\right)$$
$$\times e^{*2}\sin(\lambda - 2\pi + \Omega) + \frac{1}{8}\left(9 + (-12\gamma - 12 + 5\beta)\frac{m}{a^*}\right)$$
$$\left.\times e^{*2}\sin(3\lambda - 2\pi - \Omega) + \ldots\right] \tag{3.1.98}$$

$$\varphi = \lambda + 2e^*\sin(\lambda - \pi) + \left[\frac{5}{4} + \left(-3\gamma - 3 + \frac{5}{4}\beta\right)\frac{m}{a^*}\right] e^{*2}\sin 2(\lambda - \pi)$$
$$- \frac{1}{4}\sin^2 i \left[\sin 2(\lambda - \Omega) - 2e^*\sin(\lambda + \pi - 2\Omega) + 2e^*\sin(3\lambda - \pi - 2\Omega)\right.$$
$$- 4e^{*2}\sin 2(\lambda - \Omega) + \frac{1}{4}\left(3 + (12\gamma + 12 - 5\beta)\frac{m}{a^*}\right) e^{*2}\sin 2(\pi - \Omega)$$
$$\left.+ \frac{1}{4}\left(13 + (-12\gamma - 12 + 5\beta)\frac{m}{a^*}\right) e^{*2}\sin(4\lambda - 2\pi - 2\Omega) + \ldots\right] + \ldots. \tag{3.1.99}$$

Thus, the coordinates of the moving particle are represented by the trigonometric series in λ, π and Ω with coefficients dependent on a, e, $\sin i$ or a^*, e^*, $\sin i$. λ and π are linear functions of time. The periods $\mathcal{T}_1$ and $\mathcal{T}_2$ are directly measurable quantities (at least, at the level of the mental experiments). Therefore, the periods T_1, T_2 and the frequencies of longitudes λ and π are indirectly measurable quantities. All angular arguments λ, π, Ω as well as $\sin i$ are also indirectly measurable quantities. Since n and n^* are indirectly measurable quantities, a and a^* are unmeasurable coordinate-dependent quantities, as follows from (3.1.51). As for the eccentricity, the comparison of (3.1.97) and (3.1.99) demonstrates that e is an unmeasurable, coordinate-dependent quantity whereas e^* is a directly measurable coordinate-independent quantity. This reflects the fact that expansions (3.1.96) and (3.1.97) in powers of e depend on α and expansions (3.1.98) and (3.1.99) in powers of e^* do not depend on α. Hence, depending on the definition of the eccentricity as characterizing the form of orbit by (3.1.60) or as one half the coefficient of the leading term in the longitude of the particle by (3.1.99) there are different conclusions concerning its measurability.

These conclusions are derived here in the post-Newtonian approximation (therefore all quantities in the post-Newtonian terms have Newtonian meaning and may be regarded as measurable). But it is clear that they are valid for any approximation. The basic frequencies νn^* and n^* by (3.1.92) and (3.1.93) may be determined from the results of measurements of the periods $\mathcal{T}_1$ and $\mathcal{T}_2$ and, hence, T_1 and T_2.

3.1.6 Solution in osculating elements

Taking into account the importance of the problem of the motion of a test particle in the Schwarzschild field it is suitable to give also its solution in osculating elements. To extend the domain of application of the relevant results to other problems (two-body problem of finite masses in GRT, motion of a test particle in the spherically symmetric gravitational fields of different post-Newtonian gravitation theories, etc) it is advantageous, following the work of Chazy (1928, 1930), to consider equations (1.1.25) with disturbing force

$$\boldsymbol{F} = m\left[\left(2\sigma\frac{GM}{r} - 2\epsilon\dot{\boldsymbol{r}}^2 + 3\alpha\frac{(\boldsymbol{r}\dot{\boldsymbol{r}})^2}{r^2}\right)\frac{\boldsymbol{r}}{r^3} + 2\mu\frac{(\boldsymbol{r}\dot{\boldsymbol{r}})}{r^3}\dot{\boldsymbol{r}}\right] \qquad (3.1.100)$$

α, σ, ϵ and μ being constant parameters. It should be noted that until the advent of GRT many authors suggested different modifications of the Newton law of gravitation described by the disturbing force (3.1.100) with specific values of the constants α, σ, ϵ and μ. The GRT equations (3.1.46) of the generalized Schwarzschild problem result from this by fixing one of these constants by means of the relations

$$\sigma = \gamma + \beta - \alpha \qquad 2\epsilon = \gamma + \alpha \qquad \mu = \gamma + 1 - \alpha \tag{3.1.101}$$

with the constant α characterizing again the choice of the coordinate conditions. Solution of (1.1.25) with the disturbing force (3.1.100) might be of course obtained by using the mean elements as performed above. But to illustrate the method of variation of arbitrary constants let us apply the osculating elements. By (1.1.27) the components of the disturbing acceleration are

$$\begin{aligned}
S &= \frac{m}{r^2}\left(2\sigma\frac{GM}{r} - 2\epsilon\dot{r}^2 + (3\alpha + 2\mu)\frac{(\boldsymbol{r}\dot{\boldsymbol{r}})^2}{r^2}\right) \\
&= m\frac{n^2a^2}{r^2}\left(-(3\alpha + 2\mu)\frac{a^2}{r^2}(1 - e^2) + 2(3\alpha - 2\epsilon + 2\mu + \sigma)\frac{a}{r}\right. \\
&\qquad \left. - 3\alpha + 2\epsilon - 2\mu\right) \\
T &= m2\epsilon\frac{\sqrt{GM}}{r^4}\sqrt{a}(1 - e^2)^{1/2}(\boldsymbol{r}\dot{\boldsymbol{r}}) = m2\epsilon\frac{n^2a^3}{r^3}e\sin f \\
W &= 0.
\end{aligned}$$

Therefore, equations (1.1.26) for the osculating elements immediately imply

$$i = \text{constant} \qquad \Omega = \text{constant}$$

whereas the differential equations for other elements take the form

$$\begin{aligned}
\frac{\mathrm{d}a}{\mathrm{d}t} &= m\frac{2na^2e}{r^2(1 - e^2)^{1/2}}\sin f \\
&\quad \times\left(-3\alpha\frac{a^2}{r^2}(1 - e^2) + 2(3\alpha - 2\epsilon + 2\mu + \sigma)\frac{a}{r} - 3\alpha + 2\epsilon - 2\mu\right) \\
\frac{\mathrm{d}e}{\mathrm{d}t} &= m\frac{na(1 - e^2)^{1/2}}{r^2}\sin f \\
&\quad \times\left(-3\alpha\frac{a^2}{r^2}(1 - e^2) + 2(3\alpha - 2\epsilon + 2\mu + \sigma)\frac{a}{r} - 3\alpha + 2\epsilon - 4\mu\right) \\
\frac{\mathrm{d}\pi}{\mathrm{d}t} &= m\frac{na(1 - e^2)^{1/2}}{r^2e}\left(3\alpha\frac{a^3}{r^3}(1 - e^2)^2 + (-9\alpha + 4\epsilon - 4\mu - 2\sigma)\frac{a^2}{r^2}(1 - e^2)\right. \\
&\quad \left. + (3\alpha - 2\epsilon)\frac{a}{r}(1 - e^2) + 2(3\alpha - 2\epsilon + 4\mu + \sigma)\frac{a}{r} - 3\alpha + 2\epsilon - 4\mu\right) \\
\frac{\mathrm{d}\epsilon}{\mathrm{d}t} &= [1 - (1 - e^2)^{1/2}]\frac{\mathrm{d}\pi}{\mathrm{d}t} + m\frac{2na}{r^2}\left((3\alpha + 2\mu)\frac{a}{r}(1 - e^2)\right. \\
&\quad \left. + 2(-3\alpha + 2\epsilon - 2\mu - \sigma) + (3\alpha - 2\epsilon + 2\mu)\frac{r}{a}\right).
\end{aligned}$$

Substituting into the right-hand sides of these equations the values of the osculating elements for the initial moment of time and integrating, one obtains the first-order perturbations

$$\delta a = \frac{me}{(1-e^2)^2}\{[4(\epsilon-\mu-\sigma)+e^2(-\tfrac{9}{2}\alpha+4\epsilon-4\mu)]\cos f + (2\epsilon-2\mu-\sigma)e\cos 2f+\tfrac{1}{2}\alpha e^2\cos 3f\}|_{t_0}^{t} \tag{3.1.102}$$

$$\delta e = \frac{m}{a(1-e^2)}\{[2(\epsilon-\sigma)+e^2(-\tfrac{9}{4}\alpha+2\epsilon-4\mu)]\cos f + (\epsilon-\mu-\tfrac{1}{2}\sigma)e\cos 2f+\tfrac{1}{4}\alpha e^2\cos 3f\}|_{t_0}^{t} \tag{3.1.103}$$

$$\delta\pi = \frac{m}{a(1-e^2)}\left\{(2\epsilon+2\mu-\sigma)f+\left[2\frac{\epsilon-\sigma}{e}+\left(-\frac{3}{4}\alpha+2\epsilon\right)e\right]\sin f + \left(\epsilon-\mu-\frac{1}{2}\sigma\right)\sin 2f+\frac{1}{4}\alpha e\sin 3f\right\}\Bigg|_{t_0}^{t} \tag{3.1.104}$$

$$\delta\epsilon = [1-(1-e^2)^{1/2}]\delta\pi+\frac{2m}{a(1-e^2)^{1/2}}[(3\alpha-2\epsilon+2\mu)(1-e^2)^{1/2}E + (-3\alpha+4\epsilon-2\mu-2\sigma)f+(3\alpha+2\mu)e\sin f]|_{t_0}^{t}. \tag{3.1.105}$$

Rewriting δa as

$$\delta a = \frac{2m}{(1-e^2)^2}\left[\alpha\left(\frac{a}{r}\right)^3(1-e^2)^3+(-3\alpha+2\epsilon-2\mu-\sigma)\left(\frac{a}{r}\right)^2(1-e^2)^2 + (3\alpha-2\epsilon+2\mu)(1-e^2)^2\frac{a}{r}-\alpha+\sigma+(3\alpha-3\epsilon+3\mu+\tfrac{1}{2}\sigma)e^2\right]\Bigg|_{t_0}^{t} \tag{3.1.106}$$

one gets

$$\int_{t_0}^{t}\delta n dt = 3\frac{m}{a}\left\{(-3\alpha+2\epsilon-2\mu)E+(2\alpha-2\epsilon+2\mu+\sigma)\frac{f}{(1-e^2)^{1/2}} - \alpha\frac{e\sin f}{(1-e^2)^{1/2}}+\left[a(1-e^2)\left(\frac{a}{r_0}\right)^3+(-3\alpha+2\epsilon-2\mu-\sigma) \times\left(\frac{a}{r_0}\right)^2+(3\alpha-2\epsilon+2\mu)\frac{a}{r_0}\right]l\right\}\Bigg|_{t_0}^{t} \tag{3.1.107}$$

initial values being marked by a zero index. The variation of the mean longitude is determined by

$$\Delta\lambda = n(t - t_0) + \delta\lambda \qquad \delta\lambda = \int_{t_0}^{t} \delta n dt + \delta\epsilon$$

and for one revolution

$$[\delta\lambda]_{f_0}^{f_0+2\pi} = [\delta\pi]_{f_0}^{f_0+2\pi} + 2\pi\frac{m}{a}\left[-3\alpha + 2\epsilon - 2\mu + 3(3\alpha - 2\epsilon + 2\mu)\frac{a}{r_0} + 3(-3\alpha + 2\epsilon - 2\mu - \sigma)\left(\frac{a}{r_0}\right)^2 + 3\alpha(1 - e^2)\left(\frac{a}{r_0}\right)^3\right]. \tag{3.1.108}$$

From the Kepler equation it follows that the anomalistic period resulting in increasing the starting values of f and E by 2π is determined by the expression

$$T_1 = \frac{2\pi}{n} + \frac{1}{n}(-[\delta\lambda] + [\delta\pi])\Big|_{f_0}^{f_0+2\pi}$$

or

$$T_1 = \frac{2\pi}{n}\left\{1 + \frac{m}{a}\left[3\alpha - 2\epsilon + 2\mu + 3(-3\alpha + 2\epsilon - 2\mu)\frac{a}{r_0} + 3(3\alpha - 2\epsilon + 2\mu + \sigma)\left(\frac{a}{r_0}\right)^2 - 3\alpha(1 - e^2)\left(\frac{a}{r_0}\right)^3\right]\right\}. \tag{3.1.109}$$

The sidereal period resulting in increasing the starting value of the mean longitude by 2π is

$$T_2 = \frac{2\pi}{n} - \frac{1}{n}[\delta\lambda]_{f_0}^{f_0+2\pi}$$

or

$$T_1 = \frac{2\pi}{n}\left\{1 + \frac{m}{a}\left[\frac{3\alpha - 4\epsilon + \sigma + (-3\alpha + 2\epsilon - 2\mu)e^2}{1 - e^2} + 3(-3\alpha + 2\epsilon - 2\mu)\frac{a}{r_0} + 3(3\alpha - 2\epsilon + 2\mu + \sigma)\left(\frac{a}{r_0}\right)^2 - 3\alpha(1 - e^2)\left(\frac{a}{r_0}\right)^3\right]\right\}. \tag{3.1.110}$$

Expressions (3.1.109) and (3.1.110) with relativistic values (3.1.101) differ formally from (3.1.70) and (3.1.72). This is due to the fact that expressions (3.1.109) and (3.1.110) contain the values $a = a_0$, $e = e_0$ of the osculating elements at the initial moment of time whereas the quantities $a = a'$, $e = e'$ occurring in (3.1.70) and (3.1.72) are constants characterizing the size and the form of the orbit. It is not difficult to verify the identity of these expressions. If a_1, e_1 and a_2, e_2 are values of the osculating elements in pericentre and apocentre respectively then

$$2a' = a_2(1+e_2) + a_1(1-e_1) \qquad 2a'e' = a_2(1+e_2) - a_1(1-e_1).$$

Using (3.1.102) and (3.1.103) one may derive

$$a_i = a_0 + \delta a_i - \delta a_0 \qquad e_i = e_0 + \delta e_i - \delta e_0 \qquad (i = 1, 2)$$

with

$$\delta a_1 = \frac{me}{(1-e^2)^2}[4(\epsilon - \mu - \sigma) + (2\epsilon - 2\mu - \sigma)e + 4(-\alpha + \epsilon - \mu)e^2]$$
$$\delta a_2 = \frac{me}{(1-e^2)^2}[-4(\epsilon - \mu - \sigma) + (2\epsilon - 2\mu - \sigma)e - 4(-\alpha + \epsilon - \mu)e^2]$$
$$\delta e_1 = \frac{m}{a(1-e^2)}[2(\epsilon - \sigma) + (\epsilon - \mu - \tfrac{1}{2}\sigma)e + (-2\alpha + 2\epsilon - 4\mu)e^2]$$
$$\delta e_2 = \frac{m}{a(1-e^2)}[-2(\epsilon - \sigma) + (\epsilon - \mu - \tfrac{1}{2}\sigma)e - (-2\alpha + 2\epsilon - 4\mu)e^2].$$

δa_0 is expressed as the right-hand side of (3.1.106) with $r = r_0$. Then,

$$a' = a_0 - \delta a_0 + \tfrac{1}{2}(\delta a_1 + \delta a_2) + \tfrac{1}{2}a_0(\delta e_2 - \delta e_1) + \tfrac{1}{2}e_0(\delta a_2 - \delta a_1)$$

and therefore

$$a' = a_0 + m\left[-2\alpha(1-e^2)\left(\frac{a}{r_0}\right)^3 + 2(3\alpha - 2\epsilon + 2\mu + \sigma)\left(\frac{a}{r_0}\right)^2 \right.$$
$$\left. + 2(-3\alpha + 2\epsilon - 2\mu)\frac{a}{r_0} + 2(\alpha - \epsilon)\right].$$

The mean motion n is expressed in (3.1.109), (3.1.110) in terms of a_0 by means of the unperturbed Kepler third law whereas in (3.1.70), (3.1.72) it is expressed in terms of $a = a'$ by means of (3.1.51). Substituting into (3.1.70) and (3.1.72)

$$a'^{3/2} = a_0^{3/2}\left(1 + \frac{3}{2}\frac{a' - a_0}{a_0}\right)$$

and taking into account relations (3.1.101) one comes to (3.1.109) and (3.1.110).

As seen from (3.1.104) the advancement of the longitude of pericentre per one revolution is equal to

$$\delta\pi = \frac{2\pi m}{a(1-e^2)}(2\epsilon + 2\mu - \sigma). \tag{3.1.111}$$

For values (3.1.101) this expression results in the previous formula of the Schwarzschild advance of pericentre (3.1.66).

The equations of motion with disturbing force (3.1.100) may be put in the Lagrange form with Lagrangian

$$L = \frac{1}{2}\dot{\boldsymbol{r}}^2 + \frac{GM}{r} + \frac{2\alpha - 2\epsilon + \mu}{8c^2}(\dot{\boldsymbol{r}}^2)^2 + \frac{m}{r}\left[\left(\epsilon + \frac{1}{2}\mu - \sigma\right)\frac{GM}{r}\right.$$
$$\left. + \left(-\alpha + \epsilon + \frac{1}{2}\mu\right)\dot{\boldsymbol{r}}^2 + \alpha\frac{(\boldsymbol{r}\dot{\boldsymbol{r}})^2}{r^2}\right]. \tag{3.1.112}$$

Relations (3.1.102)–(3.1.107) yield the perturbations of the osculating elements. The same relations might be obtained on the basis of the Lagrange equations for the contact elements (1.1.39). The Lagrangian (3.1.112) contains four arbitrary constant parameters. If three of them are related by the condition

$$2\alpha - 2\epsilon + \mu = 1 \tag{3.1.113}$$

which is satisfied particularly with the values (3.1.101), then the resulting Lagrangian corresponds to the metric form with the three arbitrary parameters α, ϵ and σ equivalent to (3.1.44).

3.2 LIGHT PROPAGATION IN THE SCHWARZSCHILD PROBLEM

3.2.1 Variational principle

The geodesic variational principle for the constant gravitational field leads to the problem of constructing geodesics in the three-dimensional space. Indeed, principle (1.2.50) has the form

$$\delta\int\sqrt{f}\,\mathrm{d}x^0 = 0 \tag{3.2.1}$$

with

$$f = g_{00} + 2g_{0i}\frac{\mathrm{d}x^i}{\mathrm{d}x^0} + g_{ik}\frac{\mathrm{d}x^i}{\mathrm{d}x^0}\frac{\mathrm{d}x^k}{\mathrm{d}x^0}.$$

f being explicitly independent of x^0, there exists the integral

$$\sqrt{f} - \frac{\partial \sqrt{f}}{\partial(\mathrm{d}x^i/\mathrm{d}x^0)} \frac{\mathrm{d}x^i}{\mathrm{d}x^0} = \frac{1}{\sqrt{h}} \qquad h = \text{constant}$$

or

$$\frac{1}{\sqrt{f}} \left(g_{00} + g_{0i} \frac{\mathrm{d}x^i}{\mathrm{d}x^0} \right) = \frac{1}{\sqrt{h}}. \tag{3.2.2}$$

f may be represented in the form

$$f = \frac{1}{g_{00}} \left(g_{00} + g_{0i} \frac{\mathrm{d}x^i}{\mathrm{d}x^0} \right)^2 - \left(\frac{\mathrm{d}\ell}{\mathrm{d}x^0} \right)^2 \tag{3.2.3}$$

$\mathrm{d}\ell$ being the element of the space distance (2.3.3). Therefore, combining (3.2.2) and (3.2.3) one obtains

$$g_{00}\mathrm{d}x^0 + g_{0i}\mathrm{d}x^i = \frac{\sqrt{g_{00}}\,\mathrm{d}\ell}{(1 - hg_{00})^{1/2}}. \tag{3.2.4}$$

Again using (3.2.3), one has

$$\sqrt{f}\mathrm{d}x^0 = \frac{\sqrt{hg_{00}}\,\mathrm{d}\ell}{(1 - hg_{00})^{1/2}}. \tag{3.2.5}$$

For a fixed value of h the variational principle (3.2.1) may be rewritten in the form

$$\delta \int \left(\frac{1}{\sqrt{h}} - \sqrt{f} \right) \mathrm{d}x^0 = 0 \tag{3.2.6}$$

with arbitrary variations of x^0 at the endpoints of the interval of integration. Transformation from (3.2.1) to (3.2.6) may be easily verified taking into account the basic formula of calculus of variations:

$$\delta \int_{x_0}^{x_1} f(x, y, y')\mathrm{d}x = \int_{x_0}^{x_1} \left(\frac{\partial f}{\partial y} - \frac{\mathrm{d}}{\mathrm{d}x} \frac{\partial f}{\partial y'} \right) \delta y \, \mathrm{d}x + \left(f - y' \frac{\partial f}{\partial y'} \right) \delta x \Big|_{x_0}^{x_1} + \frac{\partial f}{\partial y'} \delta y \Big|_{x_0}^{x_1}.$$

For this case the third term is zero due to vanishing variations of the space coordinates at the endpoints, the second term is annulled by the existence of the first integral and the first term results in the same Lagrangian as defined from the original variational principle (3.2.1). Excluding $\mathrm{d}x^0$ from (3.2.6) with the aid of (3.2.4) and (3.2.5) the geodesic variational principle for the constant field finally takes the form

$$\delta \int \left(\frac{(1 - hg_{00})^{1/2}}{\sqrt{g_{00}}} \mathrm{d}\ell - \frac{g_{0i}}{g_{00}} \mathrm{d}x^i \right) = 0. \tag{3.2.7}$$

Having determined the trajectory, the dependence on time is obtained by quadrature (3.2.4). This principle is particularly adequate to study light propagation. For this case one has $h = 0$ (as a consequence of $f = 0$) and comparison of (3.2.7) and (3.2.4) leads to the conclusion that light is propagated in accordance with the Fermi principle (the principle of least time)

$$\delta \int \mathrm{d}x^0 = 0. \tag{3.2.8}$$

3.2.2 Post-Newtonian equations and their solution

For the static spherically symmetric field with metric (3.1.44) (with $b(r) = 0$) principle (3.2.8) takes the form

$$\delta \int \frac{\mathrm{d}\ell}{\sqrt{p(r)}} = 0 \tag{3.2.9}$$

with

$$\mathrm{d}\ell^2 = \left[\delta_{ik} + \frac{2m}{r}\left((\gamma - \alpha)\delta_{ik} + \alpha\frac{x^i x^k}{r^2}\right)\right] \mathrm{d}x^i \mathrm{d}x^k.$$

It is sufficient to retain in $p(r)$ only the term of order m/r

$$p(r) = 1 - \frac{2m}{r}.$$

Similar to the transformation from (1.2.50) to (1.2.55) principle (3.2.9) may be rewritten in the rational form by putting t as the canonical parameter

$$\delta \int \frac{1}{p(r)} \left(\frac{\mathrm{d}\ell}{\mathrm{d}t}\right)^2 \mathrm{d}t = 0. \tag{3.2.10}$$

For the static field time t is the canonical parameter since for this case the equations associated with the variational principle (1.2.55) admit the first integral

$$g_{00}\frac{\mathrm{d}x^0}{\mathrm{d}\lambda} = \text{constant}.$$

From this it follows that x^0 and λ are linearly related and hence are equivalent. The Lagrangian for the variational principle (3.2.10) is

$$L = \frac{1}{2}\dot{\boldsymbol{r}}^2 + \frac{m}{r}\left((\gamma + 1 - \alpha)\dot{\boldsymbol{r}}^2 + \alpha\frac{(\boldsymbol{r}\dot{\boldsymbol{r}})^2}{r^2}\right). \tag{3.2.11}$$

The corresponding equations of light propagation in rectangular coordinates are of the form

$$\ddot{\boldsymbol{r}} = \frac{m}{r^3}\left(-(\gamma+1+\alpha)(\dot{\boldsymbol{r}}^2)\boldsymbol{r} + 3\alpha\frac{(\boldsymbol{r}\dot{\boldsymbol{r}})^2}{r^2}\boldsymbol{r} + 2(\gamma+1-\alpha)(\boldsymbol{r}\dot{\boldsymbol{r}})\dot{\boldsymbol{r}}\right). \quad (3.2.12)$$

The general solution of this equation is given by the formulae

$$\boldsymbol{r}(t) = \boldsymbol{r}_0 + c(t-t_0)\boldsymbol{\sigma} + m\left[(\gamma+1)\left(\frac{\boldsymbol{\sigma}\times(\boldsymbol{r}_0\times\boldsymbol{\sigma})}{r_0-\boldsymbol{\sigma}\boldsymbol{r}_0} - \frac{\boldsymbol{\sigma}\times(\boldsymbol{r}\times\boldsymbol{\sigma})}{r-\boldsymbol{\sigma}\boldsymbol{r}} - \boldsymbol{\sigma}\ln\frac{r+\boldsymbol{\sigma}\boldsymbol{r}}{r_0+\boldsymbol{\sigma}\boldsymbol{r}_0}\right) + \alpha\left(\frac{\boldsymbol{r}}{r} - \frac{\boldsymbol{r}_0}{r_0}\right)\right] \quad (3.2.13)$$

$$\frac{1}{c}\dot{\boldsymbol{r}}(t) = \boldsymbol{\sigma} - \frac{m}{r}\left((\gamma+1)\frac{\boldsymbol{\sigma}\times(\boldsymbol{r}\times\boldsymbol{\sigma})}{r-\boldsymbol{\sigma}\boldsymbol{r}} + (\gamma+1-\alpha)\boldsymbol{\sigma} + \alpha\frac{(\boldsymbol{\sigma}\boldsymbol{r})}{r^2}\boldsymbol{r}\right). \quad (3.2.14)$$

The arbitrary constants of integration are chosen to be space coordinates $\boldsymbol{r}(t_0) = \boldsymbol{r}_0$ of the light particle at the initial moment of time t_0, the direction of the light velocity at the infinitely far distance from the primary for the infinite remote past $\boldsymbol{\sigma} = \dot{\boldsymbol{r}}(-\infty)/c$, $\boldsymbol{\sigma}^2 = 1$ and the light velocity c. In the right-hand sides of (3.2.13) and (3.2.14) $\boldsymbol{r}$ is to be replaced by its Newtonian expression

$$\boldsymbol{r}_N(t) = \boldsymbol{r}_0 + c(t-t_0)\boldsymbol{\sigma}. \quad (3.2.15)$$

From (3.2.14) for $t \to \infty$ there results

$$\frac{1}{c}\dot{\boldsymbol{r}}(\infty) \equiv \boldsymbol{\nu} = \boldsymbol{\sigma} - 2m(\gamma+1)\frac{\boldsymbol{\sigma}\times(\boldsymbol{r}_0\times\boldsymbol{\sigma})}{|\boldsymbol{r}_0\times\boldsymbol{\sigma}|^2}. \quad (3.2.16)$$

The absolute value of the vector product

$$|\boldsymbol{\nu}\times\boldsymbol{\sigma}| = 2(\gamma+1)\frac{m}{|\boldsymbol{r}_0\times\boldsymbol{\sigma}|} \quad (3.2.17)$$

determines the sine of the angle of the total deflection of light in passing in the Schwarzschild field. Within the post-Newtonian approximation one may adopt here as $\boldsymbol{r}_0$ any point of the light trajectory. In virtue of (3.2.15)

$$|\boldsymbol{r}_N(t)\times\boldsymbol{\sigma}| = |\boldsymbol{r}_0\times\boldsymbol{\sigma}| = d$$

d being the impact parameter for the light trajectory. But it may be preferable to apply (3.2.17) without using the notion of impact parameter.

For the radial light ray expressions (3.2.13) and (3.2.14) become

$$\boldsymbol{r}(t) = \boldsymbol{r}_0 + c(t-t_0)\frac{\boldsymbol{r}_0}{r_0} + m(\gamma+1)\frac{\boldsymbol{r}_0}{r_0}\ln\frac{r_0}{r} \quad (3.2.18)$$

$$\frac{1}{c}\dot{\boldsymbol{r}}(t) = \frac{\boldsymbol{r}}{r}\left(1 - (\gamma+1)\frac{m}{r}\right). \tag{3.2.19}$$

According to (3.2.14) the magnitude of the coordinate light velocity at any point $\boldsymbol{r}$ is

$$\frac{1}{c}|\dot{\boldsymbol{r}}(t)| = 1 - \frac{m}{r}\left(\gamma + 1 - \alpha\frac{|\boldsymbol{r}\times\boldsymbol{\sigma}|^2}{r^2}\right) \tag{3.2.20}$$

differing from c. But the physically measurable, chronometric invariant light velocity is, of course, always the same and is equal to c as followed from the $3+1$ metric splitting (2.3.11) and condition $ds^2 = 0$ for the light propagation.

3.2.3 Boundary value problem and Doppler displacement

Formulae (3.2.13) and (3.2.14) give the solution of the Cauchy problem for light propagation. In fact, it is easy to express $\boldsymbol{\sigma}$ from (3.2.14) in terms of the coordinate direction of the light ray at moment t_0. Then, the relation (3.2.13) determines the law of motion in terms of position and velocity of the light particle at the initial moment t_0. Consider now the boundary value problem with two given positions of the light particle $\boldsymbol{r}(t_0) = \boldsymbol{r}_0$ and $\boldsymbol{r}(t) = \boldsymbol{r}$ $(t_0 < t)$. Denoting

$$\boldsymbol{D} = \boldsymbol{r}(t) - \boldsymbol{r}_0(t_0) \qquad D = |\boldsymbol{D}| \tag{3.2.21}$$

one easily finds from (3.2.13) and (3.2.14) the light direction at $t \to -\infty$

$$\boldsymbol{\sigma} = \frac{\boldsymbol{D}}{D} + \frac{m}{D}\left[(\gamma+1)\frac{r - r_0 + D}{|\boldsymbol{r}_0\times\boldsymbol{r}|^2} + \frac{\alpha}{D^2}\left(\frac{1}{r_0} - \frac{1}{r}\right)\right][\boldsymbol{D}\times(\boldsymbol{r}_0\times\boldsymbol{r})] \tag{3.2.22}$$

and the time of the light propagation

$$t - t_0 = \frac{D}{c} + \frac{m}{c}\left((\gamma+1)\ln\frac{r_0 + r + D}{r_0 + r - D} + \frac{1}{2}\alpha\frac{(r_0 + r)[(r_0 - r)^2 - D^2]}{r_0 r D}\right). \tag{3.2.23}$$

The light signal emitted at moment t_0 from the point $\boldsymbol{r}_0$ reaches the point $\boldsymbol{r}$ at moment t having the coordinate velocity

$$\frac{1}{c}\dot{\boldsymbol{r}}(t) = \frac{\boldsymbol{D}}{D} + \frac{m}{D}\left\{-\frac{\gamma+1}{r}\boldsymbol{D} - \frac{\alpha}{r^3}[\boldsymbol{r}\times(\boldsymbol{r}_0\times\boldsymbol{r})]\right.$$
$$\left. + \left[\frac{\alpha}{D^2}\left(\frac{1}{r_0} - \frac{1}{r}\right) - \frac{\gamma+1}{r(r r_0 + \boldsymbol{r}\boldsymbol{r}_0)}\right][\boldsymbol{D}\times(\boldsymbol{r}_0\times\boldsymbol{r})]\right\} \tag{3.2.24}$$

which results from substituting (3.2.22) into (3.2.14). If the light emitter is a distant star so that $r \ll r_0$ then expanding in powers of r/r_0 and retaining only the main terms, relations (3.2.22) and (3.2.23) take the form

$$\boldsymbol{\sigma} = -\frac{\boldsymbol{r}_0}{r_0} + \frac{\boldsymbol{r}_0 \times (\boldsymbol{r} \times \boldsymbol{r}_0)}{r_0^3}\left[1 + \frac{m}{r}\left(\frac{\gamma+1}{1+\boldsymbol{r}_0\boldsymbol{r}/(r_0 r)} - \alpha\right)\right] + \ldots \quad (3.2.25)$$

$$t - t_0 = \frac{r_0}{c}\left(1 - \frac{\boldsymbol{r}_0\boldsymbol{r}}{r_0^2}\right) + \frac{m}{c}\left[(\gamma+1)\ln\frac{2r_0^2}{rr_0 + \boldsymbol{r}\boldsymbol{r}_0} + \alpha\left(\frac{\boldsymbol{r}_0\boldsymbol{r}}{r_0 r} - 1\right)\right] + \ldots . \quad (3.2.26)$$

Consider now the Doppler displacement of the light frequency. Let again the light signal be emitted at moment t_0 and received at moment t. By the previous formulae one may find the functional dependence $t = t(t_0)$ for the specific motions of the emitter and the receiver. Let the light signal be emitted with period δt_0. The corresponding interval of the proper time of the light emitter is

$$\delta\tau_0 = c^{-1}\left(\frac{\mathrm{d}s}{\mathrm{d}t}\right)_{\boldsymbol{r}_0(t_0)} \delta t_0 .$$

The receiver accepts this signal with period δt in the coordinate time or with period

$$\delta\tau = c^{-1}\left(\frac{\mathrm{d}s}{\mathrm{d}t}\right)_{\boldsymbol{r}(t)} \delta t$$

expressed in the proper time of the light receiver. Besides this one has

$$\delta t = \frac{\mathrm{d}t}{\mathrm{d}t_0}\delta t_0 .$$

The light frequency is inversely proportional to the period in the proper time. Therefore, the ratio of the frequency ν_0 of the light emitter at point $\boldsymbol{r}_0(t_0)$ to the frequency ν of the light receiver at point $\boldsymbol{r}(t)$ is

$$\frac{\nu_0}{\nu} = \frac{\delta\tau}{\delta\tau_0} = \frac{(\mathrm{d}s/\mathrm{d}t)_{\boldsymbol{r}(t)}}{(\mathrm{d}s/\mathrm{d}t)_{\boldsymbol{r}_0(t_0)}}\frac{\mathrm{d}t}{\mathrm{d}t_0} . \quad (3.2.27)$$

For the Schwarzschild field this general formula of the frequency ratio yields in the post-Newtonian approximation

$$\frac{\nu_0}{\nu} = \frac{1 - m/r - \dot{\boldsymbol{r}}^2/2c^2}{1 - m/r_0 - \dot{\boldsymbol{r}}_0^2/2c^2}\frac{\mathrm{d}t}{\mathrm{d}t_0} \quad (3.2.28)$$

with $\dot{\boldsymbol{r}}_0 = \mathrm{d}\boldsymbol{r}_0/\mathrm{d}t_0$ and $\dot{\boldsymbol{r}} = \mathrm{d}\boldsymbol{r}/\mathrm{d}t$. The derivative $\mathrm{d}t/\mathrm{d}t_0$ may be calculated by using (3.2.23) and taking into account that

$$\dot{r}_0 = \frac{\boldsymbol{r}_0\dot{\boldsymbol{r}}_0}{r_0} \qquad \dot{r} = \frac{\boldsymbol{r}\dot{\boldsymbol{r}}}{r} \qquad \frac{\mathrm{d}r}{\mathrm{d}t_0} = \frac{\boldsymbol{r}\dot{\boldsymbol{r}}}{r}\frac{\mathrm{d}t}{\mathrm{d}t_0} \qquad \frac{\mathrm{d}D}{\mathrm{d}t_0} = \frac{1}{D}\left(\boldsymbol{D}\dot{\boldsymbol{r}}\frac{\mathrm{d}t}{\mathrm{d}t_0} - \boldsymbol{D}\dot{\boldsymbol{r}}_0\right).$$

Then

$$\frac{dt}{dt_0} = \frac{1 - (\boldsymbol{D}\dot{\boldsymbol{r}}_0/cD)[1 + mS(r_0, r, D)] + m(\dot{r}_0/c)T(r, r_0, D)}{1 - (\boldsymbol{D}\dot{\boldsymbol{r}}/cD)[1 + mS(r_0, r, D)] - m(\dot{r}/c)T(r, r_0, D)} \qquad (3.2.29)$$

with

$$S(r_0, r, D) = 2(\gamma + 1)\frac{r_0 + r}{(r_0 + r)^2 - D^2} - \frac{1}{2}\alpha(r_0 + r)\frac{(r_0 - r)^2 + D^2}{r_0 r D^2} \qquad (3.2.30)$$

$$T(r_0, r, D) = -2(\gamma + 1)\frac{D}{(r_0 + r)^2 - D^2} + \alpha\frac{r^2 - r_0^2 + D^2}{2r^2 D} + \alpha\frac{r}{D}\left(\frac{1}{r_0} - \frac{1}{r}\right). \qquad (3.2.31)$$

In contrast to the S function T is not symmetric with respect to r_0 and r and enters into numerator and denominator of (3.2.29) with different order of its arguments. Relations (3.2.28) and (3.2.29) with (3.2.30) and (3.2.31) are sufficient for solving all problems concerning Doppler observations in the Schwarzschild field. The Doppler displacement (3.2.28) includes the gravitational shift of frequency due to the difference of the gravitational potentials at points $\boldsymbol{r}_0$ and $\boldsymbol{r}$. The Doppler effect is treated in detail, for example, in Tausner (1966) and Kislik (1985). The different cases of the mutual position of observer, light emitter and the primary are considered in the latter paper.

3.2.4 Post-post-Newtonian equations and their solution

At present the discussion of high precision measurements of the light deflection and time radio ranging has confirmed the GRT post-Newtonian effects at the level of 0.1 to 1.5% (Will 1986). In this regard some specific programmes for investigating the post-post-Newtonian effects are coming into use. In particular, much attention is paid to the project of an astrometric optical interferometer to be put in orbit around the Earth (Reasenberg and Shapiro 1986). From preliminary estimates one may hope to gain a precision of $1 \times 10^{-6''}$ for measurement of the light deflection whereas the post-post-Newtonian effect of the light deflection in the vicinity of the Sun is $11 \times 10^{-6''}$. Therefore, investigation of the post-post-Newtonian approximation of the light propagation is of interest for future practical applications. This problem is treated here within the framework of GRT taking into account coordinate parameters (Brumberg 1987a). Within the framework of the PPN formalism with a fixed choice of coordinate conditions the post-post-Newtonian effects in the light deflection and the time of flight of radio signal in the Schwarzschild field have been investigated by Richter and Matzner (1982, 1983).

Let us start from the general expression of the static Schwarzschild metric (3.1.2) with (3.1.22) and $b(r) = 0$. The equations of light propagation admit the Lagrangian obtained from (3.2.10)

$$L = \frac{1}{2[1-2m/a(r)]}\left[\frac{a^2(r)}{r^2}\dot{\boldsymbol{r}}^2 + \frac{1}{r^2}\left(\frac{a'^2(r)}{1-2m/a(r)} - \frac{a^2(r)}{r^2}\right)(\boldsymbol{r}\dot{\boldsymbol{r}})^2\right]. \tag{3.2.32}$$

With $a(r)$ given by expansion (3.1.43) the Lagrangian in the post-post-Newtonian approximation is

$$L = \frac{1}{2}\dot{\boldsymbol{r}}^2 + \frac{m}{r}\left[(2-\alpha)\dot{\boldsymbol{r}}^2 + \alpha\left(\frac{\boldsymbol{r}\dot{\boldsymbol{r}}}{r}\right)^2\right] + \frac{m^2}{r^2}\left[\left(\frac{7}{2} - 2\alpha + \frac{1}{2}\alpha^2 + \epsilon\right)\dot{\boldsymbol{r}}^2 + \left(\frac{1}{2} + 4\alpha - \frac{1}{2}\alpha^2 - 2\epsilon\right)\left(\frac{\boldsymbol{r}\dot{\boldsymbol{r}}}{r}\right)^2\right] + \ldots. \tag{3.2.33}$$

The post-post-Newtonian equations of light propagation generated by this Lagrangian take the form

$$\begin{aligned}\ddot{\boldsymbol{r}} = {}& \frac{m}{r^3}\left[(4-2\alpha)(\boldsymbol{r}\dot{\boldsymbol{r}})\dot{\boldsymbol{r}} - (2+\alpha)\dot{\boldsymbol{r}}^2\boldsymbol{r} + 3\alpha\left(\frac{\boldsymbol{r}\dot{\boldsymbol{r}}}{r}\right)^2\boldsymbol{r}\right] \\ & + \frac{m^2}{r^4}\left[(-2+8\alpha-2\alpha^2+4\epsilon)(\boldsymbol{r}\dot{\boldsymbol{r}})\dot{\boldsymbol{r}} + 2\epsilon\dot{\boldsymbol{r}}^2\boldsymbol{r}\right. \\ & \left. + (2-4\alpha+2\alpha^2-8\epsilon)\left(\frac{\boldsymbol{r}\dot{\boldsymbol{r}}}{r}\right)^2\boldsymbol{r}\right] + \ldots. \end{aligned} \tag{3.2.34}$$

Choosing again six arbitrary constants of the general solution of equations (3.2.34) the triplet of the spatial coordinates of the light particle $\boldsymbol{r}_0 = \boldsymbol{r}(t_0)$ at moment t_0, the unit vector $\boldsymbol{\sigma} = \dot{\boldsymbol{r}}(-\infty)/c$ of the light direction in the remote past and light velocity c at infinity and denoting post-Newtonian and post-post-Newtonian values of the desired quantities by indices pN and ppN respectively one gets

$$c^{-1}\dot{\boldsymbol{r}}_{pN}(t) = \boldsymbol{\sigma} + m\boldsymbol{A}_1(\boldsymbol{r}_N) \tag{3.2.35}$$

$$\boldsymbol{r}_{pN}(t) = \boldsymbol{r}_N(t) + m[\boldsymbol{B}_1(\boldsymbol{r}_N) - \boldsymbol{B}_1(\boldsymbol{r}_0)] \tag{3.2.36}$$

$$c^{-1}\dot{\boldsymbol{r}}_{ppN}(t) = \boldsymbol{\sigma} + m\boldsymbol{A}_1(\boldsymbol{r}_{pN}) + m^2\boldsymbol{A}_2(\boldsymbol{r}_N) \tag{3.2.37}$$

$$\boldsymbol{r}_{ppN}(t) = \boldsymbol{r}_N(t) + m[\boldsymbol{B}_1(\boldsymbol{r}_{pN}) - \boldsymbol{B}_1(\boldsymbol{r}_0)] + m^2[\boldsymbol{B}_2(\boldsymbol{r}_N) - \boldsymbol{B}_2(\boldsymbol{r}_0)]. \tag{3.2.38}$$

$\boldsymbol{r}_N(t)$ stands here for the Newtonian value (3.2.15) and

$$\boldsymbol{A}_1(\boldsymbol{r}) = -\alpha\frac{\boldsymbol{\sigma}\boldsymbol{r}}{r^3}\boldsymbol{r} + (\alpha-2)\frac{\boldsymbol{\sigma}}{r} - 2\frac{\boldsymbol{\sigma}\times(\boldsymbol{r}\times\boldsymbol{\sigma})}{r(r-\boldsymbol{\sigma}\boldsymbol{r})} \tag{3.2.39}$$

$$A_2(r) = \frac{2\alpha}{r^3}r + \left(-\frac{1}{2} + 4\alpha - \alpha^2 + 2\epsilon\right)\frac{\sigma r}{r^4}r - \frac{4r}{r^2(r - \sigma r)}$$
$$+ \left(\frac{1}{2} - 4\alpha + \alpha^2 - \epsilon\right)\frac{\sigma}{r^2} + (12 - 4\alpha)\frac{\sigma \times (r \times \sigma)}{r^2(r - \sigma r)} + 4\frac{\sigma \times (r \times \sigma)}{r(r - \sigma r)^2}$$
$$- \frac{15}{4}(\sigma r)\frac{\sigma \times (r \times \sigma)}{r^2 |r \times \sigma|^2} - \frac{15}{4}\frac{\sigma \times (r \times \sigma)}{|r \times \sigma|^3}\left(\tan^{-1}\frac{\sigma r}{|r \times \sigma|} + \frac{\pi}{2}\right)$$
(3.2.40)

$$B_1(r) = \frac{\alpha}{r}r - 2\frac{\sigma \times (r \times \sigma)}{r - \sigma r} - 2\sigma \ln(r + \sigma r) \qquad (3.2.41)$$

$$B_2(r) = \left(\frac{1}{4} - \epsilon\right)\frac{r}{r^2} + 2\alpha\frac{\sigma}{r} - \frac{4\sigma}{r - \sigma r} + 4\frac{\sigma \times (r \times \sigma)}{(r - \sigma r)^2}$$
$$- \frac{15}{4}\frac{\sigma}{|r \times \sigma|}\tan^{-1}\frac{\sigma r}{|r \times \sigma|}$$
$$- \frac{15}{4}(\sigma r)\frac{\sigma \times (r \times \sigma)}{|r \times \sigma|^3}\left(\tan^{-1}\frac{\sigma r}{|r \times \sigma|} + \frac{\pi}{2}\right) \qquad (3.2.42)$$

$|r \times \sigma|$ is to be meant as the arithmetic value of the vector product. Immediate differentiation of expressions (3.2.39)–(3.2.42) demonstrates that functions (3.2.37) and (3.2.38) represent the post-post-Newtonian solution of (3.2.34). It should be mentioned that on the right-hand sides of (3.2.34) r and $\dot{r}$ have their post-Newtonian values in the terms $O(m)$ and Newtonian values in the terms $O(m^2)$.

Taking in (3.2.37) the limit $t \to \infty$ and substituting, in particular, expression (3.2.36) into (3.2.39) one finds the post-post-Newtonian generalization of relation (3.2.16)

$$c^{-1}\dot{r}_{ppN}(\infty) = \nu = \sigma - 4m\frac{\sigma \times (r_0 \times \sigma)}{|r_0 \times \sigma|^2}$$
$$+ m^2\left(-\frac{15}{4}\pi\frac{\sigma \times (r_0 \times \sigma)}{|r_0 \times \sigma|^3} - \frac{8\sigma}{|r_0 \times \sigma|^2}\right.$$
$$\left. + 8(r_0 + \sigma r_0)\frac{\sigma \times (r_0 \times \sigma)}{|r_0 \times \sigma|^4} - 4\alpha\frac{\sigma \times (r_0 \times \sigma)}{r_0 |r_0 \times \sigma|^2}\right).$$
(3.2.43)

From this

$$|\sigma \times \nu| = \frac{4m}{|r_0 \times \sigma|} + \frac{m^2}{|r_0 \times \sigma|^2}\left(\frac{15}{4}\pi - 8\frac{r_0 + \sigma r}{|r_0 \times \sigma|} + \frac{4\alpha}{r_0}|r_0 \times \sigma|\right). \qquad (3.2.44)$$

This relation defines the sine of the angle of the total deflection of light in the post-post-Newtonian approximation of the Schwarzschild problem.

3.2.5 Boundary value problem in the post-post-Newtonian approximation

Retaining designations (3.2.21) for the boundary value problem $\boldsymbol{r}(t_0) = \boldsymbol{r}_0$, $\boldsymbol{r}(t) = \boldsymbol{r}$ $(t_0 < t)$ one has

$$c(t - t_0) = D \tag{3.2.45}$$

$$\boldsymbol{\sigma} = \boldsymbol{D}/D \tag{3.2.46}$$

in the Newtonian approximation,

$$c(t - t_0) = D - \frac{m}{D}\boldsymbol{D}[\boldsymbol{B}_1(\boldsymbol{r}) - \boldsymbol{B}_1(\boldsymbol{r}_0)] \tag{3.2.47}$$

$$\boldsymbol{\sigma} = \frac{\boldsymbol{D}}{D} + \frac{m}{D^3}\{\boldsymbol{D} \times [\boldsymbol{D} \times (\boldsymbol{B}_1(\boldsymbol{r}) - \boldsymbol{B}_1(\boldsymbol{r}_0))]\} \tag{3.2.48}$$

in the post-Newtonian approximation and finally

$$\begin{aligned} c(t - t_0) = D &- \frac{m}{D}\boldsymbol{D}[\boldsymbol{B}_1(\boldsymbol{r}) - \boldsymbol{B}_1(\boldsymbol{r}_0)] - \frac{m^2}{D}\boldsymbol{D}[\boldsymbol{B}_2(\boldsymbol{r}) - \boldsymbol{B}_2(\boldsymbol{r}_0)] \\ &+ \frac{m^2}{2D^3}|\boldsymbol{D} \times (\boldsymbol{B}_1(\boldsymbol{r}) - \boldsymbol{B}_1(\boldsymbol{r}_0))|^2 \end{aligned} \tag{3.2.49}$$

$$\begin{aligned} \boldsymbol{\sigma} = \frac{\boldsymbol{D}}{D} &+ \frac{m}{D^3}\{\boldsymbol{D} \times [\boldsymbol{D} \times (\boldsymbol{B}_1(\boldsymbol{r}) - \boldsymbol{B}_1(\boldsymbol{r}_0))]\} \\ &+ \frac{m^2}{D^3}\{\boldsymbol{D} \times [\boldsymbol{D} \times (\boldsymbol{B}_2(\boldsymbol{r}) - \boldsymbol{B}_2(\boldsymbol{r}_0))]\} \\ &+ \frac{m^2}{D^3}\{(\boldsymbol{B}_1(\boldsymbol{r}) - \boldsymbol{B}_1(\boldsymbol{r}_0)) \times [\boldsymbol{D} \times (\boldsymbol{B}_1(\boldsymbol{r}) - \boldsymbol{B}_1(\boldsymbol{r}_0))]\} \\ &- \frac{3}{2}\frac{m^2}{D^5}\boldsymbol{D}|\boldsymbol{D} \times (\boldsymbol{B}_1(\boldsymbol{r}) - \boldsymbol{B}_1(\boldsymbol{r}_0))|^2 \end{aligned} \tag{3.2.50}$$

in the post-post-Newtonian approximation. In the right-hand sides of (3.2.47) and (3.2.48) and in the terms $O(m^2)$ of (3.2.49) and (3.2.50) one may use for $\boldsymbol{\sigma}$ the Newtonian expression (3.2.46). In the terms $O(m)$ of (3.2.49) and (3.2.50) one should use the post-Newtonian value (3.2.48).

Substitution of (3.2.41) and (3.2.42) into (3.2.49) and (3.2.50) gives

$$c(t - t_0) = D + m\left[2\ln\frac{r_0 + r + D}{r_0 + r - D}\right.$$

$$+\frac{\alpha}{2D}\left(\frac{r_0^2-r^2-D^2}{r}+\frac{r^2-r_0^2-D^2}{r_0}\right)\Bigg]$$
$$+m^2\left[\frac{15}{4}\frac{D}{|\boldsymbol{r}_0\times\boldsymbol{r}|}\left(\tan^{-1}\frac{r^2-r_0^2-D^2}{2|\boldsymbol{r}_0\times\boldsymbol{r}|}-\tan^{-1}\frac{r^2-r_0^2-D^2}{2|\boldsymbol{r}_0\times\boldsymbol{r}|}\right)\right.$$
$$+\frac{2D}{|\boldsymbol{r}_0\times\boldsymbol{r}|^2}(r-r_0+D)(r-r_0-D)$$
$$+\frac{1-4\epsilon}{8D}\left(\frac{r_0^2-r^2-D^2}{r^2}+\frac{r^2-r_0^2-D^2}{r_0^2}\right)$$
$$\left.+2\alpha\frac{(r-r_0)^2}{rr_0D}+\frac{\alpha^2}{2D^3}|\boldsymbol{r}_0\times\boldsymbol{r}|^2\left(\frac{1}{r_0}-\frac{1}{r}\right)^2\right] \tag{3.2.51}$$

$$\boldsymbol{\sigma}=\frac{\boldsymbol{D}}{D}+m\left[2\frac{r-r_0+D}{D|\boldsymbol{r}_0\times\boldsymbol{r}|^2}+\frac{\alpha}{D^3}\left(\frac{1}{r_0}-\frac{1}{r}\right)\right][\boldsymbol{D}\times(\boldsymbol{r}_0\times\boldsymbol{r})]$$
$$+m^2\left\{-\frac{|\boldsymbol{r}_0\times\boldsymbol{r}|^2}{2D^5}\left[\alpha\left(\frac{1}{r_0}-\frac{1}{r}\right)+\frac{2D^2}{|\boldsymbol{r}_0\times\boldsymbol{r}|^2}(r-r_0+D)\right]^2\boldsymbol{D}\right.$$
$$+\left[\frac{15}{8}\frac{\pi D}{|\boldsymbol{r}_0\times\boldsymbol{r}|^3}+\frac{15}{8}\frac{r^2-r_0^2+D^2}{D|\boldsymbol{r}_0\times\boldsymbol{r}|^3}\tan^{-1}\frac{r^2-r_0^2+D^2}{2|\boldsymbol{r}_0\times\boldsymbol{r}|}\right.$$
$$-\frac{15}{8}\frac{r^2-r_0^2-D^2}{D|\boldsymbol{r}_0\times\boldsymbol{r}|^3}\tan^{-1}\frac{r^2-r_0^2-D^2}{2|\boldsymbol{r}_0\times\boldsymbol{r}|}$$
$$+2\frac{(r+r_0)(r-r_0-D)}{D|\boldsymbol{r}_0\times\boldsymbol{r}|^4}(r-r_0+D)^2$$
$$+\left(\frac{1}{4}+\frac{1}{2}\alpha^2-\epsilon\right)\frac{1}{D^3}\left(\frac{1}{r_0^2}-\frac{1}{r^2}\right)$$
$$\left.\left.+\alpha\frac{(r+r_0)(r-r_0+D)^2}{D^2rr_0|\boldsymbol{r}_0\times\boldsymbol{r}|^2}-\alpha^2\frac{(r+r_0)(r-r_0)^3}{2D^5r_0^2r^2}\right][\boldsymbol{D}\times(\boldsymbol{r}_0\times\boldsymbol{r})]\right\} \tag{3.2.52}$$

with

$$4|\boldsymbol{r}_0\times\boldsymbol{r}|^2=[D^2-(r-r_0)^2][(r+r_0)^2-D^2]. \tag{3.2.53}$$

Thus, $t-t_0$ and $\boldsymbol{\sigma}$ are expressed in terms of the boundary values. Actually, it is not necessary to have an analytical expression for each effect related to the light propagation and one may merely use (3.2.51) and (3.2.52) for numerical calculation.

3.3 FIELD OF A ROTATING SPHEROID

3.3.1 Kerr metric

Next to the Schwarzschild solution, among other exact solutions of the

Einstein field equations the most important solution is the Kerr solution for the gravitational field of a rotating spherical body. This solution belongs to the class of axially symmetric fields described by the metric

$$\begin{aligned} ds^2 = {} & p(r,\theta)c^2\, dt^2 + 2b(r,\theta)c\, dt dr + 2d(r,\theta)\sin^2\theta c\, dt d\varphi - q(r,\theta)\, dr^2 \\ & - a^2(r,\theta)\, d\theta^2 - f^2(r,\theta)\sin^2\theta\, d\varphi^2 - 2g(r,\theta)\sin^2\theta\, dr d\varphi. \end{aligned} \tag{3.3.1}$$

All the functions a, b, d, f, g, p, q are even functions of θ. A characteristic of the Kerr metric is the fact that b and g may be annulled by a suitable coordinate transformation but that d is always different from zero. The Kerr metric is most commonly used in one of the following three forms:

$$\begin{array}{llll} & \text{I} & \text{II} & \text{III} \\ b(r,\theta) & \frac{2mr}{R^2} & 0 & \frac{2mr}{r^2+A^2} \\ q(r,\theta) & 1+\frac{2mr}{R^2} & \frac{R^2}{r^2-2mr+A^2} & \frac{R^2}{r^2+A^2}\left(1+\frac{2mr}{r^2+A^2}\right) \\ g(r,\theta) & A\left(1+\frac{2mr}{R^2}\right) & 0 & \frac{2mrA}{r^2+A^2} \end{array} \tag{3.3.2}$$

The remaining coefficients are the same for all three forms:

$$p(r,\theta) = 1 - \frac{2mr}{R^2} \qquad d(r,\theta) = \frac{2mrA}{R^2} \qquad a^2(r,\theta) = R^2$$

$$f^2(r,\theta) = r^2 + A^2 + \frac{2mr}{R^2}A^2\sin^2\theta \tag{3.3.3}$$

with

$$R^2 = r^2 + A^2\cos^2\theta.$$

m and A are two arbitrary constants of the Kerr solution . m has the same meaning as in the Schwarzschild solution. Systems I and III reduce under $A = 0$ to system V (3.1.23) of the Schwarzschild problem whereas system II reduces to system I (3.1.23). By comparison with the case of the weak field one obtains

$$A = 2I\omega/cM \tag{3.3.4}$$

where ω is the angular velocity of rotation of the body, I is the moment of inertia (for the homogeneous sphere $I = ML^2/5$, L being the radius of the sphere). Systems I and II have been derived in Boyer and Lindquist (1967) and system III is studied in Carter (1966).

3.3.2 Weyl–Levi-Civita metric

Another physically important axially symmetric solution of type (3.3.1) is the Weyl–Levi-Civita solution describing the gravitational field of a fixed spheroid. The form of this solution most often employed is given in Young and Coulter (1969). The coefficients of the metric (3.3.1) for this solution have the values

$$p(r,\theta) = e^{2\psi} \qquad q(r,\theta) = e^{2\gamma-2\psi}\left(1+\frac{m^2\sin^2\theta}{r^2-2mr}\right)$$

$$a^2(r,\theta) = e^{2\gamma-2\psi}(r^2-2mr+m^2\sin^2\theta) \qquad f^2(r,\theta) = e^{-2\psi}(r^2-2mr)$$

$$b(r,\theta) = d(r,\theta) = g(r,\theta) = 0 \tag{3.3.5}$$

with

$$\psi = \frac{1}{2}\ln\left(1-\frac{2m}{r}\right) + \frac{1}{2}q\frac{r^2}{m^2}(3\cos^2\theta-1) \times \left[\left(\frac{3}{4}-\frac{3}{2}\frac{m}{r}+\frac{1}{2}\frac{m^2}{r^2}\right)\ln\left(1-\frac{2m}{r}\right)+\frac{3}{2}\frac{m}{r}-\frac{3}{2}\frac{m^2}{r^2}\right] \tag{3.3.6}$$

$$\begin{aligned}
\gamma = {} & \frac{1}{2}\ln\frac{r^2-2mr}{r^2-2mr+m^2\sin^2\theta} + q\left\{\ln\frac{r^2-2mr}{r^2-2mr+m^2\sin^2\theta}\right. \\
& \left. -3\frac{r}{m}\sin^2\theta\left[\frac{1}{2}\left(1-\frac{m}{r}\right)\ln\left(1-\frac{2m}{r}\right)+\frac{m}{r}\right]\right\} \\
& + q^2\left\{\frac{1}{2}\ln\frac{r^2-2mr}{r^2-2mr+m^2\sin^2\theta}\right. \\
& + \frac{r^4}{m^4}\sin^2\theta\left[\left(-\frac{9}{8}+\frac{9}{2}\frac{m}{r}-\frac{45}{8}\frac{m^2}{r^2}+\frac{9}{4}\frac{m^3}{r^3}\right)\ln^2\left(1-\frac{2m}{r}\right)\right. \\
& + \left(-\frac{9}{2}\frac{m}{r}+\frac{27}{2}\frac{m^2}{r^2}-\frac{21}{2}\frac{m^3}{r^3}+\frac{3}{2}\frac{m^4}{r^4}\right) \\
& \left.\times\ln\left(1-\frac{2m}{r}\right)-\frac{9}{2}\frac{m^2}{r^2}+9\frac{m^3}{r^3}-3\frac{m^4}{r^4}\right] \\
& + \frac{r^4}{m^4}\sin^4\theta\left[\left(\frac{81}{64}-\frac{81}{16}\frac{m}{r}+\frac{99}{16}\frac{m^2}{r^2}-\frac{9}{4}\frac{m^3}{r^3}\right)\ln^2\left(1-\frac{2m}{r}\right)\right. \\
& + \left(\frac{81}{16}\frac{m}{r}-\frac{243}{16}\frac{m^2}{r^2}+\frac{45}{4}\frac{m^3}{r^3}-\frac{9}{8}\frac{m^4}{r^4}\right) \\
& \left.\left.\times\ln\left(1-\frac{2m}{r}\right)+\frac{81}{16}\frac{m^2}{r^2}-\frac{81}{8}\frac{m^3}{r^3}+\frac{45}{16}\frac{m^4}{r^4}\right]\right\}.
\end{aligned} \tag{3.3.7}$$

m and q are two physical parameters of this metric. m has the same meaning as before. For $q = 0$ this metric reduces to the Schwarzschild metric I (3.1.23). Expansion in powers of m/r yields

$$\psi = \frac{1}{2}\ln\left(1 - \frac{2m}{r}\right) - \frac{1}{2}q(3\cos^2\theta - 1)\left(\frac{2}{15}\frac{m^3}{r^3} + \frac{2}{5}\frac{m^4}{r^4} + \ldots\right) \quad (3.3.8)$$

$$\gamma = \frac{1}{2}\ln\frac{r^2 - 2mr}{r^2 - 2mr + m^2\sin^2\theta} + q\left[\left(-\frac{2}{5}\sin^2\theta + \frac{1}{2}\sin^4\theta\right)\frac{m^4}{r^4} + \ldots\right]$$
$$+ q^2\left[\left(-\frac{2}{25}\sin^2\theta + \frac{6}{25}\sin^4\theta - \frac{1}{6}\sin^6\theta\right)\frac{m^6}{r^6} + \ldots\right]. \quad (3.3.9)$$

Therefore, the expansion of $p(r,\theta)$ starts with the terms

$$p(r,\theta) = 1 - \frac{2m}{r} + \frac{2}{15}q\frac{m^3}{r^3}(1 - 3\cos^2\theta) + \ldots. \quad (3.3.10)$$

On the other hand, the Newtonian potential of spheroid is expressed as

$$U = \frac{GM}{r} - \frac{Q}{2r^3}(1 - 3\cos^2\theta) \quad (3.3.11)$$

with the quadrupole moment

$$Q = G(A - C) \quad (3.3.12)$$

A and C being equatorial and polar moments of inertia of the spheroid. Comparison with the weak field metric gives

$$q = \frac{15}{2}\frac{Q}{c^2m^3}. \quad (3.3.13)$$

Detailed expansions for all coefficients of the metric of a spheroid have been derived in the paper cited above.

Celestial mechanics analysis of the motion in an axially symmetrical field (3.3.1) for the cases of rotating spheres and fixed spheroids has not been performed as thoroughly as for the Schwarzschild field. In fact, only the cases of radial motion and motion in the equatorial plane have been investigated adequately. However, for practical purposes it is not necessary to use the rigorous solution (3.3.1) with coefficients (3.3.2), (3.3.3) or (3.3.5). Instead, it is sufficient to have metric of a slowly rotating spheroid.

3.3.3 Metric of slowly rotating spheroid

The gravitational field of the slowly rotating spheroid is characterized by the Newtonian potential (3.3.11) and the vector potential U^i determined by (2.2.27) with velocity components

$$v^i = \epsilon_{ijk}\omega^j x^k \quad (3.3.14)$$

of the points of the spheroid (ϵ_{ijk} is the completely antisymmetric Levi-Civita tensor, $\epsilon_{123} = +1$) and moments of inertia

$$I^{km} = \int \rho x^k x^m \, \mathrm{d}^3x. \tag{3.3.15}$$

Choosing the plane $x^3 = z = 0$ as the equatorial plane one has

$$\omega^1 = \omega^2 = 0 \qquad \omega^3 = \omega \qquad I^{11} = I^{22} = \tfrac{1}{2}C \quad I^{33} = A - \tfrac{1}{2}C$$

and the components of the vector potential become

$$U^i = G\epsilon_{ijk}\omega^j I^{km} x^m / r^3$$

or

$$U^1 = -\frac{1}{2}GC\omega\frac{y}{r^3} \qquad U^2 = \frac{1}{2}GC\omega\frac{x}{r^3} \qquad U^3 = 0. \tag{3.3.16}$$

As far as the additive term in h_{00} is concerned, by neglecting the terms $O(\omega^2)$ and using (2.2.33)–(2.2.35), this term reduces to the correction $2U^2/c^4$ alone. Indeed, the function χ in this case does not depend explicitly on t and the contribution from $\tilde{U}$ is

$$\tilde{U} = \frac{G\xi}{r} \qquad \xi = \int (-\rho U + \rho\Pi + 3p) \, \mathrm{d}^3x. \tag{3.3.17}$$

As was shown by Fock (1955), one may assume the validity of the condition

$$\rho\Pi - \rho U + p = \rho\Omega \tag{3.3.18}$$

inside the body. The integral of pressure may be evaluated as

$$\int p \, \mathrm{d}^3x = \tfrac{1}{3}\epsilon - \tfrac{2}{3}T \tag{3.3.19}$$

Ω is the potential of centrifugal force

$$\Omega = \tfrac{1}{2}\omega^2(x^2 + y^2) \tag{3.3.20}$$

T is the kinetic energy of rotation of the body

$$T = \int \rho\Omega \, \mathrm{d}^3x = \tfrac{1}{2}\omega^2 C \tag{3.3.21}$$

and

$$\epsilon = \tfrac{1}{2}\int \rho U \, \mathrm{d}^3x. \tag{3.3.22}$$

Therefore,

$$\xi = \tfrac{2}{3}\epsilon - \tfrac{1}{3}T.$$

The term with T vanishes in the linear approximation. Then the additive term (3.3.17) may be included into the Newtonian potential by re-defining the mass of the spheroid. Finally, from (2.2.32) and (2.2.37) the metric of the slowly rotating spheroid is determined by

$$\begin{aligned}
h_{00} &= -\frac{2m}{r} + \frac{Q}{c^2 r^3}\left(1 - \frac{3z^2}{r^2}\right) + \frac{2m^2}{r^2} - \frac{2mQ}{c^2 r^4}\left(1 - \frac{3z^2}{r^2}\right) + 2a_{0,0} \\
&\quad + \left[-\frac{2m}{r^3}x^s + \frac{Q}{c^2}\left(\frac{3x^s}{r^5} - 15\frac{z^2}{r^7}x^s + 6\frac{z}{r^5}\delta_{3s}\right)\right] a_s \\
h_{01} &= -\frac{2GC\omega}{c^3}\frac{y}{r^3} + a_{0,1} + a_{1,0} \\
h_{02} &= \frac{2GC\omega}{c^3}\frac{x}{r^3} + a_{0,2} + a_{2,0} \\
h_{03} &= a_{0,3} + a_{3,0} \\
h_{ik} &= -\frac{2m}{r}\delta_{ik} + \frac{Q}{c^2 r^3}\left(1 - \frac{3z^2}{r^2}\right)\delta_{ik} + a_{i,k} + a_{k,i}.
\end{aligned} \tag{3.3.23}$$

These expressions are valid in the linear approximation with respect to angular velocity ω and quadrupole moment Q. Let arbitrary functions a_0, a_s be chosen in the same manner as for the Schwarzschild problem, i.e.

$$a_0 = 0 \qquad a_i = \alpha\frac{m}{r}x^i \tag{3.3.24}$$

where α is again the coordinate parameter . With values (3.3.24) the potentials of gravitation of the slowly rotating spheroid take the form

$$\begin{aligned}
h_{00} &= -\frac{2m}{r} + \frac{Q}{c^2 r^3}\left(1 - \frac{3z^2}{r^2}\right) + 2(1-\alpha)\frac{m^2}{r^2} - (-2+3\alpha)\frac{mQ}{c^2 r^4}\left(1 - \frac{3z^2}{r^2}\right) \\
h_{01} &= -\frac{2GC\omega}{c^3}\frac{y}{r^3} \qquad h_{02} = \frac{2GC\omega}{c^3}\frac{x}{r^3} \qquad h_{03} = 0 \\
h_{ik} &= -2(1-\alpha)\frac{m}{r}\delta_{ik} - 2\alpha\frac{m}{r^3}x^i x^k + \frac{Q}{c^2 r^3}\left(1 - \frac{3z^2}{r^2}\right)\delta_{ik}.
\end{aligned} \tag{3.3.25}$$

By using (2.2.49) and (2.2.53) the motion of a test particle in the field (3.3.25) can be described by the Lagrangian

$$L = \frac{1}{2}\dot{r}^2 + \frac{GM}{r} + R_1 + R_2 + R_3 + R_4 \tag{3.3.26}$$

$$R_1 = -\frac{Q}{2r^3}\left(1 - \frac{3z^2}{r^2}\right)$$

$$R_2 = \frac{1}{8c^2}(\dot{\boldsymbol{r}}^2)^2 + \frac{m}{r}\left[\left(\frac{3}{2} - \alpha\right)\dot{\boldsymbol{r}}^2 + \left(-\frac{1}{2} + \alpha\right)\frac{GM}{r} + \alpha\frac{(\boldsymbol{r}\dot{\boldsymbol{r}})^2}{r^2}\right]$$

$$R_3 = -\frac{2GC\omega}{c^2 r^3}(x\dot{y} - y\dot{x})$$

$$R_4 = \frac{Q}{c^2}\frac{1}{2r^3}\left(1 - \frac{3z^2}{r^2}\right)\left((1 - 3\alpha)\frac{GM}{r} - \frac{3}{2}\dot{\boldsymbol{r}}^2\right).$$

The corresponding equations of motion take the form (1.1.25) with disturbing force

$$\boldsymbol{F} = \boldsymbol{F}_1 + \boldsymbol{F}_2 + \boldsymbol{F}_3 + \boldsymbol{F}_4 \tag{3.3.27}$$

$$\boldsymbol{F}_1 = \frac{3Q}{r^5}\left[\frac{1}{2}\left(1 - 5\frac{z^2}{r^2}\right)\boldsymbol{r} + z\boldsymbol{s}\right]$$

$$\boldsymbol{F}_2 = \frac{m}{r^3}\left[\left((4 - 2\alpha)\frac{GM}{r} - (1 + \alpha)\dot{\boldsymbol{r}}^2 + 3\alpha\frac{(\boldsymbol{r}\dot{\boldsymbol{r}})^2}{r^2}\right)\boldsymbol{r} + (4 - 2\alpha)(\boldsymbol{r}\dot{\boldsymbol{r}})\dot{\boldsymbol{r}}\right]$$

$$\boldsymbol{F}_3 = \frac{2GC\omega}{c^2 r^3}\left(\frac{3z}{r^2}(\boldsymbol{r} \times \dot{\boldsymbol{r}}) + (\dot{\boldsymbol{r}} \times \boldsymbol{s})\right)$$

$$\boldsymbol{F}_4 = \frac{Q}{c^2 r^5}\left\{\left[\left(-8 + 6\alpha + 3(12 - 11\alpha)\frac{z^2}{r^2}\right)\frac{GM}{r} + \frac{3}{2}\left(1 - 5\frac{z^2}{r^2}\right)\dot{\boldsymbol{r}}^2\right]\boldsymbol{r}\right.$$
$$\left. + 3\left((-4 + 5\alpha)\frac{GM}{r} + \dot{\boldsymbol{r}}^2\right)z\boldsymbol{s} - 6\left[\left(1 - 5\frac{z^2}{r^2}\right)(\boldsymbol{r}\dot{\boldsymbol{r}}) + 2z\dot{z}\right]\dot{\boldsymbol{r}}\right\}.$$

Here $\boldsymbol{s}$ denotes the unit vector $(0, 0, 1)$ directed along the z axis. The perturbing functions R_k and perturbing forces $\boldsymbol{F}_k$ (k =1,2,3,4) have the following meanings: $k = 1$—Newtonian perturbations due to the quadrupole moment; $k = 2$—relativistic Schwarzschild perturbations; $k = 3$—relativistic perturbations caused by rotation of the primary; $k = 4$—direct relativistic perturbations due to the quadrupole moment. In solving the problem (1.1.25), (3.3.27) the latter class of perturbations should be extended by adding within the same level of accuracy the indirect second-order perturbations from the interaction of the perturbations of $k = 1$ and $k = 2$. In spite of the large number of papers devoted to relativistic quadrupole perturbations (see, for example, Krause 1963, Barker and O'Connell 1976, Soffel *et al* 1988), this problem still needs detailed investigation. For example, the second-order perturbations just mentioned have been obtained only quite recently (Heimberger *et al* 1990). Only the main secular perturbations of problem (3.3.27) are indicated here. Using expressions (1.1.29) and averaging over the mean anomaly one obtains the equations for the secular variations of the area vector and the Laplace vector:

$$\dot{c} = \frac{3}{4}Q\frac{\sin 2i}{a^3(1-e^2)^{3/2}}\boldsymbol{l} + \frac{2GC\omega}{c^2}\frac{n\sin i}{a(1-e^2)}\boldsymbol{l} + \frac{3}{4}Qm\frac{\sin i}{a^4(1-e^2)^{5/2}}$$
$$\times \{\boldsymbol{l}[-6+10\alpha+5\alpha e^2+(1-\tfrac{5}{2}\alpha)e^2\cos 2\varpi]\cos i$$
$$+(\boldsymbol{s}\times\boldsymbol{l})(1-\tfrac{5}{2}\alpha)e^2\sin 2\varpi\} \quad (3.3.28)$$

$$\dot{\boldsymbol{f}} = \frac{3}{2}Q\frac{ne}{a^2(1-e^2)^2}\Bigg[-\frac{1}{8}\boldsymbol{P}\sin^2 i\sin 2\varpi$$
$$+\boldsymbol{Q}\left(-1+\frac{11}{8}\sin^2 i-\frac{1}{8}\sin^2 i\cos 2\varpi\right)$$
$$-\frac{5}{8}\boldsymbol{k}\sin 2i\cos\varpi+\frac{1}{4}\boldsymbol{s}\sin i\cos\varpi\Bigg]+3m\frac{n^3a^2e}{1-e^2}\boldsymbol{Q}$$
$$+\frac{2GC\varpi}{c^2}\frac{n^2e}{(1-e^2)^{3/2}}[-3\boldsymbol{Q}\cos i+(\boldsymbol{s}\times\boldsymbol{P})]$$
$$+3Qm\frac{ne}{a^3(1-e^2)^3}\{\tfrac{1}{8}\boldsymbol{P}(3-13\alpha-\tfrac{7}{2}e^2+2\alpha e^2)\sin^2 i\sin 2\varpi$$
$$+\boldsymbol{Q}[1-3\alpha+(-\tfrac{17}{4}+\tfrac{19}{2}\alpha)\sin^2 i+(-\tfrac{3}{8}-\tfrac{3}{4}\alpha)e^2$$
$$+(\tfrac{13}{16}+\tfrac{19}{8}\alpha)e^2\sin^2 i+(\tfrac{3}{8}-\tfrac{13}{8}\alpha-\tfrac{17}{16}\alpha e^2)\sin^2 i\cos 2\varpi]$$
$$+(\boldsymbol{s}\times\boldsymbol{k})[-4+\tfrac{15}{2}\alpha+(\tfrac{1}{4}+\tfrac{15}{8}\alpha)e^2]\sin i\sin\varpi$$
$$+\boldsymbol{s}[\tfrac{3}{2}-\tfrac{5}{2}\alpha-(\tfrac{1}{4}+\tfrac{5}{8}\alpha)e^2]\sin i\cos\varpi\}. \quad (3.3.29)$$

Keplerian elements and unit vectors $\boldsymbol{P}$, $\boldsymbol{Q}$, $\boldsymbol{l}$, $\boldsymbol{m}$, $\boldsymbol{k}$ are denoted here as earlier in accordance with (1.1.6) and (1.1.7) (we use here the symbol ϖ for the argument of pericentre to distinguish from ω for the angular velocity of rotation). For a spherical body with $Q = 0$ equations (3.3.28) and (3.3.29), using $\boldsymbol{s}\times\boldsymbol{k} = \boldsymbol{l}\sin i$, $\boldsymbol{sk} = \cos i$, may be transformed to

$$\dot{\boldsymbol{c}} = \boldsymbol{\Omega}\times\boldsymbol{c} \qquad \dot{\boldsymbol{f}} = \boldsymbol{\Omega}\times\boldsymbol{f} \quad (3.3.30)$$

demonstrating the precession law of changing vectors $\boldsymbol{c}$ and $\boldsymbol{f}$ with the rate

$$\boldsymbol{\Omega} = \frac{3mn}{a(1-e^2)}\boldsymbol{k} + \frac{2GC\omega}{c^2}\frac{1}{a^3(1-e^2)^{3/2}}[\boldsymbol{s}-3(\boldsymbol{sk})\boldsymbol{k}]. \quad (3.3.31)$$

The terms of $\boldsymbol{\Omega}$ proportional to $\boldsymbol{k}$ evidently give no contribution to the expression for $\dot{c}$. The first term of (3.3.31) is the Schwarzschild advance. The second term, proportional to the angular velocity ω, describes the Lense–Thirring precession due to rotation of the primary. Taking scalar products of (3.3.28) and (3.3.29) with $\boldsymbol{l}$, $\boldsymbol{m}$, $\boldsymbol{k}$, $\boldsymbol{P}$ and $\boldsymbol{Q}$ one obtains the equations for the secular variations of the Keplerian elements:

$$\frac{\mathrm{d}p}{\mathrm{d}t} = \frac{3}{2}Qm\frac{e^2}{na^5(1-e^2)^2}(-1+\tfrac{5}{2}\alpha)\sin^2 i\sin 2\varpi \quad (3.3.32)$$

$$\frac{de}{dt} = \frac{21}{8}Qm\frac{e}{na^6(1-e^2)^3}(-1+\alpha-\tfrac{1}{2}e^2+\alpha e^2)\sin^2 i \sin 2\varpi \quad (3.3.33)$$

$$\frac{di}{dt} = \frac{3}{8}Qm\frac{e^2}{na^6(1-e^2)^3}(-1+\tfrac{5}{2}\alpha)\sin 2i \sin 2\varpi \quad (3.3.34)$$

$$\frac{d\Omega}{dt} = \frac{3}{2}Q\frac{\cos i}{na^5(1-e^2)^2} + \frac{2GC\omega}{c^2}\frac{1}{a^3(1-e^2)^{3/2}} + \frac{3}{4}Qm\frac{\cos i}{na^6(1-e^2)^3}[-6+10\alpha+5\alpha e^2+(1-\tfrac{5}{2}\alpha)e^2\cos 2\varpi] \quad (3.3.35)$$

$$\frac{d\varpi}{dt} + \cos i\frac{d\Omega}{dt} = \frac{3}{2}\frac{Q}{na^5(1-e^2)^2}\left(\frac{3}{2}\sin^2 i - 1\right) + \frac{3mn}{a(1-e^2)} - \frac{4GC\omega}{c^2}\frac{\cos i}{a^3(1-e^2)^{3/2}} + \frac{3}{4}\frac{Qm}{na^6(1-e^2)^3} \times [(4-12\alpha-\tfrac{3}{2}(1+2\alpha)e^2)(1-\tfrac{3}{2}\sin^2 i) + \tfrac{7}{2}(-1+\alpha-\tfrac{1}{2}\alpha e^2)\sin^2 i\cos 2\varpi]. \quad (3.3.36)$$

Equations (3.3.32) and (3.3.36) may be replaced by two equivalent ones

$$\frac{da}{dt} = \frac{3Qm}{na^5}\frac{e^2}{(1-e^2)^4}\left[-\frac{9}{4}+3\alpha+\left(-\frac{3}{8}+\frac{1}{2}\alpha\right)e^2\right]\sin^2 i\sin 2\varpi \quad (3.3.37)$$

$$\frac{d\varpi}{dt} = \frac{3}{4}Q\frac{1-5\cos^2 i}{na^5(1-e^2)^2} + \frac{3mn}{a(1-e^2)} - \frac{6GC\omega}{c^2}\frac{\cos i}{a^3(1-e^2)^{3/2}} + \frac{3Qm}{4na^6(1-e^2)^3}(10-22\alpha+(-12+28\alpha)\sin^2 i-(\tfrac{3}{2}+8\alpha)e^2 + (\tfrac{9}{4}+\tfrac{19}{2}\alpha)e^2\sin^2 i+\{(-1+\tfrac{5}{2}\alpha)e^2 + [-\tfrac{7}{2}+\tfrac{7}{2}\alpha+(1-\tfrac{17}{4}\alpha)e^2]\sin^2 i\}\cos 2\varpi). \quad (3.3.38)$$

An equation for the variation of the sixth element, the mean anomaly at the epoch, is not given here. The possible contribution into terms of the order Qm due to combination of the first-order terms in Q and m separately is not included. These equations demonstrate that in the first approximation elements Ω and ω contain secular terms whereas all other elements contain long period terms of order Qm. Indeed, integration of these equations in the first approximation yields

$$\Delta a = \frac{3}{2}Qm\frac{e^2\sin^2 i}{\omega_1 na^5(1-e^2)^4}[\tfrac{9}{4}-3\alpha+(\tfrac{3}{8}-\tfrac{1}{2}\alpha)e^2]\cos 2\varpi \quad (3.3.39)$$

$$\Delta e = \frac{21}{16} Qm \frac{e \sin^2 i}{\omega_1 n a^6 (1-e^2)^3} [1 - \alpha + (\tfrac{1}{2} - \alpha) e^2] \cos 2\varpi \tag{3.3.40}$$

$$\Delta i = \frac{3}{16} Qm \frac{e^2 \sin 2i}{\omega_1 n a^6 (1-e^2)^3} (1 - \tfrac{5}{2}\alpha) \cos 2\varpi \tag{3.3.41}$$

$$\Delta \Omega = \Omega_1 t + \frac{3}{8} Qm \frac{e^2 \cos i}{\omega_1 n a^6 (1-e^2)^3} (1 - \tfrac{5}{2}\alpha) \sin 2\varpi \tag{3.3.42}$$

$$\Delta \varpi = \omega_1 t + \frac{3}{8} Qm \frac{1}{\omega_1 n a^6 (1-e^2)^3} \{(-1 + \tfrac{5}{2}\alpha) e^2 + [-\tfrac{7}{2} + \tfrac{7}{2}\alpha + (1 - \tfrac{17}{4}\alpha) e^2] \sin^2 i\} \sin 2\varpi. \tag{3.3.43}$$

The frequencies Ω_1 and ω_1 have values

$$\Omega_1 = \frac{3Q \cos i}{2na^5(1-e^2)^2} + \frac{2GC\omega}{c^2 a^3 (1-e^2)^{3/2}} + \frac{3Qm \cos i}{4na^6(1-e^2)^3} (-6 + 10\alpha + 5\alpha e^2) \tag{3.3.44}$$

$$\omega_1 = \frac{3}{4} Q \frac{1 - 5\cos^2 i}{na^5(1-e^2)^2} + \frac{3mn}{a(1-e^2)} - \frac{6GC\omega \cos i}{c^2 a^3 (1-e^2)^{3/2}} + \frac{3Qm}{4na^6(1-e^2)^3} \times [10 - 22\alpha + (-12 + 28\alpha) \sin^2 i - (\tfrac{3}{2} + 8\alpha) e^2 + (\tfrac{9}{4} + \tfrac{19}{2}\alpha) e^2 \sin^2 i]. \tag{3.3.45}$$

On the right-hand sides of (3.3.39)–(3.3.43) ϖ is to be meant as a linear function of time with the frequency ω_1.

In application to the theory of motion of Earth satellites of particular interest is the case of the critical inclination when $\cos^2 i = 1/5$ and ω_1 is of the relativistic order of smallness (for the model problem considered here). This means that amplitudes of relativistic perturbations in (3.3.39)–(3.3.43) are of Newtonian order of smallness $O(Q)$. Thus, the case of the critical inclination corresponds to relativistic resonance leading to relativistic solutions to be expanded in fractional powers of the relativistic small parameter. This case deserves further investigation.

Concluding this chapter devoted to the one-body problems let us note that one may find in the literature descriptions of numerous effects in the motion of particles and light propagation due to rotation of the primary and its non-spherical form. All these effects are not observable as yet and, excluding the Lense–Thirring precession, can hardly have a great chance of being detected in the near future. A detailed bibliography on these effects may be found in the monograph by Ivanitskaya (1979).

4

Approximate Solutions of the Field Equations and Approximate Equations of Motion

4.1 *N*-BODY PROBLEM

4.1.1 History of the question

Exact solutions of the field equations and the form of the rigorous equations of motion of the N-body problem are not known in GRT even for $N = 2$. To solve this problem one has to apply the approximate methods and construct a solution in the form of series in powers of small parameters. The N-body problem applied to Solar System bodies is characterized by three main simplifying factors:

(1) the slowness of motion: $v \ll c$, v being the characteristic velocity of the bodies;

(2) the weak gravitational field (everywhere including the interiors of the bodies): $U \ll c^2$, U being the Newtonian potential;

(3) quasi-point structure of the bodies: $L \ll R$, L is the characteristic linear size of the bodies, R is the characteristic distance between the bodies.

This enables us to look for the solution of the field equations (2.1.11) in the form (2.1.3) taking Galilean values (2.1.2) as the initial approximation and expanding the components of the metric tensor in the series in powers v/c and U/c^2

$$\begin{aligned} h_{00} &= c^{-2}h_{00}^{(2)} + c^{-4}h_{00}^{(4)} + c^{-5}h_{00}^{(5)} + c^{-6}h_{00}^{(6)} + \dots \\ h_{0i} &= c^{-3}h_{0i}^{(3)} + c^{-5}h_{0i}^{(5)} + c^{-6}h_{0i}^{(6)} + \dots \\ h_{ik} &= c^{-2}h_{ik}^{(2)} + c^{-4}h_{ik}^{(4)} + c^{-5}h_{ik}^{(5)} + c^{-6}h_{ik}^{(6)} + \dots . \end{aligned} \tag{4.1.1}$$

Coefficients of these series are expanded in series with respect to parameters of the order L/R. First of all, it is to be noted that the quantities $h_{00}^{(5)}$, $h_{0i}^{(6)}$ and $h_{ik}^{(5)}$ are due to gravitational radiation from the bodies. Not so long ago some authors considered it possible to construct the expansions (4.1.1) without terms of odd degree in h_{00}, h_{ik} and without terms of even degree in h_{0i}. This would allow the possibility of annulling the gravitational radiation by a suitable choice of coordinate system. More recent investigations showed that this point of view was erroneous. A culminating contribution to the problem of gravitational radiation was made by Damour (1983, 1987a) and in a series of papers by Grishchuk and Kopejkin (Grishchuk and Kopejkin 1983, 1986, Kopejkin 1985). In the practical application of GRT to the motion of Solar System bodies one may not take into account the gravitational radiation and in most cases the radiation terms in (4.1.1) are omitted below.

Historically, the equations of the motion of the problem of N bodies considered as point masses were first obtained by generalizing the geodesic principle. If the metric tensor $g_{\mu\nu}$ is known then the equations of motion of a test particle result from the geodesic variational principle. Taking into account in a suitable manner the influence of the body on the surrounding field this principle may be generalized and made applicable to the motion of finite masses. It was by this method that de Sitter (1916) first derived the relativistic equations of motion of the N-body problem. Some arithmetic errors occurred in these equations and are reproduced in the the encyclopaedic paper of Kottler (1922) and the treatise by Chazy (1928,1930), but were corrected by Eddington and Clark (1938). Not claiming to give an extensive bibliography, let us add that the correct equations of motion of the N-body problem were also derived by means of the geodesic principle by Bertotti (1954).

A more refined method enables us to derive the equations of motion directly from the field equations avoiding the use of the geodesic principle. In contrast to the Newtonian theory of gravitation, the GRT equations of motion of the gravitating masses as the sources of the field are intimately related to the field equations as was demonstrated by Einstein and Grommer (1927). This is due to the non-linear form of the field equations and the existence of the four Bianchi identities. In the linear Newtonian field theory determined by the linear Poisson equations the equations of motion are not related to the field equations. The field equations admit solutions under arbitrary given motion of masses and the equations of motion are to be postulated separately. In GRT, for the solvability of the field equations or the fulfillment of the coordinate conditions one should impose some definite relations on the coordinates of masses. These relations turn out to be the differential equations of motion.

The equations of motion of the N-body problem were derived for the first time from the field equations by Einstein *et al* (1938). Their method, called the EIH technique, was finally completed in the paper by Einstein and Infeld (1949). The specific feature of this technique is the use of the vacuum field equations outside the masses and the treatment of masses as the field singularities. The EIH technique is sometimes regarded as rather arduous. In deriving the post-Newtonian equations of motion by this technique one needs to determine $h_{00}^{(2)}$, $h_{ik}^{(2)}$, $h_{0i}^{(3)}$, $h_{00}^{(4)}$, $h_{ik}^{(4)}$ and $h_{0i}^{(5)}$ whereas in most other methods $h_{ik}^{(4)}$ and $h_{0i}^{(5)}$ are not needed. However the mathematical elegance of the EIH technique implying in particular the use of the vacuum field equations alone compensates to a large measure its practical complexity. In distinction from the original papers of the EIH technique the book by Brumberg (1972) contains the determination of the gravitation potentials and the derivation of the equations of motion by this technique under harmonic coordinate conditions resulting in some simplification of the necessary manipulation. Independently and a little later than the appearance of the EIH technique the equations of motion were derived by Fock using the mass tensor of the extended bodies. This method is exposed in its final form in Fock (1955). A set of other methods of somewhat intermediate nature have been developed since including, for example, the Infeld technique (Infeld 1954, 1957, Infeld and Plebanski 1960) involving the use of the mass tensors with δ-functions and the derivation of the equations of motion from the variational principle (2.1.38).

Beginning in the middle of the 1970s interest in the problem of motion in GRT greatly increased. The new approaches have led to more efficient mathematical techniques and more physically meaningful solutions although the technical side of the methods has changed little (Thorne and Hartle 1985). The mathematical structure of expansions (4.1.1) and the chances of finding, even in principle, the terms of arbitrary high order have become clearer. The technique of taking into account the internal structure and shape of the bodies, their rotation, density distribution, etc, has been rigorously improved. Of particular importance is the description of the physical characteristics of the body (sphericity, rigid body rotation, etc) in its proper reference system co-moving the body (Kopejkin 1987). Ignoring this condition might lead in expansions (4.1.1) to non-physical terms of the order L^2/R^2 (Brumberg 1972). A detailed review and discussion of the new approaches to the GRT problem of motion may be found in Damour (1987b). This paper is an excellent introduction to the present state of the problem of motion in GRT.

4.1.2 Gravitational field of the N-body problem

For the practical purposes of relativistic celestial mechanics it is sufficient to use the results obtained using the Fock method. In accordance with the designations employed in (2.2.38)–(2.2.40) the metric of the field of N bodies is determined by

$$g_{00} = 1 - c^{-2}2U + c^{-4}2(U^2 - W) + 2\left(a_{0,0} + c^{-2}U_{,k}a_k + c^2\sum_A \frac{\partial U}{\partial x_A^k}\tilde{a}_k(\boldsymbol{x}_A)\right) + \ldots \quad (4.1.2)$$

$$g_{0i} = c^{-3}4U^i + a_{0,i} + a_{i,0} + \ldots \quad (4.1.3)$$

$$g_{ik} = -\delta_{ik} - c^{-2}2U\delta_{ik} + a_{i,k} + a_{k,i} + \ldots \quad (4.1.4)$$

$$W = \tfrac{3}{2}\Phi_1 - \Phi_2 + \Phi_3 + 3\Phi_4 + \frac{\partial^2\chi}{\partial t^2}. \quad (4.1.5)$$

The functions U, U^i and W dependent on $\boldsymbol{x} = (x^1, x^2, x^3)$ and t are the Newtonian potential, vector potential and additive potential, respectively. The bodies are enumerated by capital latin letters $A, B, \ldots$. a_i and a_0 are arbitrary functions of the second and the third order of smallness respectively. The coordinates of the bodies $\boldsymbol{x}_A = \boldsymbol{x}_A(t)$ are functions of time to be determined from the differential equations of motion. The bodies have masses M_A. In addition, they possess dipole and quadrupole moments I_A^k and I_A^{km}, respectively. They are assumed to be in approximately rigid body rotation with angular velocities ω_A^k so that the velocity distribution inside body A is

$$v^i = v_A^i + \epsilon_{ijk}\omega_A^j r_A^k \quad (4.1.6)$$

$v_A^i = \dot{x}_A^i(t)$ being the translatory velocity of body A. ϵ_{ijk} is the three-dimensional antisymmetric Levi-Civita symbol ($\epsilon_{123} = +1$). r_A^k stands for $r_A^k = x^k - x_A^k$ and the designation $r_{AB}^k = x_A^k - x_B^k$ is used below. The assumption (4.1.6) is valid only for the Newtonian approximation and may be used in the relativistic terms. Only linear terms with respect to I_A^k, I_A^{km} and ω_A^k are further retained. The mass of the body and its moments are defined by the expressions

$$M_A = \int_{(A)} \rho' \,\mathrm{d}^3x' \quad (4.1.7)$$

$$I_A^k = \int_{(A)} \rho'(x'^k - x_A^k)\,\mathrm{d}^3x' \qquad I_A^{km} = \int_{(A)} \rho'(x'^k - x_A^k)(x'^m - x_A^m)\,\mathrm{d}^3x' \quad (4.1.8)$$

where ρ' is the mass density at the point $\boldsymbol{x}'$ and the integration is performed over the hypersurface t = constant corresponding to the volume of the body A. The moments of the body are functions of time and their derivatives satisfy the relations

$$\dot{I}_A^k = \epsilon_{kjn}\omega_A^j I_A^n \qquad \dot{I}_A^{km} = (\epsilon_{kjn}I_A^{mn} + \epsilon_{mjn}I_A^{kn})\omega_A^j. \quad (4.1.9)$$

The state of matter inside the body is determined by the mass tensor (2.1.37). Using (2.2.27), (2.2.34), (2.2.35) and the expansion

$$|\boldsymbol{x}-\boldsymbol{x}'|^n = (r_A)^n \left[1 - \frac{n}{r_A^2} r_A^k r_A'^k + \frac{n}{2r_A^2}\left(r_A'^k r_A'^k + \frac{n-2}{r_A^2}(r_A^k r_A'^k)^2\right) + \ldots\right] \tag{4.1.10}$$

valid for the point $\boldsymbol{x}$ outside the body one finds (Fock 1955, Brumberg 1972)

$$U = \sum_A G \int_{(A)} \frac{\rho' \mathrm{d}^3 x'}{|\boldsymbol{x}-\boldsymbol{x}'|} = \sum_A G \left[\frac{M_A}{r_A} + I_A^k \frac{r_A^k}{r_A^3} + \frac{1}{2r_A^3} I_A^{km} \times \left(-\delta_{km} + \frac{3}{r_A^2} r_A^k r_A^m\right) + \ldots\right] \tag{4.1.11}$$

$$U^i = \sum_A G \int_{(A)} \frac{\rho' v'^i \mathrm{d}^3 x'}{|\boldsymbol{x}-\boldsymbol{x}'|} = \sum_A G \left[\frac{M_A}{r_A} v_A^i + \frac{1}{2r_A^3} I_A^{km} \times \left(-\delta_{km} + \frac{3}{r_A^2} r_A^k r_A^m\right) v_A^i + \epsilon_{ijk} \omega_A^j I_A^{km} \frac{r_A^m}{r_A^3} + \ldots\right] \tag{4.1.12}$$

$$\begin{aligned} W &= \sum_A G \int_{(A)} \left(\tfrac{3}{2}\rho v^2 - \rho U + \rho \Pi + 3p\right)' \frac{\mathrm{d}^3 x'}{|\boldsymbol{x}-\boldsymbol{x}'|} + \frac{\partial^2 \chi}{\partial t^2} \\ &= \sum_A G \left[\frac{\xi_A}{r_A} + \frac{1}{r_A^3} r_A^k \xi_A^k + \frac{1}{2r_A^3}\left(-\delta_{km} + \frac{3}{r_A^2} r_A^k r_A^m\right) \xi_A^{km} + \ldots\right] \\ &+ \sum_A G \left[\frac{3}{2}\frac{M_A}{r_A} v_A^2 + \frac{3}{4r_A^3} I_A^{km}\left(-\delta_{km} + \frac{3}{r_A^2} r_A^k r_A^m\right) v_A^2 \right. \\ &+ 3\epsilon_{kjn} \omega_A^j v_A^k I_A^{mn} \frac{r_A^m}{r_A^3} - \frac{M_A}{r_A} \bar{U}_A(\boldsymbol{x}_A) \\ &- \frac{1}{2r_A^3} I_A^{km}\left(-\delta_{km} + \frac{3}{r_A^2} r_A^k r_A^m\right) \bar{U}_A(\boldsymbol{x}_A) \\ &\left. - I_A^{km} \frac{r_A^m}{r_A^3} \bar{U}_{A,k}(\boldsymbol{x}_A) - \frac{1}{2r_A} I_A^{km} \bar{U}_{A,km}(\boldsymbol{x}_A) + \ldots\right] + \frac{\partial^2 \chi}{\partial t^2} \end{aligned} \tag{4.1.13}$$

$$\begin{aligned} \chi &= \tfrac{1}{2} \sum_A G \int_A \rho' |\boldsymbol{x}-\boldsymbol{x}'| \mathrm{d}^3 x' \\ &= \tfrac{1}{2} \sum_A G \left[M_A r_A + \frac{1}{2r_A} I_A^{km}\left(\delta_{km} - \frac{1}{r_A^2} r_A^k r_A^m\right) + \ldots\right]. \end{aligned} \tag{4.1.14}$$

The accent in the integrand for W reminds us that the appropriate function is to be evaluated at the point $\boldsymbol{x}'$. The potential U as well as the potentials U^i, W and χ may be divided into an 'internal' part U_A due to body A and an 'external' part $\bar{U}_A$ due to other bodies and obtained by summation in (4.1.11) over all bodies excluding body A. If the value of U is taken inside body A as in the integrand for W then U_A is replaced by u_A, the 'internal' part of the potential inside the body. In evaluating (4.1.13) one uses the following integrals dependent only on the internal structure of the relevant body:

$$\xi_A = \int_{(A)} (-\rho u_A + \rho\Pi + 3p)'\, \mathrm{d}^3x' \tag{4.1.15}$$

$$\xi_A^k = \int_{(A)} (-\rho u_A + \rho\Pi + 3p)'(x'^k - x_A^k)\, \mathrm{d}^3x' \tag{4.1.16}$$

$$\xi_A^{km} = \int_{(A)} (-\rho u_A + \rho\Pi + 3p)'(x'^k - x_A^k)(x'^m - x^m)\, \mathrm{d}^3x'. \tag{4.1.17}$$

These quantities enter into function W in the same manner as the quantities (4.1.7) and (4.1.8) enter into U. This means that in re-defining the mass density of body A by means of

$$\tilde{\rho} = \rho + c^{-2}(-\rho u_A + \rho\Pi + 3p) \tag{4.1.18}$$

the mass and the moments of the body occurring in U become

$$\tilde{M}_A = M_A + c^{-2}\xi_A \qquad \tilde{I}_A^k = I_A^k + c^{-2}\xi_A^k \qquad \tilde{I}_A^{km} = I_A^{km} + c^{-2}\xi_A^{km}. \tag{4.1.19}$$

Then the terms of W dependent on internal structure may be omitted. This process is called by Damour (1987b) as 'the effacing of internal structure in the external problem'.

The reference system t, x^i with metric tensor (4.1.2)–(4.1.4) and (4.1.11)–(4.1.13) is valid for the space domain for which the system of bodies at hand may be regarded as isolated and its gravitational radiation may be neglected. In this sense this reference system is global. In a global system the coordinates of the centre of mass of body A are defined by the relation

$$M_A x_A^i = \int_{(A)} \rho' x'^i\, \mathrm{d}^3x'. \tag{4.1.20}$$

Then the dipole moments I_A^k vanish. If one demands the validity of (4.1.20) only in the Newtonian approximation then the moments I_A^k are of post-Newtonian order. Therefore, they are retained in U in the Newtonian term of (4.1.2) and are omitted in U^i and W entering only into $h_{0i}^{(3)}$ and $h_{00}^{(4)}$.

In what follows the harmonic coordinates are used omitting in (4.1.2)–(4.1.4) arbitrary functions a_0 and a_i. As mentioned in Chapter 2, the choice (2.2.41) for a_0 corresponds to the coordinate system of the PPN formalism. But a_0 does not affect the form of the equations of the motion. In most papers the equations of motion in harmonic coordinates are used and the coordinate transformation (2.2.30) enables one to change to any other coordinates.

4.1.3 Equations of motion of a test particle in the *N*-body field

Let us start with the equations of motion of the restricted problem of $N+1$ bodies, i.e. with the equations of motion of a test particle in the N-body gravitational field determined by the harmonic metric (4.2.2)–(4.2.4). The equations of motion have the form (2.2.49), using (4.1.2)–(4.2.4)

$$\ddot{x}^i = U_{,i} + c^{-2} G^i \tag{4.1.21}$$

$$G^i = -4UU_{,i} - U_{,k}\dot{x}^k\dot{x}^i + U_{,i}\dot{x}^k\dot{x}^k - 3\dot{U}\dot{x}^i + 4\dot{U}^i - 4U^k_{,i}\dot{x}^k + W_{,i}.$$

The total derivative with respect to time is denoted here by a dot; for example

$$\dot{U} = cU_{,0} + U_{,k}\dot{x}^k.$$

This equation may be rewritten in the Lagrange form (2.2.51) with Lagrangian (2.2.53) or

$$L = \tfrac{1}{2}\dot{x}^k\dot{x}^k + U + c^{-2}\left[\tfrac{1}{8}(\dot{x}^k\dot{x}^k)^2 + \tfrac{3}{2}U\dot{x}^k\dot{x}^k - 4U^k\dot{x}^k - \tfrac{1}{2}U^2 + W\right]. \tag{4.1.22}$$

4.1.4 Geodesic principle in the *N*-body problem

To apply the geodesic principle to the motion of a given body (more specifically, for the Earth E) in the N-body problem let us separate again in the potentials U, U^i, W, and χ their internal and external parts (with respect to the substitution $\boldsymbol{x} = \boldsymbol{x}_E$)

$$U = U_E + \bar{U}_E \qquad U^i = U^i_E + \bar{U}^i_E \qquad W = W_E + \bar{W}_E \qquad \chi = \chi_E + \bar{\chi}_E. \tag{4.1.23}$$

The internal parts U_E, U^i_E, W_E, and χ_E result from (4.1.11)–(4.1.14) retaining in the summation over A only one term for $A = E$. The external parts $\bar{U}_E$, $\bar{U}^i_E$, $\bar{W}_E$, and $\bar{\chi}_E$ are given by (4.1.11)–(4.1.14) excluding in the summation over A the value $A = E$. Unlike $\bar{U}_E$, $\bar{U}^i_E$ and $\bar{\chi}_E$ the function $\bar{W}_E$ depends on the characteristics of the body E but all four functions are regular with respect to the substitution $\boldsymbol{x} = \boldsymbol{x}_E$. Substituting into the equations of geodesic motion (4.1.21) the coordinates of the body E, $\boldsymbol{x} = \boldsymbol{x}_E$, and removing the singular parts one obtains the equations of the motion of the body E:

$$a_E^i = \bar{U}_{E,i}(\boldsymbol{x}_E) + H_E^i + c^{-2}\bar{G}_E^i \tag{4.1.24}$$

$$\begin{aligned}\bar{G}_E^i = & - 4\bar{U}_E(\boldsymbol{x}_E)\bar{U}_{E,i}(\boldsymbol{x}_E) - \bar{U}_{E,k}(\boldsymbol{x}_E)v_E^k v_E^i + \bar{U}_{E,i}(\boldsymbol{x}_E)v_E^2 \\ & - 3\dot{\bar{U}}_E(\boldsymbol{x}_E)v_E^i + 4\dot{\bar{U}}{}_E^i(\boldsymbol{x}_E) - 4\bar{U}_{E,i}^k(\boldsymbol{x}_E)v_E^k + \bar{W}_{E,i}(\boldsymbol{x}_E).\end{aligned} \tag{4.1.25}$$

$v_E^i = \dot{x}_E^i$ and $a_E^i = \ddot{x}_E^i$ are components of velocity and acceleration of the body E. $\bar{U}_E(\boldsymbol{x}_E)$, $\bar{U}_E^i(\boldsymbol{x}_E)$ and $\bar{W}_E(\boldsymbol{x}_E)$ result from substituting $\boldsymbol{x} = \boldsymbol{x}_E$ into the regular parts $\bar{U}_E$, $\bar{U}_E^i$, and $\bar{W}_E$. This substitution is to be performed after partial differentiation with respect to $\boldsymbol{x}$ and t. H_E^i are corrections for the deviation of motion of the body from the geodesic motion. They are due to rotation of the body E and its quadrupole moments. For all bodies of the Solar System these corrections are extremely small enabling us to take into account only their Newtonian values. Evidently, these corrections cannot be obtained on the basis of the geodesic principle.

Equations (4.1.24) may be written in the Lagrange form with Lagrangian

$$\begin{aligned}L_E = & \tfrac{1}{2}v_E^2 + \bar{U}_E(\boldsymbol{x}_E) + H_E + c^{-2}[\tfrac{1}{8}(v_E^2)^2 + \tfrac{3}{2}\bar{U}_E(\boldsymbol{x}_E)v_E^2 \\ & - 4\bar{U}_E^k(\boldsymbol{x}_E)v_E^k - \tfrac{1}{2}\bar{U}_E^2(\boldsymbol{x}_E) + \bar{W}_E(\boldsymbol{x}_E)].\end{aligned} \tag{4.1.26}$$

The correction H_E corresponds to the term H_E^i in equations (4.1.24). Corrections H_E^i and H_E are derived in the following section. Here, we will just indicate their Newtonian values:

$$H_E^i = \tfrac{1}{2}M_E^{-1}I_E^{km}\bar{U}_{E,ikm}(\boldsymbol{x}_E) + \dots \tag{4.1.27}$$

$$H_E = \tfrac{1}{2}M_E^{-1}I_E^{km}\bar{U}_{E,km}(\boldsymbol{x}_E) + \dots. \tag{4.1.28}$$

4.1.5 Equations of motion of the N-body problem

Equations (4.1.24) with the correction (4.1.27) from Newtonian theory are quite adequate for all present practical requirements. Let us derive them now by the Fock method. By this method (Fock 1955) the equations of the translatory motion of the N-body problem are described in integral form as

$$\int_{(A)} g\nabla_\alpha T^{\alpha i}\, \mathrm{d}^3x = 0 \tag{4.1.29}$$

∇_α being the covariant derivative. On the basis of (2.1.37) and components (4.1.2)–(4.1.4) of the metric tensor the components of the mass tensor admit expansions

$$T^{00} = \rho[1 + c^{-2}(\tfrac{1}{2}v^2 + \Pi - U)] + \ldots$$
$$cT^{0i} = \rho v^i[1 + c^{-2}(\tfrac{1}{2}v^2 + \Pi - U)] + c^{-2}pv^i + \ldots \tag{4.1.30}$$
$$c^2T^{ik} = \rho v^i v^k + p\delta_{ik} + \ldots .$$

Inside the body one may assume the validity of the equation of continuity

$$c\rho_{,0} + (\rho v^k)_{,k} = 0 \tag{4.1.31}$$

and the equations of motion of continuous matter

$$\rho\dot{v}^i = \rho U_{,i} - p_{,i}. \tag{4.1.32}$$

Due to (4.1.31) the integrals taken over the volume of the body satisfy the relation

$$\begin{aligned}\frac{\mathrm{d}}{\mathrm{d}t}\int_{(A)} \rho f(t, x^i)\mathrm{d}^3x &= \int_{(A)} \left(\frac{\partial \rho}{\partial t} f + \rho\frac{\partial f}{\partial t}\right) \mathrm{d}^3x \\ &= \int_{(A)} \left(-(\rho v^k)_{,k} f + \rho\frac{\partial f}{\partial t}\right) \mathrm{d}^3x \\ &= \int_{(A)} \rho \left(v^k\frac{\partial f}{\partial x^k} + \frac{\partial f}{\partial t}\right) \mathrm{d}^3x = \int_{(A)} \rho\frac{\mathrm{d}f}{\mathrm{d}t}\, \mathrm{d}^3x\end{aligned} \tag{4.1.33}$$

often used below. Using equations (4.1.31) and (4.1.32) and neglecting the square of the angular velocity of rotation of the body one may derive the following approximation relation already met with in (3.3.19):

$$\int_{(E)} p\,\mathrm{d}^3x = \tfrac{1}{3}\epsilon_E \qquad \epsilon_E = \tfrac{1}{2}\int_{(E)} \rho u_E\, \mathrm{d}^3x. \tag{4.1.34}$$

ϵ_E represents the contraction of the more general integral

$$B_E^{km} = \tfrac{1}{2}G \iint_{(E)} \frac{\rho\rho'}{|\boldsymbol{x} - \boldsymbol{x}'|^3}(x^k - x'^k)(x^m - x'^m)\, \mathrm{d}^3x\mathrm{d}^3x' \qquad \epsilon_E = B_E^{kk}.$$

For quasi-spherical bodies one has approximately

$$B_E^{km} = \tfrac{1}{3}\delta_{km}\epsilon_E. \tag{4.1.35}$$

Substitution of (4.1.2)–(4.1.4) and (4.1.30) enables one to transform the equations of motion to the form

$$\dot{P}^i_E = F^i_E \tag{4.1.36}$$

with

$$P^i_E = \int_{(E)} \{[\rho + c^{-2}\rho(\tfrac{1}{2}v^2 + 3U + \Pi) + c^{-2}p]v^i - 4c^{-2}\rho U^i - c^{-1}\rho\chi_{,0i}\}\,\mathrm{d}^3x \tag{4.1.37}$$

$$F^i_E = \int_{(E)} (\sigma U_{,i} + c^{-2}\rho W_{,i} - 4c^{-2}\rho v^k U^k_{,i} - c^{-1}\rho\dot{\chi}_{,0i})\,\mathrm{d}^3x \tag{4.1.38}$$

and

$$\sigma = \rho + c^{-2}[\rho(\tfrac{3}{2}v^2 - U + \Pi) + 3p]. \tag{4.1.39}$$

These expressions are given in Fock (1955) and Brumberg (1972) but using slightly different notations. Description in the form of (4.1.37) and (4.1.38) has the advantage of making clear, with the aid of (4.1.33), that the function χ does not affect the equations of motion and may be omitted in (4.1.37) and (4.1.38). Evaluation of integrals (4.1.37) and (4.1.38) presents no difficulties. As before, the functions U, U^i and W are represented by (4.1.23) as the sum of internal and external parts. The integrals of the internal parts reduce to typical integrals (4.1.15)–(4.1.17), (4.1.34), and (4.1.35). The integrals of the external parts are evaluated by expanding integrals in the vicinity of the body E. The internal parts U_E, U^i_E, and W_E are determined by the relevant integral expressions (4.1.11)–(4.1.14) for the value $A = E$ of the summation index. In particular,

$$U^i_E = v^i U_E - 2\epsilon_{ijk}\omega^j_E \chi_{E,k} \tag{4.1.40}$$

$$W_E = G\int_{(E)} \frac{c^2(\sigma' - \rho')}{|\boldsymbol{x} - \boldsymbol{x}'|}\,\mathrm{d}^3x' + c^2\chi_{E,00} \tag{4.1.41}$$

with

$$\chi_E = \tfrac{1}{2}G\int_{(E)} \rho'|\boldsymbol{x} - \boldsymbol{x}'|\,\mathrm{d}^3x'. \tag{4.1.42}$$

Substituting (4.1.40) into (4.1.37) and using (4.1.33) one finds the separate contributions in (4.1.37) (retaining again in equations (4.1.36) only linear terms in angular velocities and rejecting terms dependent only on the internal structure and having no effect on the equations of motion)

$$\int_{(E)} \rho v^i\,\mathrm{d}^3x = \frac{\mathrm{d}}{\mathrm{d}t}\int_{(E)} \rho x^i\,\mathrm{d}^3x = M_E v^i_E + \ldots \tag{4.1.43}$$

$$c^{-2}\int_{(E)}(\tfrac{1}{2}\rho v^2-\rho U_E+\rho\Pi+p)v^i\,\mathrm{d}^3x = c^{-2}[\tfrac{1}{2}M_E v_E^2 v_E^i+(\xi_E-\tfrac{2}{3}\epsilon_E)v_E^i+\ldots) \tag{4.1.44}$$

$$\int_{(E)}\rho\chi_{E,k}\,\mathrm{d}^3x = \tfrac{1}{2}G\iint_{(E)}\rho\rho'\frac{x^k-x'^k}{|\boldsymbol{x}-\boldsymbol{x}'|}\,\mathrm{d}^3x\mathrm{d}^3x' = 0 \tag{4.1.45}$$

$$c^{-2}\int_{(E)}3\rho\bar{U}_E v^i\,\mathrm{d}^3x = c^{-2}[3M_E v_E^i\bar{U}_E(\boldsymbol{x}_E)+3\epsilon_{ijm}\omega_E^j I_E^{km}\bar{U}_{E,k}(\boldsymbol{x}_E) + \tfrac{3}{2}v_E^i I_E^{km}\bar{U}_{E,km}(\boldsymbol{x}_E)+\ldots] \tag{4.1.46}$$

$$-c^{-2}\int_{(E)}4\rho\bar{U}_E^i\,\mathrm{d}^3x = c^{-2}[-4M_E\bar{U}_E^i(\boldsymbol{x}_E)-2I_E^{km}\bar{U}_{E,km}^i(\boldsymbol{x}_E)+\ldots]. \tag{4.1.47}$$

Thus, integral (4.1.37) has the value (without taking into account the last term of the integrand)

$$\begin{aligned}P_E^i = &\{\tilde{M}_E+c^{-2}[-\tfrac{2}{3}\epsilon_E+\tfrac{1}{2}M_E v_E^2+3M_E\bar{U}_E(\boldsymbol{x}_E)]\}v_E^i\\ &+c^{-2}[-4M_E\bar{U}_E^i(\boldsymbol{x}_E)-2I_E^{km}\bar{U}_{E,km}^i(\boldsymbol{x}_E)+\tfrac{3}{2}v_E^i I_E^{km}\bar{U}_{E,km}(\boldsymbol{x}_E)\\ &+3\epsilon_{ijm}\omega_E^j I_E^{km}\bar{U}_{E,k}(\boldsymbol{x}_E)].\end{aligned} \tag{4.1.48}$$

Turning to the evaluation of (4.1.38) let us start with the internal parts of the integrand terms. The third term makes no contribution due to the antisymmetric structure of the integrand

$$\int_{(E)}\rho v^k U_{E,i}^k\,\mathrm{d}^3x = -G\iint_{(E)}\rho\rho' v^k v'^k\frac{x^i-x'^i}{|\boldsymbol{x}-\boldsymbol{x}'|^3}\,\mathrm{d}^3x\mathrm{d}^3x' = 0. \tag{4.1.49}$$

The contribution from the first two terms as seen from (4.1.41) reduces only to the effect due to χ_E. Indeed,

$$\int_{(E)}(\sigma U_{E,i}+c^{-2}\sigma W_{E,i})\,\mathrm{d}^3x = -G\iint_{(E)}\sigma\sigma'\frac{x^i-x'^i}{|\boldsymbol{x}-\boldsymbol{x}'|^3}\,\mathrm{d}^3x\mathrm{d}^3x' + \int_{(E)}\rho\chi_{E,00i}\,\mathrm{d}^3x$$

and the first integral vanishes as in (4.1.45). Using (4.1.31), (4.1.35) and (4.1.45) the second integral in its main part may be reduced to

$$\begin{aligned}\int_{(E)}\rho\chi_{E,00i}\,\mathrm{d}^3x &= \int_{(E)}(\rho\chi_{E,0i})_{,0}\,\mathrm{d}^3x = -\int_{(E)}(\rho_{,0}\chi_{E,i})_{,0}\,\mathrm{d}^3x\\ &= c^{-2}\frac{\mathrm{d}}{\mathrm{d}t}\int_{(E)}(\rho v^k)_{,k}\chi_{E,i}\,\mathrm{d}^3x = -c^{-2}\frac{\mathrm{d}}{\mathrm{d}t}\int_{(E)}\rho v^k\chi_{E,ik}\,\mathrm{d}^3x\\ &= -c^{-2}\frac{\mathrm{d}}{\mathrm{d}t}(v_E^i\epsilon_E-v_E^k B_E^{ik}) = -\frac{2}{3}c^{-2}\frac{\mathrm{d}}{\mathrm{d}t}(\epsilon_E v_E^i).\end{aligned} \tag{4.1.50}$$

In evaluating the contributions from the external parts $\bar{U}_E$, $\bar{U}_E^k$, and $\bar{W}_E$ in (4.1.38) it is assumed that U has been already subjected to re-definition (4.1.15)–(4.1.19) resulting in the absence of the internal structure terms in the expression (4.1.13) for W. Then

$$\int_{(E)} \tilde{\rho}\bar{U}_{E,i}\,\mathrm{d}^3x = \tilde{M}_E\bar{U}_{E,i}(\boldsymbol{x}_E)+\tilde{I}_E^k\bar{U}_{E,ik}(\boldsymbol{x}_E)+\tfrac{1}{2}\tilde{I}_E^{km}\bar{U}_{E,ikm}(\boldsymbol{x}_E) \quad (4.1.51)$$

$$\begin{aligned}\int_{(E)} (\sigma-\tilde{\rho})\bar{U}_{E,i}\,\mathrm{d}^3x = c^{-2}[&\tfrac{3}{2}M_E v_E^2\bar{U}_{E,i}(\boldsymbol{x}_E) + \tfrac{3}{4}v_E^2 I_E^{km}\bar{U}_{E,ikm}(\boldsymbol{x}_E) \\ &+ 3\epsilon_{knj}v_E^n\omega_E^j I_E^{km}\bar{U}_{E,im}(\boldsymbol{x}_E) - M_E\bar{U}_E(\boldsymbol{x}_E)\bar{U}_{E,i}(\boldsymbol{x}_E) \\ &- \tfrac{1}{4}I_E^{km}(\bar{U}_E^2(\boldsymbol{x}_E))_{,ikm}] \end{aligned} \quad (4.1.52)$$

$$c^{-2}\int_{(E)} \rho\bar{W}_{E,i}\,\mathrm{d}^3x = c^{-2}[M_E\bar{W}_{E,i}(\boldsymbol{x}_E) + \tfrac{1}{2}I_E^{km}\bar{W}_{E,ikm}(\boldsymbol{x}_E)] \quad (4.1.53)$$

$$\begin{aligned}-4c^{-2}\int_{(E)} \rho v^k\bar{U}_{E,i}^k\,\mathrm{d}^3x = c^{-2}[&-4M_E v_E^k\bar{U}_{E,i}^k(\boldsymbol{x}_E) - 2I_E^{mn}v_E^k\bar{U}_{E,imn}^k(\boldsymbol{x}_E) \\ &- 4\epsilon_{kjm}\omega_E^j\bar{U}_{E,in}^k(\boldsymbol{x}_E)I^{mn}]. \end{aligned} \quad (4.1.54)$$

Combining all contributions (4.1.49)–(4.1.54) one obtains

$$\begin{aligned} F_E^i = {}& \tilde{M}_E\bar{U}_{E,i}(\boldsymbol{x}_E) + \tilde{I}_E^k\bar{U}_{E,ik}(\boldsymbol{x}_E) + \tfrac{1}{2}\tilde{I}_E^{km}\bar{U}_{E,ikm}(\boldsymbol{x}_E) \\ &+ c^{-2}\Big(-\frac{2}{3}\frac{\mathrm{d}}{\mathrm{d}t}(\epsilon_E v_E^i) + \tfrac{3}{2}M_E v_E^2\bar{U}_{E,i}(\boldsymbol{x}_E) - M_E\bar{U}_E(\boldsymbol{x}_E)\bar{U}_{E,i}(\boldsymbol{x}_E) \\ &+ M_E\bar{W}_{E,i}(\boldsymbol{x}_E) - 4M_E v_E^k\bar{U}_{E,i}^k(\boldsymbol{x}_E) + \tfrac{3}{4}v_E^2 I_E^{km}\bar{U}_{E,ikm}(\boldsymbol{x}_E) \\ &- \tfrac{1}{4}I_E^{km}(\bar{U}_E^2(\boldsymbol{x}_E))_{,ikm} + \tfrac{1}{2}I_E^{km}\bar{W}_{E,ikm}(\boldsymbol{x}_E) - 2I_E^{km}v_E^n\bar{U}_{E,ikm}^n(\boldsymbol{x}_E) \\ &+ 3\epsilon_{knj}v_E^n\omega_E^j I_E^{km}\bar{U}_{E,im}(\boldsymbol{x}_E) - 4\epsilon_{knj}\omega_E^j I_E^{km}\bar{U}_{E,im}^n(\boldsymbol{x}_E)\Big). \end{aligned} \quad (4.1.55)$$

Substitution of (4.1.48) and (4.1.55) into (4.1.36) gives the equations of translatory motion of the N-body problem. Comparison of (4.1.48) and (4.1.55) immediately demonstrates that the terms with ϵ_E cancel out and the internal structure of bodies manifests itself only in the re-definition (4.1.18) and (4.1.19). It is assumed further that such re-definition has been made both in the field metric and the equations of motion and the tilde in designations of mass and multipole moments will be omitted.

Expressed explicitly, equations (4.1.36), (4.1.48), and (4.1.55) yield again equations (4.1.24). This time it is possible to refine the correction (4.1.27) adding the relativistic terms

$$
\begin{aligned}
H_E^i = {} & \tfrac{1}{2}M_E^{-1}I_E^{km}\bar{U}_{E,ikm}(\boldsymbol{x}_E) + M_E^{-1}I_E^k\bar{U}_{E,ik}(\boldsymbol{x}_E) \\
& + c^{-2}M_E^{-1}I_E^{km}\{\omega_E^j[-3\epsilon_{mnj}v_E^i\bar{U}_{E,kn}(\boldsymbol{x}_E) - 3\epsilon_{ijm}\dot{\bar{U}}_{E,k}(\boldsymbol{x}_E) \\
& + 4\epsilon_{mnj}\bar{U}_{E,kn}^i(\boldsymbol{x}_E) + 3\epsilon_{mnj}v_E^n\bar{U}_{E,ik}(\boldsymbol{x}_E) - 4\epsilon_{mnj}\bar{U}_{E,ik}^n(\boldsymbol{x}_E)] \\
& + \tfrac{1}{2}v_E^2\bar{U}_{E,ikm}(\boldsymbol{x}_E) - \tfrac{1}{2}v_E^i v_E^n\bar{U}_{E,kmn}(\boldsymbol{x}_E) - \tfrac{3}{2}v_E^i\dot{\bar{U}}_{E,km}(\boldsymbol{x}_E) \\
& - 2v_E^n\bar{U}_{E,ikm}^n(\boldsymbol{x}_E) - 2\bar{U}_E(\boldsymbol{x}_E)\bar{U}_{E,ikm}(\boldsymbol{x}_E) - 2\bar{U}_{E,i}(\boldsymbol{x}_E)\bar{U}_{E,km}(\boldsymbol{x}_E) \\
& - \bar{U}_{E,k}(\boldsymbol{x}_E)\bar{U}_{E,im}(\boldsymbol{x}_E) + 2\dot{\bar{U}}_{E,km}(\boldsymbol{x}_E) + \tfrac{1}{2}\bar{W}_{E,ikm}(\boldsymbol{x}_E)\}. \qquad (4.1.56)
\end{aligned}
$$

The first term gives the Newtonian correction for the non-geodesic motion of the body caused by its quadrupole moment. The second term is due to the possible non-zero dipole moment of the relativistic order of smallness. Other corrections are of relativistic origin caused by rotation of the body and its quadrupole moments. As already stated, only the first Newtonian term is of practical importance.

4.1.6 Lagrangian of the equations of motion

The Lagrangian for the equations of motion of body E has already been determined (equation (4.1.26)). Now it is possible to refine the value H_E

$$
\begin{aligned}
H_E = {} & \tfrac{1}{2}M_E^{-1}I_E^{km}\bar{U}_{E,km}(\boldsymbol{x}_E) + M_E^{-1}I_E^k\bar{U}_{E,k}(\boldsymbol{x}_E) + c^{-2}M_E^{-1}I_E^{km} \\
& \times \{\epsilon_{mnj}\omega_E^j[3v_E^n\bar{U}_{E,k}(\boldsymbol{x}_E) - 4\bar{U}_{E,k}^n(\boldsymbol{x}_E)] + \tfrac{3}{4}v_E^2\bar{U}_{E,km}(\boldsymbol{x}_E) \\
& - 2v_E^n\bar{U}_{E,km}^n(\boldsymbol{x}_E) - \tfrac{1}{4}(\bar{U}_E^2(\boldsymbol{x}_E))_{,km} + \tfrac{1}{2}\bar{W}_{E,km}(\boldsymbol{x}_E)\}. \qquad (4.1.57)
\end{aligned}
$$

Corrections H_E^i and H_E demonstrate to what extent the motion of a body possessing dipole and quadrupole moments differs from the geodesic motion. Equations (4.1.24) with (4.1.25), (4.1.56) or equations (2.2.51) with Lagrangian (4.1.26), (4.1.57) enable one to solve all questions of the motion of the body E taking into account the first degree terms with respect to the mass moments of all bodies. The equations take into account the first degree terms relative to the angular velocities as well as the mixed terms containing the products of the angular velocities of the body E and any body A. However, all these terms are of importance only in theoretical investigations. For example, the series of papers by Barker and O'Connell (the last being Barker *et al* 1986) deal with the relativistic problem of two bodies with spins and quadrupole moments.

For pure theoretical investigations it may be convenient to have the Lagrangian valid for all bodies. But one loses thereby the representation of the Lagrangian in terms of the coefficients of expansions (4.1.2)–(4.1.4) and one has to use the explicit expressions of the potentials. Such a global

Lagrangian is derived for example in Fock (1955) and Brumberg (1972). The Lagrange form of the equations of motion enables one to find easily all ten first integrals. Only the integrals of the linear momentum and the motion of the centre of mass are reproduced here. By applying the transformation (4.1.50) to the external potential $\tilde{\chi}_E$ and expanding in the vicinity of the body E one has

$$c^{-2}\int_{(E)} \rho v^k \bar{\chi}_{E,ik}\,\mathrm{d}^3x = c^{-2}[M_E v_E^k \bar{\chi}_{E,ik}(\boldsymbol{x}_E) + \tfrac{1}{2} v_E^n I_E^{km} \bar{\chi}_{E,ikmn}(\boldsymbol{x}_E) + \epsilon_{njk}\omega_E^j I_E^{km} \bar{\chi}_{E,imn}(\boldsymbol{x}_E) + \ldots]. \tag{4.1.58}$$

Combining this result with (4.1.48) one finds the integral of momentum

$$\begin{aligned}\sum_A [&M_A(1 + \tfrac{1}{2}c^{-2}v_A^2 + 3c^{-2}\bar{U}_A(\boldsymbol{x}_A))v_A^i \\ &+ c^{-2}M_A\left(-4\bar{U}_A^i(\boldsymbol{x}_A) + v_A^k\bar{\chi}_{A,ik}(\boldsymbol{x}_A)\right) \\ &+ c^{-2}I_A^{km}(-2\bar{U}^i_{A,km}(\boldsymbol{x}_A) + \tfrac{3}{2}v_A^i\bar{U}_{A,km}(\boldsymbol{x}_A) + \tfrac{1}{2}\bar{\chi}_{A,ikmn}(\boldsymbol{x}_A)v_A^n) \\ &+ c^{-2}\omega_A^j I_A^{km}(3\epsilon_{ijm}\bar{U}_{A,k}(\boldsymbol{x}_A) + \epsilon_{njk}\bar{\chi}_{A,imn}(\boldsymbol{x}_A))] = K^i. \end{aligned} \tag{4.1.59}$$

For point non-rotating bodies this integral takes the simple form

$$\sum_A M_A\left[v_A^i + \tfrac{1}{2}c^{-2}v_A^2 v_A^i - \tfrac{1}{2}c^{-2}\sum_{B\neq A}\frac{GM_B}{r_{AB}}\left(v_A^i + \frac{1}{r_{AB}^2}r_{AB}^i r_{AB}^k v_A^k\right)\right] = K^i. \tag{4.1.60}$$

Integration of this relation yields the integral of the motion of the centre of mass

$$\sum_A \tilde{M}_A x_A^i = K^i t + N^i \tag{4.1.61}$$

where $\tilde{M}_A$ is the Tolman mass of body A,

$$\tilde{M}_A = M_A\left(1 + \tfrac{1}{2}c^{-2}v_A^2 - \tfrac{1}{2}c^{-2}\sum_{B\neq A}\frac{GM_B}{r_{AB}}\right). \tag{4.1.62}$$

The Tolman mass of the whole system of bodies for which $\dot{\tilde{M}} = 0$ is determined as

$$\tilde{M} = \sum_A \tilde{M}_A \tag{4.1.63}$$

and the coordinates $\tilde{\chi}^i$ of the centre of mass of the system of bodies are

$$\tilde{M}\tilde{\chi}^i = \sum_A \tilde{M}_A x_A^i. \tag{4.1.64}$$

For the barycentric reference system defined by $K^i = N^i = 0$ the centre of mass of the system of bodies is at rest at the origin $\tilde{X}^i = 0$.

These relations can also be obtained from the integral relations (2.1.51)–(2.1.53).

4.2 GEOCENTRIC REFERENCE FRAME

4.2.1 Local and global coordinates

The solution of the preceding subsection is valid for an isolated and gravitationally non-radiating system of bodies. Both these assumptions do not comply with reality but in the case of the Solar System they are satisfied with an accuracy adequate for all practical problems. Therefore, the solution obtained above is correct in the sufficiently large space domain where all linear sizes are small as compared with the length of gravitational waves and the gravitational influence of stars is not yet noticeable. In this sense the coordinates used above may be considered as global ones. The reference system defined by (4.1.2)–(4.1.4) is not rotating since expansion (4.1.3) for g_{0i} starts with the third-order terms and does not contain the first-order term $c^{-1}\epsilon_{ijk}\omega^j x^k$ characteristic of rotating systems. The specific choice of coordinates resulting in vanishing expressions (4.1.60) and (4.1.61) leads to the barycentric reference system (BRS). It is assumed thereby that coordinates t, and x^i of BRS are harmonic.

Global BRS is adequate to study the motion of major and minor planets and comets as well as for the reduction of observations. For investigating the motion of satellites of the planets the planetocentric reference system is preferred. One would think that it is sufficient to introduce the relative coordinates $r^k_E = x^k - x^k_E$, x^k being the BRS coordinates of a satellite and x^k_E being the BRS coordinates of a planet (the Earth to be more specific). Then the difference between (4.1.21) and (4.1.24) would result in the geocentric equations of satellite motion. But, in fact, one remains thereby within the BRS framework retaining the BRS coordinate time t as the independent argument and regarding the space geocentric coordinates only as the differences of the BRS coordinates of the satellite and the Earth. Similarly, introducing the relative heliocentric Earth coordinates $r^k_{ES} = x^k_E - x^k_S$ and using (4.1.24) for the Earth E and the Sun S one obtains the heliocentric equations of the motion of the Earth in BRS but not in the dynamically adequate heliocentric reference system. Of course, considering that it is impossible to introduce in GRT physically meaningful inertial coordinates, one may use any coordinates in solving the relativity problems. In practical astronomical problems solving the differential equations of motion represents the first dynamical step. To compare with observations one should introduce the observable quantities. This second kinematic step of solving the astronomical problem requires the analysis of the light propagation and the description of the observational procedure in the same coordinates that have been used for solving the dynamical problem. If a reference system is not dynamically adequate for a given problem (the description of motion of an Earth satellite or the Earth in relative coordinates r^k_E or r^k_{ES}, for example) then both steps will contain extra large terms due only to the inadequate choice of reference system. These terms cancel out in the expressions of the actually measurable quantities and the real relativistic effects turn out to be much smaller than the relativistic perturbations obtained at the first step. In the examples at hand these terms are caused by ignoring Lorentz and gravitation space-time terms in transforming from one coordinate system to another. On the other hand, if the reference system is dynamically adequate for a given problem then the solution of the dynamical problem does not contain large terms of kinematic origin and the transformation to observable quantities demands only insignificant corrections. Different methods have been suggested to construct physically adequate planetocentric systems. Some information about these methods will be given in Chapter 6. Here the geocentric reference system (GRS) constructed in Brumberg and Kopejkin (1989a) is used. Changing the reference body one can derive heliocentric or any planetocentric reference system. Similar to BRS, GRS is built in harmonic coordinates. The use of one and the same type of the coordinate conditions simplifies mathematical interrelations between different systems and facilitates the comparison of results obtained in different systems.

The Earth, like any other body in the Solar System, has its own gravitational field characterized by the multipole moments. The gravitational field generated by all other bodies of the Solar System may be regarded as external with respect to the Earth field. The external gravitational field has three characteristic sizes: an inhomogeneity scale $\mathcal{L}_e$ (the characteristic distance between bodies), a radius of curvature $\mathcal{R}_e$ and a time scale $\mathcal{T}_e$ (the average time for a significant change of curvature). Each body of the Solar System is isolated in the sense that its characteristic size L satisfies the inequalities

$$L \ll \mathcal{L}_e \qquad L \ll \mathcal{R}_e \qquad L \ll c\mathcal{T}_e. \tag{4.2.1}$$

These conditions are satisfied for each body in the Solar System.

One of the basic principles of modern approaches to the problem of motion in GRT is to split up the space-time in the vicinity of the isolated body into three regions (Thorne and Hartle 1985): an internal region $L < \rho < r_I$, a buffer region $r_I < \rho < r_0 \ll \mathcal{L}_e, \mathcal{R}_e, c\mathcal{T}_e$ and an external region $r_0 < \rho$, ρ being the characteristic distance from the body. In the body's internal region its own gravitational field dominates, in the external region the gravitational field of other bodies dominates and in the buffer region both fields have a comparable effect. The internal region solution corresponds to the one-body problem considered in Chapter 3. The external region solution describes the tidal gravitational influence of the external bodies on the body at hand and may be represented by series in powers of $\rho/\mathcal{R}_e$, $\rho/\mathcal{L}_e$ and $\rho/(c\mathcal{T}_e)$. The buffer region solution reflects the gravitational interaction of the considered body and the external bodies and is described by combination of the series for internal and external regions.

GRS as any other planetocentric reference system furnishes an example of a reference system valid for the vicinity of the reference body. The solution in local coordinates is dynamically more compact than the BRS solution. Besides this, it is very important that the physical characteristics of the body (its angular velocity of rotation, the multipole moments, etc) should be defined in local coordinates. As noted for the first time by Kopejkin (1987), such important characteristics of celestial bodies as rigid body rotation, sphericity, etc, should be formulated just in local coordinates. Matching of the global BRS solution with the local GRS solution enables one afterwards to re-formulate these characteristics and to express the BRS quantities in terms of the physically more reliable internal characteristics of the body.

4.2.2 Geocentric metric

The GRS time coordinate is denoted by $w^0 = cu$. The space coordinates are denoted by $\boldsymbol{w} = (w^1, w^2, w^3)$ with the special designation $\rho = |\boldsymbol{w}|$ for the absolute magnitude of the GRS position vector. The GRS components of the metric tensor are designated by $\hat{g}_{\alpha,\beta}(u, \boldsymbol{w})$. GRS is used below directly only for solving kinematic questions related with the light propagation. Hence, the terms $O(c^{-4})$ in $\hat{g}_{00}$ are not needed. Such terms are necessary in solving dynamical problems dealing with the motion of the Moon or Earth satellites. These problems are examined here in GRS but based on the initial BRS equations. The terms $O(c^{-4})$ in $\hat{g}_{00}$ may be found in Kopejkin (1988) and Brumberg and Kopejkin (1989b). In ignoring these terms the GRS metric represents the linear superposition of the proper Earth terms and the terms due to the tidal action of the external masses, namely (Thorne and Hartle 1985)

$$\hat{g}_{00}(u, \boldsymbol{w}) = 1 - c^{-2}(2\hat{U}_E + 2Q_k w^k + 3Q_{km} w^k w^m + 5Q_{kmn} w^k w^m w^n + \ldots) + \ldots \tag{4.2.2}$$

$$\hat{g}_{0i}(u, \boldsymbol{w}) = 4c^{-3}(\hat{U}_E^i + \epsilon_{ijk} C_{jm} w^k w^m - \tfrac{3}{10}\dot{Q}_k w^k w^i + \tfrac{1}{10}\dot{Q}_i w^k w^k + \ldots) + \ldots \tag{4.2.3}$$

$$\begin{aligned} \hat{g}_{ij}(u, \boldsymbol{w}) = & -\delta_{ij} - c^{-2}(2\hat{U}_E + 2Q_k w^k + 3Q_{km} w^k w^m \\ & + 5Q_{kmn} w^k w^m w^n + \ldots)\delta_{ij} + \ldots . \end{aligned} \tag{4.2.4}$$

All coefficients of these expansions are determined from matching with the global BRS metric but the proper terrestrial terms may be written immediately on the basis of the solution of the one-body problem

$$\hat{U}_E = \frac{G\hat{M}_E}{\rho} + \frac{1}{2\rho^3} G\hat{I}_E^{km} \left(-\delta_{km} + \frac{3}{\rho^2} w^k w^m \right) + \ldots \tag{4.2.5}$$

$$\hat{U}_E^i = G\epsilon_{ijk}\hat{\omega}_E^j \hat{I}_E^{km} \frac{w^m}{\rho^3} + \ldots . \tag{4.2.6}$$

$\hat{M}_E$, $\hat{\omega}_E^i$, and $\hat{I}_E^{km}$ are the mass, angular velocity of rotation and quadrupole moments, respectively, of the Earth in GRS. It is easy to verify that components (4.2.2)–(4.2.4) satisfy the harmonic conditions (2.2.17) and (2.2.18) (in doing this one should take into account the symmetry of C_{km}). Only the initial terms of expansions in powers of w^k are given in (4.2.2)–(4.2.4). The general form of these expansions is indicated in Kopejkin (1988). The GRS metric tensor may also be given in closed form without expanding in powers of w^k (see the Postscript). For our purposes the approximate expressions (4.2.2)–(4.2.4) are quite adequate.

The terms with acceleration Q_k in (4.2.2) and (4.2.4) deserve particular attention. By definition the centre of mass of the Earth is at rest at the GRS origin and acceleration Q_k is due only by the deviation of the Earth motion from the geodesic motion. This deviation, even its Newtonian part (4.1.27), may be often neglected.

4.2.3 Matching of BRS and GRS

Transformation from BRS to GRS generalizes the Lorentz transformation of special relativity and has the form (Kopejkin 1987, 1988, Brumberg and Kopejkin 1989a)

$$\begin{aligned} u = t &- c^{-2}(S(t) + v_E^k r_E^k) + c^{-4}(\tfrac{3}{8} v_E^4 t - \tfrac{1}{2} v_E^2 v_E^k x^k + B \\ &+ B^k r_E^k + B^{km} r_E^k r_E^k + B^{kmn} r_E^k r_E^m r_E^n + \ldots) + \ldots \end{aligned} \tag{4.2.7}$$

$$w^i = r_E^i + c^{-2}[(\tfrac{1}{2} v_E^i v_E^k + F^{ik} + D^{ik}) r_E^k + D^{ikm} r_E^k r_E^m] + \ldots \tag{4.2.8}$$

where S, $F^{ik}(= -F^{ki})$, $D^{ik}(= D^{ki})$, $D^{ikm}(= D^{imk})$, B, B^k, $B^{km}(= B^{mk})$ and $B^{kmn}(= B^{mnk} = B^{nkm})$ are functions of the barycentric time t. These functions as well as the metric coefficients (4.2.2)–(4.2.4) are determined by matching BRS and GRS metrics with the aid of the fundamental tensor relation

$$g_{\alpha\beta}(t, x) = \hat{g}_{\mu\nu}(u, w) \frac{\partial w^\mu}{\partial x^\alpha} \frac{\partial w^\nu}{\partial x^\beta} . \tag{4.2.9}$$

The BRS potentials are again represented in form (4.1.23) and their external parts are expanded in powers of $r_E^k = x^k - x_E^k$. Along with the coefficients of the transformation (4.2.7), (4.2.8) and metric (4.2.2)–(4.2.4) this matching determines the BRS acceleration a^i_E of the Earth which gives another technique for deriving the equations of motion. The GRS metric is determined here only in the approximation (4.2.2)–(4.2.4) without considering the terms $O(c^{-4})$ in $\hat{g}_{00}$. In this approximation one cannot determine coefficient B in (4.2.7) and the post-Newtonian terms in a^i_E. All these terms may be found in Kopejkin (1988). It is of importance that transformation (4.2.8) within the post-Newtonian accuracy is rigorous in w^i.

The presentation of $g_{00}(t, \boldsymbol{x})$ in terms of $\hat{g}_{\mu\nu}(u, \boldsymbol{w})$ enables one to determine functions $S(t)$, a^i_E and coefficients of (4.2.2). In fact, the identity (4.2.9) for $\alpha = \beta = 0$ and for order $O(c^{-2})$ results in

$$\begin{aligned} 2U_E + 2\bar{U}_E = {} & 2\hat{U}_E + 2\dot{S} - v_E^2 + 2(a^k_E + Q_k)w^k \\ & + 3Q_{km}w^k w^m + 5Q_{kmn}w^k w^m w^n + \ldots . \end{aligned}$$

Comparing terms with equal powers of w^k one has

$$\hat{U}_E = U_E \tag{4.2.10}$$

$$\dot{S} = \tfrac{1}{2}v_E^2 + \bar{U}_E(\boldsymbol{x}_E) \tag{4.2.11}$$

$$a^i_E = \bar{U}_{E,i}(\boldsymbol{x}_E) - Q_i \tag{4.2.12}$$

$$Q_{km} = \tfrac{1}{3}\bar{U}_{E,km}(\boldsymbol{x}_E) = \sum_{A \neq E} \frac{GM_A}{r^5_{EA}}(r^k_{EA} r^m_{EA} - \tfrac{1}{3} r^2_{EA}\delta_{km}) \tag{4.2.13}$$

$$\begin{aligned} Q_{kmn} = \tfrac{1}{15}\bar{U}_{E,kmn}(\boldsymbol{x}_E) = \sum_{A \neq E} \frac{GM_A}{r^5_{EA}} \Big(& \tfrac{1}{5}\delta_{mn} r^k_{EA} + \tfrac{1}{5}\delta_{nk} r^m_{EA} \\ & + \tfrac{1}{5}\delta_{km} r^n_{EA} - \frac{1}{r^2_{EA}} r^k_{EA} r^m_{EA} r^n_{EA} \Big). \end{aligned} \tag{4.2.14}$$

Comparison of (4.2.12) and (4.1.24) yields

$$Q_i = -H^i_E \tag{4.2.15}$$

with the Newtonian value (4.1.27). Applying (4.2.9) for the spatial components of the left-hand side one has for the order $O(c^{-2})$

$$\begin{aligned} (2U_E + 2\bar{U}_E)\delta_{ij} = {} & (2\hat{U}_E + 2Q_k w^k + 3Q_{km}w^k w^m \\ & + 5Q_{kmn}w^k w^m w^n + \ldots)\delta_{ij} + 2D^{ij} + 2(D^{ijk} + D^{jik})r^k_E \end{aligned} \tag{4.2.16}$$

resulting again in (4.2.10), (4.2.13), (4.2.14) and the relations

$$D^{ij} = \delta_{ij}\bar{U}_E(\boldsymbol{x}_E) = \delta_{ij}\sum_{A\neq E}\frac{GM_A}{r_{EA}} \tag{4.2.17}$$

$$D^{ijk} = \tfrac{1}{2}(\delta_{ij}a_E^k + \delta_{ik}a_E^j - \delta_{jk}a_E^i) = \tfrac{1}{2}\sum_{A\neq E}\frac{GM_A}{r_{EA}^3}(\delta_{jk}r_{EA}^i - \delta_{ij}r_{EA}^k - \delta_{ik}r_{EA}^j). \tag{4.2.18}$$

Finally, matching (4.2.9) for the mixed components of the BRS metric gives for the order $O(c^{-3})$

$$\begin{aligned}4(U_E^i + \bar{U}_E^i) = {} & 4\hat{U}_E + 2v_E^i(2\hat{U}_E + 2Q_k w^k + 3Q_{km}w^k w^m + \ldots) \\ & + v_E^i(\dot{S} - \tfrac{1}{2}v_E^2 + \tfrac{1}{2}a_E^k r_E^k) \\ & + v_E^k(2D^{ik} - \tfrac{1}{2}a_E^i r_E^k + 2D^{ikm}r_E^m + 2D^{kim}r_E^m) \\ & + B^i + (2B^{ik} - \dot{F}^{ik} - \dot{D}^{ik})r_E^k + (3B^{ikm} - \dot{D}^{ikm})r_E^k r_E^m \\ & + 4\epsilon_{ijk}C_{jm}w^k w^m - \tfrac{6}{5}\dot{Q}_k w^i w^k + \tfrac{2}{5}\dot{Q}_i w^k w^k. \end{aligned} \tag{4.2.19}$$

Using the values obtained previously and equating coefficients with the same powers of w^k one finds

$$\hat{U}_E^i = U_E^i - v_E^i\hat{U}_E \tag{4.2.20}$$

$$B^i = 4\bar{U}_E^i(\boldsymbol{x}_E) - 3v_E^i\bar{U}_E(\boldsymbol{x}_E) = \sum_{A\neq E}\frac{GM_A}{r_{EA}}(4v_A^i - 3v_E^i) \tag{4.2.21}$$

$$2B^{ik} - \dot{F}^{ik} = 4\bar{U}_{E,k}^i(\boldsymbol{x}_E) - 4v_E^i Q_k + \dot{D}^{ik} - \tfrac{5}{2}a_E^k v_E^i + \tfrac{1}{2}a_E^i v_E^k \tag{4.2.22}$$

$$\begin{aligned}3B^{ikm} + 2\epsilon_{ijk}C_{jm} + 2\epsilon_{ijm}C_{jk} = {} & 2\bar{U}_{E,km}^i(\boldsymbol{x}_E) - 6v_E^i Q_{km} + \dot{D}^{ikm} \\ & + \tfrac{3}{5}\delta_{im}\dot{Q}_k + \tfrac{3}{5}\delta_{ik}\dot{Q}_m - \tfrac{2}{5}\dot{Q}_i\delta_{km}. \end{aligned} \tag{4.2.23}$$

To solve (4.2.22) and (4.2.23) one has to take into account the conditions of symmetry and antisymmetry of the functions occurring in the left-hand sides. Therefore,

$$\begin{aligned}B^{ik} & = \tfrac{1}{2}\dot{D}^{ik} + \bar{U}_{E,k}^i(\boldsymbol{x}_E) + \bar{U}_{E,i}^k(\boldsymbol{x}_E) - (v_E^i Q_k + v_E^k Q_i) - \tfrac{1}{2}(v_E^i a_E^k + v_E^k a_E^i) \\ & = \sum_{A\neq E}\frac{GM_A}{r_{EA}^3}[(\tfrac{1}{2}v_E^i - v_A^i)r_{EA}^k + (\tfrac{1}{2}v_E^k - v_A^k)r_{EA}^i - \tfrac{1}{2}\delta_{ik}(v_E^m - v_A^m)r_{EA}^m] \end{aligned} \tag{4.2.24}$$

$$\begin{aligned}\dot F^{ik} &= -2[\bar U^i_{E,k}(\boldsymbol{x}_E) - \bar U^k_{E,i}(\boldsymbol{x}_E)] + 2(v^i_E Q_k - v^k_E Q_i) + \tfrac{3}{2}(v^i_E a^k_E - v^k_E a^i_E)\\ &= \sum_{A\neq E} \frac{GM_A}{r^3_{EA}}[\tfrac{3}{2}(v^k_E r^i_{EA} - v^i_E r^k_{EA}) + 2(v^k_A r^i_{EA} - v^i_A r^k_{EA})] \qquad (4.2.25)\end{aligned}$$

$$\begin{aligned}B^{ikm} &= \tfrac{2}{9}[\bar U^i_{E,km}(\boldsymbol{x}_E) + \bar U^k_{E,mi}(\boldsymbol{x}_E) + \bar U^m_{E,ik}(\boldsymbol{x}_E)]\\ &\quad - \tfrac{2}{3}(v^i_E Q_{km} + v^k_E Q_{mi} + v^m_E Q_{ik}) + \tfrac{1}{9}(\dot D^{ikm} + \dot D^{kmi} + \dot D^{mik})\\ &\quad + \tfrac{4}{45}(\delta_{ik}\dot Q_m + \delta_{im}\dot Q_k + \delta_{km}\dot Q_i) \qquad (4.2.26)\end{aligned}$$

$$\begin{aligned}\epsilon_{ijk}C_{jm} &= \tfrac{1}{3}[\bar U^i_{E,km}(\boldsymbol{x}_E) - \bar U^k_{E,im}(\boldsymbol{x}_E)] - (v^i_E Q_{km} - v^k_E Q_{im})\\ &\quad + \tfrac{1}{6}(\dot D^{ikm} - \dot D^{kim}) + \tfrac{1}{6}(\delta_{im}\dot Q_k - \delta_{km}\dot Q_i). \qquad (4.2.27)\end{aligned}$$

By using the identity

$$\epsilon_{kij}\epsilon_{kmn} = \delta_{im}\delta_{jn} - \delta_{in}\delta_{jm} \qquad (4.2.28)$$

the last relation may be rewritten in the form

$$\begin{aligned}C_{ij} &= \epsilon_{ikm}[-\tfrac{1}{3}\bar U^k_{E,jm}(\boldsymbol{x}_E) + v^k_E Q_{jm} - \tfrac{1}{6}\dot D^{kjm} + \tfrac{1}{6}\delta_{jm}\dot Q_k]\\ &= \tfrac{1}{3}\epsilon_{ikm}[v^k_E \bar U_{E,jm}(\boldsymbol{x}_E) - \bar U^k_{E,jm}(\boldsymbol{x}_E)] - \tfrac{1}{6}\epsilon_{ijk}\dot a^k_E - \tfrac{1}{6}\epsilon_{ijk}\dot Q_k.\end{aligned}$$

$$(4.2.29)$$

The antisymmetric part of C_{ij} is equal to

$$\begin{aligned}\tfrac{1}{2}(C_{ij} - C_{ji}) &= \tfrac{1}{6}\{-\epsilon_{ijk}(\dot a^k_E + \dot Q_k) + \epsilon_{ikm}[v^k_E \bar U_{E,jm}(\boldsymbol{x}_E) - \bar U^k_{E,jm}(\boldsymbol{x}_E)]\\ &\quad - \epsilon_{jkm}[v^k_E \bar U_{E,im}(\boldsymbol{x}_E) - \bar U^k_{E,im}(\boldsymbol{x}_E)]\}.\end{aligned}$$

Omitting as in all explicit expressions (4.2.13), (4.2.14), (4.2.17), (4.2.18), (4.2.24), (4.2.25) the terms with quadrupole moments one obtains

$$\begin{aligned}\tfrac{1}{2}(C_{ij} - C_{ji}) &= \tfrac{1}{2}\sum_{A\neq E}\frac{GM_A}{r^3_{EA}}\Big(\epsilon_{ijk}(v^k_E - v^k_A)\\ &\quad + \frac{1}{r^2_{EA}} r^m_{EA}[-\epsilon_{ijk} r^k_{EA}(v^m_E - v^m_A)\\ &\quad + (\epsilon_{ikm} r^j_{EA} - \epsilon_{jkm} r^i_{EA})(v^k_E - v^k_A)]\Big).\end{aligned}$$

It may be easily verified that this expression vanishes identically. Hence, C_{ij} reduces to the symmetric part alone

$$\begin{aligned}\tfrac{1}{2}(C_{ij} + C_{ji}) &= \tfrac{1}{6}\epsilon_{ikm}[v^k_E \bar U_{E,jm}(\boldsymbol{x}_E) - \bar U^k_{E,jm}(\boldsymbol{x}_E)]\\ &\quad + \tfrac{1}{6}\epsilon_{jkm}[v^k_E \bar U_{E,im}(\boldsymbol{x}_E) - \bar U^k_{E,im}(\boldsymbol{x}_E)]\\ &= \tfrac{1}{2}\sum_{A\neq E}\frac{GM_A}{r^5_{EA}}(v^k_E - v^k_A)(\epsilon_{ikm} r^j_{EA} + \epsilon_{jkm} r^i_{EA}) r^m_{EA}.\end{aligned}$$

$$(4.2.30)$$

Thus all coefficients occurring in metric (4.2.2)–(4.2.4) and transformation (4.2.7) and (4.2.8) become known. As already mentioned, with the greater accuracy involving particularly $\hat{g}_{00}^{(4)}$ in (4.2.2), B in (4.2.7) and relativistic terms in (4.2.12), the matching of the BRS and GRS metrics has been performed in Kopejkin (1988).

4.2.4 Physical characteristics of a body

With the aid of transformations (4.2.7), (4.2.8) extending the Lorentz transformation for the presence of the gravitational field one may obtain the interrelation between the BRS and GRS expressions of physical quantities. First of all, assuming x^i to be the coordinates of a moving point and putting $v^i = \dot{x}^i$ one has from (4.2.7)

$$\frac{\mathrm{d}u}{\mathrm{d}t} = 1 - c^{-2}(v^k v_E^k - \tfrac{1}{2}v_E^2 + \bar{U}_E(\boldsymbol{x}_E) + a_E^k r_E^k). \tag{4.2.31}$$

Let x^i be a point of the body at hand (the Earth for the given case). Its GRS coordinates are w^i and its GRS velocity is

$$\begin{aligned}\frac{\mathrm{d}w^i}{\mathrm{d}u} &= v^i - v_E^i + c^{-2}\{[v^k v_E^k - \tfrac{1}{2}v_E^2 + \bar{U}_E(\boldsymbol{x}_E)](v^i - v_E^i) \\ &\quad + (\tfrac{1}{2}v_E^i v_E^k + F^{ik} + D^{ik})(v^k - v_E^k) + [(v^i - v_E^i)a_E^k \\ &\quad + 2D^{ikm}(v^m - v_E^m) + \tfrac{1}{2}a_E^i v_E^k + \tfrac{1}{2}v_E^i a_E^k \\ &\quad + \dot{F}^{ik} + \dot{D}^{ik}]r_E^k + \dot{D}^{ikm} r_E^k r_E^m\}. \end{aligned} \tag{4.2.32}$$

If the body is rigid and non-rotating then, evidently, $\mathrm{d}w^i/\mathrm{d}u = 0$ but the BRS velocities v^i of the points of the body differ from its centre of mass velocity v_E^i. Such a body cannot be regarded in BRS as moving translatory and the presence of the term $\dot{F}^{ik} r_E^k$ is characteristic of its rotation (Kopejkin 1987). For a rigid rotating body its rigid body velocity distribution should be given just in GRS

$$\frac{\mathrm{d}w^i}{\mathrm{d}u} = \epsilon_{ijk}\hat{\omega}_E^j w^k \tag{4.2.33}$$

and the BRS description of the same rotation is, as seen from (4.2.32), much more complicated.

Consider now the relationship between dipole and quadrupole moments in GRS and BRS. The GRS 'proper' dipole and quadrupole moments are generally functions of time u and may be defined by the integrals

$$\hat{I}_E^k = \int_{(E)} \hat{\rho}' w'^k \,\mathrm{d}^3 w' \qquad \hat{I}_E^{km} = \int_{(E)} \hat{\rho}' w'^k w'^m \,\mathrm{d}^3 w' \tag{4.2.34}$$

taken over the hypersurface u = constant. $\hat{\rho}'$ is the proper mass density at the point $\boldsymbol{w}', u$. The GRS origin being coincident with the Earth centre of mass, one has $\hat{I}^k_E = 0$. Evidently,

$$\rho' \, \mathrm{d}^3x' = \hat{\rho}' \, \mathrm{d}^3w' \tag{4.2.35}$$

and hence $M_E = \hat{M}_E$. To set the relationship of GRS moments (4.2.34) and BRS moments (4.1.8) one has to perform a Lie transfer between the hypersurfaces u = constant and t = constant. These hypersurfaces intersect in the matching point of BRS and GRS where relations (4.2.7), (4.2.8) are valid. Consider an arbitrary point t', $\boldsymbol{x}'$. Its GRS coordinates are u', $\boldsymbol{w}'$. If this point and the matching point belong to the hypersurface u = constant then from (4.2.7) and the condition $u' = u$ there results

$$\Delta t = t' - t = c^{-2} v^n_E (x'^n - x^n).$$

If any quantity A is defined on the hypersurface t = constant then its value $\tilde{A}$ on the hypersurface u = constant is determined by the Lie transfer

$$\tilde{A} = A + v'^i A_{,i} \Delta t$$

v'^i being the velocity of the point of the hypersurface. In evaluating the integrals (4.1.8) and (4.2.34) the Earth is considered as a rotating rigid body. Hence, the velocity of each point in it is approximately

$$v'^i = \epsilon_{ikm} \omega^k_E r'^m_E.$$

From this, it follows that

$$\tilde{A} = A + c^{-2} \epsilon_{ikm} \omega^k_E v^n_E r'^m_E (x'^n - x^n) A_{,i}. \tag{4.2.36}$$

By applying this formula to (4.2.8) it is easy to find that the required functions occurring in (4.2.34) are determined by the relations

$$\begin{aligned} w'^i &= r'^i_E + c^{-2}[(\tfrac{1}{2} v^i_E v^k_E + F^{ik} + D^{ik}) r'^k_E + D^{ikm} r'^k_E r'^m_E] \\ &\quad + c^{-2} \epsilon_{ikm} \omega^k_E v^n_E r'^m_E (r'^n_E - r^n_E) \end{aligned} \tag{4.2.37}$$

differing from (4.2.8) by an extra term dependent on the matching point and caused by the Lie transfer. Using (4.2.35) and (4.2.37) one finds the relationship of moments (4.1.8) and (4.2.34) as in Brumberg and Kopejkin (1989b)

$$I^i_E = -c^{-2} \epsilon_{ikm} \omega^k_E v^n_E \hat{I}^{mn}_E - c^{-2} a^k_E \hat{I}^{ik}_E + \tfrac{1}{2} c^{-2} a^i_E \hat{I}^{kk}_E \tag{4.2.38}$$

$$I_E^{ik} = \hat{I}_E^{ik} - c^{-2}(\tfrac{1}{2}v_E^i v_E^m + F^{im} + D^{im})\hat{I}_E^{km} - c^{-2}(\tfrac{1}{2}v_E^k v_E^m + F^{km} + D^{km})\hat{I}_E^{im} + c^{-2}\omega_E^j v_E^n w^n(\epsilon_{ijm}\hat{I}_E^{km} + \epsilon_{kjm}\hat{I}_E^{im}). \quad (4.2.39)$$

Expression (4.2.39) relating the BRS and GRS quadrupole moments involves coordinates w^n of the matching point. At first sight it might seem strange. Considering that the BRS quadrupole moments are functions of t and GRS moments depend on u and taking into account that the relationship (4.2.7) of t and u includes the spatial coordinates of the matching point at hand this fact becomes clear.

The spherical body may be defined as a body with the diagonal matrix of the quadrupole moments, i.e.

$$\hat{I}_E^{km} = \delta_{km}\hat{I}_E. \quad (4.2.40)$$

For such a body its BRS moments by (4.2.38) and (4.2.39) are

$$I_E^i = c^{-2}\hat{I}_E(-\epsilon_{ikm}\omega_E^k v_E^m + \tfrac{1}{2}a_E^i) \quad (4.2.41)$$

$$I_E^{km} = [1 - 2c^{-2}\bar{U}_E(\boldsymbol{x}_E)]\hat{I}_E\delta_{km} - c^{-2}v_E^k v_E^m \hat{I}_E \quad (4.2.42)$$

demonstrating that the BRS matrix of the quadrupole moments is not reduced to the diagonal unit matrix (Kopejkin 1987).

Transformations (4.2.7), (4.2.8) and (4.2.38), (4.2.39) enable one to derive the GRS equations of motion of celestial bodies starting from the known BRS equations. Equations of motion of the major planets, the Moon and Earth satellites based on these techniques are given in the next chapter.

4.3 EQUATIONS IN VARIATIONS FOR THE SPHERICALLY SYMMETRICAL METRIC

4.3.1 Generating metric

In all methods of solving the problem of motion in GRT based on the expansions in powers of v/c and U/c^2 the masses of the bodies involved are treated as being of one and the same order. In the Solar System the ratio of the total mass of all planets to the mass of the Sun is of order 10^{-3}. Hence, it may be reasonable to take the Schwarzschild solution for the Sun's field as the intermediary and to look for the solution as series in powers of the ratio of the total planetary mass to that of Sun. Such an approach is of interest enabling one in principle to avoid expansions in powers of v/c which is of importance for the rigorous treatment of the problem of gravitational radiation (although the practical realization of such an approach still remains vague). A similar problem arises in studying the motion of the macroscopic bodies in cosmology. As the intermediary one may again take the spherically symmetric metric, for example, the Friedman metric. The problem of perturbations both for Schwarzschild and Friedman metrics has been repeatedly examined in the literature (see, for example, Peters 1966, Irvine 1965), but not for the case of perturbations due to the gravitating masses. The results derived below are based on Brumberg and Tarasevich 1983).

The GRT field equations (2.1.7) rewritten in the form

$$G^{\mu\nu} + \Lambda g^{\mu\nu} = -\kappa(\mathcal{T}^{\mu\nu} + T^{\mu\nu}) \tag{4.3.1}$$

are considered under the assumption that in vanishing perturbations $T^{\mu\nu} = 0$ they admit the spherically symmetric isotropic solution

$$\eta_{00} = A \qquad \eta_{0m} = 0 \qquad \eta_{mn} = -B\delta_{mn} \tag{4.3.2}$$

A, B being some function of radial coordinate r and, possibly, time t. With $T^{\mu\nu} \neq 0$ the solution of (4.3.1) is presented in the form

$$g_{\mu\nu} = \eta_{\mu\nu} + h_{\mu\nu} \tag{4.3.3}$$

$h_{\mu\nu}$ being small corrections to the metric tensor caused by the perturbing mass tensor $T^{\mu\nu}$ of the macroscopic bodies. Along with equations (4.3.1) one may use the equivalent equations involving the Ricci tensor

$$R_{\mu\nu} = -\kappa(\mathcal{T}^*_{\mu\nu} + T^*_{\mu\nu}) + \Lambda g_{\mu\nu} \tag{4.3.4}$$

with

$$\mathcal{T}^*_{\mu\nu} = \mathcal{T}_{\mu\nu} - \tfrac{1}{2} g_{\mu\nu}\mathcal{T} \qquad T^*_{\mu\nu} = T_{\mu\nu} - \tfrac{1}{2} g_{\mu\nu} T. \tag{4.3.5}$$

For the Schwarzschild field in the whole external space outside the central mass M one gets

$$\mathcal{T}^{\mu\nu} = 0 \qquad \Lambda = 0 \qquad A_{,0} = B_{,0} = 0 \tag{4.3.6}$$

and the solution is given by (3.1.10)–(3.1.20). In the cosmological solution for the homogeneous isotropic model describing uniformly and continuously distributed matter with density ρ and pressure p one has

$$\mathcal{T}^{\mu\nu} = (\rho + p)\frac{\mathrm{d}x^\mu}{\mathrm{d}s}\frac{\mathrm{d}x^\nu}{\mathrm{d}s} - p g^{\mu\nu} \tag{4.3.7}$$

with

$$\frac{\mathrm{d}x^m}{\mathrm{d}s} = 0 \qquad \frac{\mathrm{d}x^0}{\mathrm{d}s} = 1 \qquad A = 1 \tag{4.3.8}$$

in co-moving coordinates (with the choice $c = 1$). From (4.3.4) and (3.1.14) we obtain

$$\frac{3}{2B}\left(B_{,00} - \frac{1}{2B}B_{,0}B_{,0}\right) = -\frac{\kappa}{2}(\rho + 3p) + \Lambda \tag{4.3.9}$$

$$B_{,0m} - \frac{1}{B} B_{,0} B_{,m} = 0 \tag{4.3.10}$$

$$\frac{1}{2B}\Big(-B_{,mn} - \delta_{mn} B_{,ss} + \frac{1}{2}\delta_{mn} B_{,0} B_{,0} + \frac{1}{2B}(3B_{,m}B_{,n} + \delta_{mn} B_{,s} B_{,s}) + \delta_{mn} B B_{,00}\Big) = B\left(\frac{\kappa}{2}(\rho - p) + \Lambda\right)\delta_{mn}. \tag{4.3.11}$$

These equations may be satisfied by putting

$$B = \frac{R^2}{(1 + kr^2/4)^2} \tag{4.3.12}$$

with $k =$ constant and functions $R = R(t)$, ρ, p being determined from the equations

$$3\frac{\ddot{R}}{R} = \Lambda - \frac{\kappa}{2}(\rho + 3p) \tag{4.3.13}$$

$$R\ddot{R} + 2\dot{R}^2 + 2k = R^2\left(\Lambda + \frac{\kappa}{2}(\rho - p)\right). \tag{4.3.14}$$

These equations should be complemented by the equation of state relating the density ρ and the pressure p. Different particular solutions for $R(t)$ are well known. In addition, it is to be noted that in virtue of the Bianchi identities one should have

$$\nabla_\mu T^{\mu\nu} = 0. \tag{4.3.15}$$

For the space components these relations are satisfied identically. For the time component they lead to the relation

$$\dot{\rho} + 3(\rho + p)\frac{\dot{R}}{R} = 0. \tag{4.3.16}$$

Thus, for the Schwarzschild problem A and B are functions of r alone determined by (3.1.17). For the cosmological model $A = 1$ and B is a function (4.3.12) of r and t. For the particular case $k = 0$ (the open flat model) B becomes a function of t alone.

4.3.2 Equations in variations

Taking the difference of equations (4.3.4) and the background field equations one has

$$\delta R_{\mu\nu} = -\kappa T^*_{\mu\nu} - \kappa\delta T^*_{\mu\nu} + \Lambda h_{\mu\nu} \tag{4.3.17}$$

with

$$\delta T^*_{\mu\nu} = T^*_{\mu\nu}(g_{\alpha\beta}) - T^*_{\mu\nu}(\eta_{\alpha\beta}) \tag{4.3.18}$$

$\delta R_{\mu\nu}$ being the correction terms of the Ricci tensor due to the perturbations $h_{\mu\nu}$. Christoffel symbols of the first and the second kind and the Ricci tensor for metric (4.3.2) are given in (3.1.12)–(3.1.14). For the disturbed metric (4.3.3) the linear corrections to these quantities are listed below.

Variations of the Christoffel symbols of the first kind:

$$\delta\Gamma_{000} = \tfrac{1}{2}h_{00,0} \qquad \delta\Gamma_{00m} = \tfrac{1}{2}h_{00,m} \qquad \delta\Gamma_{0mn} = \tfrac{1}{2}(h_{0m,n} + h_{0n,m} - h_{mn,0})$$

$$\delta\Gamma_{s00} = h_{0s,0} - \tfrac{1}{2}h_{00,s} \qquad \delta\Gamma_{s0m} = \tfrac{1}{2}(h_{0s,m} + h_{ms,0} - h_{0m,s})$$

$$\delta\Gamma_{smn} = \tfrac{1}{2}(h_{ms,n} + h_{ns,m} - h_{mn,s}). \tag{4.3.19}$$

Variations of the Christoffel symbols of the second kind:

$$\begin{aligned}
\delta\Gamma^0_{00} &= \frac{1}{2A}\left(h_{00,0} - \frac{1}{A}A_{,0}h_{00} - \frac{1}{B}A_{,s}h_{0s}\right)\\
\delta\Gamma^m_{00} &= \frac{1}{2B}\left(h_{00,m} - 2h_{0m,0} + \frac{1}{A}A_{,0}h_{0m} + \frac{1}{B}A_{,s}h_{ms}\right)\\
\delta\Gamma^0_{0m} &= \frac{1}{2A}\left(h_{00,m} - \frac{1}{A}A_{,m}h_{00} - \frac{1}{B}B_{,0}h_{0m}\right)\\
\delta\Gamma^s_{0m} &= \frac{1}{2B}\left(h_{0m,s} - h_{0s,m} - h_{ms,0} + \frac{1}{A}A_{,m}h_{0s} + \frac{1}{B}B_{,0}h_{ms}\right)\\
\delta\Gamma^0_{mn} &= \frac{1}{2A}\Big(h_{0m,n} + h_{0n,m} - h_{mn,0} - \frac{1}{A}B_{,0}h_{00}\delta_{mn} + \frac{1}{B}B_{,s}h_{0s}\delta_{mn}\\
&\quad - \frac{1}{B}B_{,n}h_{0m} - \frac{1}{B}B_{,m}h_{0n}\Big)\\
\delta\Gamma^s_{mn} &= \frac{1}{2B}\Big(h_{mn,s} - h_{ms,n} - h_{ns,m} + \frac{1}{A}B_{,0}h_{0s}\delta_{mn} + \frac{1}{B}B_{,n}h_{ms}\\
&\quad + \frac{1}{B}B_{,m}h_{ns} - \frac{1}{B}B_{,r}h_{rs}\delta_{mn}\Big).
\end{aligned} \tag{4.3.20}$$

Variations of the Ricci tensor:

$$\delta R_{00} = -\frac{1}{2B}(h_{00,ss} - 2h_{0s,0s} + h_{ss,00}) + Q_{00} \tag{4.3.21}$$

$$\delta R_{0m} = -\frac{1}{2B}(h_{0m,ss} - h_{0s,ms} + h_{ss,0m} - h_{ms,0s}) + Q_{0m} \tag{4.3.22}$$

$$\begin{aligned}
\delta R_{mn} = &-\frac{1}{2B}\Big[h_{mn,ss} + h_{ss,mn} - h_{ms,ns} - h_{ns,ms} + \frac{B}{A}(-h_{00,mn}\\
&- h_{mn,00} + h_{0m,0n} + h_{0n,0m})\Big] + Q_{mn}
\end{aligned} \tag{4.3.23}$$

with

$$
\begin{aligned}
Q_{00} = &-\frac{1}{2B}\Big[\left(\frac{1}{2B}B_{,s}-\frac{1}{A}A_{,s}\right)h_{00,s}-\frac{1}{B}B_{,s}h_{0s,0}+\frac{1}{A}A_{,0}h_{0s,s}\\
&-\left(\frac{1}{B}B_{,0}+\frac{1}{2A}A_{,0}\right)h_{ss,0}-\frac{1}{2B}A_{,s}h_{rr,s}+\frac{1}{B}A_{,r}h_{rs,s}\\
&+\frac{1}{2A^2}\left(A_{,s}A_{,s}-3A_{,0}B_{,0}\right)h_{00}+\frac{1}{2AB}(A_{,0}B_{,s}-A_{,s}B_{,0})h_{0s}\\
&+\left(-\frac{1}{B}B_{,00}+\frac{1}{2AB}A_{,0}B_{,0}+\frac{1}{B^2}B_{,0}B_{,0}+\frac{1}{2B^2}A_{,s}B_{,s}\right)h_{rr}\\
&+\left(\frac{1}{B}A_{,rs}-\frac{1}{2AB}A_{,r}A_{,s}-\frac{1}{2B^2}A_{,r}B_{,s}\right)h_{rs}\Big]
\end{aligned}
\tag{4.3.24}
$$

$$
\begin{aligned}
Q_{0m} = &-\frac{1}{2B}\Big[\frac{1}{A}B_{,0}h_{00,m}+\frac{1}{A}A_{,m}h_{0s,s}\\
&+\left(\frac{1}{2B}B_{,s}-\frac{1}{2A}A_{,s}\right)h_{0s,m}-\left(\frac{1}{2B}B_{,s}+\frac{1}{2A}A_{,s}\right)h_{0m,s}\\
&+\left(\frac{1}{2A}A_{,s}-\frac{1}{2B}B_{,s}\right)h_{ms,0}-\left(\frac{1}{2B}B_{,m}+\frac{1}{2A}A_{,m}\right)h_{ss,0}\\
&-\frac{1}{B}B_{,0}h_{ss,m}+\frac{1}{B}B_{,0}h_{ms,s}-\frac{1}{A^2}A_{,m}B_{,0}h_{00}\\
&+\left(-\frac{1}{A}B_{,00}+\frac{1}{2A^2}A_{,0}B_{,0}-\frac{1}{2AB}B_{,0}B_{,0}+\frac{1}{AB}A_{,s}B_{,s}\right)h_{0m}\\
&+\left(\frac{1}{A}A_{,ms}-\frac{1}{2A^2}A_{,m}A_{,s}-\frac{1}{2AB}A_{,m}B_{,s}\right)h_{0s}\\
&+\left(-\frac{1}{B}B_{,0m}+\frac{1}{2AB}A_{,m}B_{,0}+\frac{3}{2B^2}B_{,m}B_{,0}\right)h_{ss}\\
&+\left(\frac{1}{B}B_{,0s}-\frac{1}{2AB}A_{,s}B_{,0}-\frac{1}{2B^2}B_{,s}B_{,0}\right)h_{ms}\Big]
\end{aligned}
\tag{4.3.25}
$$

$$
\begin{aligned}
Q_{mn} = &-\frac{1}{2B}\Big[-\frac{B}{2A^2}B_{,0}h_{00,0}\delta_{mn}+\frac{1}{2A}\left(\frac{B}{A}A_{,n}+B_{,n}\right)h_{00,m}\\
&+\frac{1}{2A}\left(\frac{B}{A}A_{,m}+B_{,m}\right)h_{00,n}-\frac{1}{2A}B_{,s}h_{00,s}\delta_{mn}\\
&-\frac{1}{A}B_{,n}h_{0m,0}-\frac{1}{A}B_{,m}h_{0n,0}+\frac{1}{A}B_{,s}h_{0s,0}\delta_{mn}\\
&+\frac{1}{2A}\left(B_{,0}-\frac{B}{A}A_{,0}\right)h_{0m,n}+\frac{1}{2A}\left(B_{,0}-\frac{B}{A}A_{,0}\right)h_{0n,m}\\
&+\frac{1}{A}B_{,0}h_{0s,s}\delta_{mn}+\frac{1}{2A}\left(\frac{B}{A}A_{,0}+B_{,0}\right)h_{mn,0}-\frac{1}{2A}B_{,0}h_{ss,0}\delta_{mn}
\end{aligned}
$$

$$
\begin{aligned}
&+\left(\frac{1}{2A}A_{,s}-\frac{3}{2B}B_{,s}\right)h_{mn,s}-\frac{1}{B}B_{,n}h_{ss,m}\\
&-\frac{1}{B}B_{,m}h_{ss,n}+\frac{1}{2B}B_{,s}h_{rr,s}\delta_{mn}\\
&+\left(\frac{1}{2B}B_{,s}-\frac{1}{2A}A_{,s}\right)h_{ms,n}+\left(\frac{1}{2B}B_{,s}-\frac{1}{2A}A_{,s}\right)h_{ns,m}\\
&+\frac{1}{B}B_{,n}h_{ms,s}+\frac{1}{B}B_{,m}h_{ns,s}-\frac{1}{B}B_{,r}h_{rs,s}\delta_{mn}\\
&+\frac{B}{A^2}\Big(A_{,mn}-B_{,00}\delta_{mn}+\frac{1}{A}A_{,0}B_{,0}\delta_{mn}\\
&-\frac{1}{A}A_{,m}A_{,n}-\frac{1}{2B}B_{,0}B_{,0}\delta_{mn}\\
&-\frac{1}{2B}A_{,m}B_{,n}-\frac{1}{2B}A_{,n}B_{,m}+\frac{1}{2B}A_{,s}B_{,s}\delta_{mn}\Big)h_{00}\\
&+\frac{1}{A}\left(-B_{,0n}+\frac{1}{2A}A_{,0}B_{,n}+\frac{1}{2B}B_{,0}B_{,n}\right)h_{0m}\\
&+\frac{1}{A}\left(-B_{,0m}+\frac{1}{2A}A_{,0}B_{,m}+\frac{1}{2B}B_{,0}B_{,m}\right)h_{0n}\\
&+\frac{1}{A}\left(2B_{,0s}-\frac{1}{2A}A_{,0}B_{,s}-\frac{1}{2A}A_{,s}B_{,0}-\frac{1}{B}B_{,0}B_{,s}\right)h_{0s}\delta_{mn}\\
&+\frac{1}{B}\left(B_{,ns}+\frac{1}{2A}A_{,s}B_{,n}-\frac{3}{2B}B_{,n}B_{,s}\right)h_{ms}\\
&+\frac{1}{B}\left(B_{,ms}+\frac{1}{2A}A_{,s}B_{,m}-\frac{3}{2B}B_{,m}B_{,s}\right)h_{ns}\\
&+\frac{1}{B}\left(-B_{,rs}-\frac{1}{2A}A_{,r}B_{,s}+\frac{3}{2B}B_{,r}B_{,s}\right)h_{rs}\delta_{mn}\\
&+\frac{1}{B}\left(\frac{1}{B}B_{,s}B_{,s}-\frac{1}{A}B_{,0}B_{,0}\right)h_{mn}+\frac{1}{B}\Big(-B_{,mn}+\frac{1}{2A}B_{,0}B_{,0}\delta_{mn}\\
&-\frac{1}{2B}B_{,s}B_{,s}\delta_{mn}+\frac{2}{B}B_{,m}B_{,n}\Big)h_{rr}\Big].
\end{aligned}
\tag{4.3.26}
$$

Denoting now

$$L_{\mu\nu}=B(\kappa T^*_{\mu\nu}+\kappa\delta T^*_{\mu\nu}-\Lambda h_{\mu\nu}+Q_{\mu\nu}) \tag{4.3.27}$$

one obtains from (4.3.17) the equations to determine $h_{\mu\nu}$:

$$h_{00,ss}-2h_{0s,0s}+h_{ss,00}=2L_{00} \tag{4.3.28}$$

$$h_{0m,ss}-h_{0s,ms}+h_{ss,0m}-h_{ms,0s}=2L_{0m} \tag{4.3.29}$$

$$h_{mn,ss} + h_{ss,mn} - h_{ms,ns} - h_{ns,ms} + \frac{B}{A}(-h_{00,mn} - h_{mn,00} + h_{0m,0n} + h_{0n,0m}) = 2L_{mn}. \tag{4.3.30}$$

By contracting (4.3.30) and using (4.3.28), we obtain

$$h_{rr,ss} - h_{rs,rs} = L_{ss} + \frac{B}{A}L_{00}. \tag{4.3.31}$$

Equations (4.3.28)–(4.3.30) may be solved by iterations with respect to $h_{\mu\nu}$. At each step of iteration the right-hand members $2L_{\mu\nu}$ are known, and equations (4.3.28)–(4.3.30) represent linear partial differential equations. To facilitate their examination one may introduce the coordinate conditions

$$h_{00,0} + h_{ss,0} - 2h_{0s,s} = 0 \tag{4.3.32}$$

$$h_{00,m} - h_{ss,m} + 2h_{ms,s} = 0. \tag{4.3.33}$$

Under these conditions equaitons (4.3.28)–(4.3.30) take the form

$$h_{00,ss} - h_{00,00} = 2L_{00} \tag{4.3.34}$$

$$h_{0m,ss} = 2L_{0m} \tag{4.3.35}$$

$$h_{mn,ss} - \frac{B}{A}h_{mn,00} = 2L_{mn} + \left(\frac{B}{A} - 1\right)h_{00,mn} - \frac{B}{A}(h_{0m,0n} + h_{0n,0m}) \tag{4.3.36}$$

with the equation for the contraction h_{rr} resulting from (4.3.36)

$$h_{rr,ss} = 2L_{ss} + 2\frac{B}{A}L_{00} - h_{00,ss}. \tag{4.3.37}$$

First, one has to solve (4.3.34) for the time component h_{00}. This is the wave equation with constant coefficients. Then one solves the Poisson equation (4.3.35) for the mixed components h_{0m}. Equation (4.3.36) for each space component h_{mn} has the form of the wave equation with variable coefficient B/A in the second derivative with respect to time. For the Schwarzschild field this coefficient is determined by (3.1.17), (3.1.20) and depends only on r. In the case of the cosmological background this coefficient is determined by (4.3.12) and depends on t and r (only on t for the open flat model). Solution of the wave equation (4.3.36) may be constructed, for example, by the Sobolev (1950) method but this remains to be actually done. Equation (4.3.37) for the space contraction represents the Poisson equation.

4.3.3 Equations of motion

The equations of motion of bodies in the field (4.3.2), (4.3.3) may be derived either from the modified geodesic principle or from the Bianchi identities (Infeld and Plebanski 1960). In the first case it is sufficient to substitute (3.1.13) and (4.3.20) into equations (2.2.48). In so doing the components $h_{\mu\nu}$ entering into the Christoffel symbols are evaluated along the body's trajectory with a suitable re-normalization. There results

$$\begin{aligned}\ddot{x}^m = &-\frac{1}{2B}A_{,m} + \left(\frac{1}{2A}A_{,0} - \frac{1}{B}B_{,0}\right)\dot{x}^m + \left(\frac{1}{A}A_{,s} - \frac{1}{B}B_{,s}\right)\dot{x}^s\dot{x}^m \\ &+ \frac{1}{2B}B_{,m}\dot{x}^s\dot{x}^s + \frac{1}{2A}B_{,0}\dot{x}^s\dot{x}^s\dot{x}^m + \frac{1}{2A}\dot{x}^m h_{00,0} - \frac{1}{2B}h_{00,m} \\ &+ \frac{1}{A}\dot{x}^m\dot{x}^s h_{00,s} + \frac{1}{B}h_{0m,0} - \frac{1}{B}\dot{x}^s h_{0s,m} + \frac{1}{B}\dot{x}^s h_{0m,s} + \frac{1}{A}\dot{x}^m\dot{x}^r\dot{x}^s h_{0r,s} \\ &+ \frac{1}{B}\dot{x}^s h_{ms,0} - \frac{1}{2A}\dot{x}^m\dot{x}^r\dot{x}^s h_{rs,0} - \frac{1}{2B}\dot{x}^r\dot{x}^s h_{rs,m} \\ &+ \frac{1}{B}\dot{x}^r\dot{x}^s h_{mr,s} - \frac{1}{A^2}(\tfrac{1}{2}A_{,0} + A_{,s}\dot{x}^s + \tfrac{1}{2}B_{,0}\dot{x}^s\dot{x}^s)\dot{x}^m h_{00} \\ &- \frac{1}{AB}(\tfrac{1}{2}A_{,0} + A_{,s}\dot{x}^s + \tfrac{1}{2}B_{,0}\dot{x}^s\dot{x}^s)h_{0m} \\ &+ \frac{1}{AB}(-\tfrac{1}{2}A_{,s} - B_{,0}\dot{x}^s + \tfrac{1}{2}B_{,s}\dot{x}^r\dot{x}^r - B_{,r}\dot{x}^r\dot{x}^s)\dot{x}^m h_{0s} \\ &+ \frac{1}{B^2}(-\tfrac{1}{2}A_{,s} - B_{,0}\dot{x}^s - B_{,r}\dot{x}^r\dot{x}^s + \tfrac{1}{2}B_{,s}\dot{x}^r\dot{x}^r)h_{ms}.\end{aligned} \tag{4.3.38}$$

The second way is more complicated but enables one to gain a more penetrating insight into the relationship of the field equations and the equations of motion. Direct calculation gives the left-hand members of the Bianchi identities in the form

$$\nabla_\mu G^{0\mu} = -\frac{1}{4AB^2}(h_{00,0} + h_{rr,0} - 2h_{0r,r})_{,ss} + S^0 \tag{4.3.39}$$

$$\begin{aligned}\nabla_\mu G^{m\mu} = &-\frac{1}{4B^3}\Big((h_{00,m} - h_{rr,m} + 2h_{mr,r})_{,ss} \\ &- \frac{B}{A}(h_{00,m} - h_{rr,m} + 2h_{mr,r})_{,00} \\ &- \frac{B}{A}(h_{00,0} + h_{rr,0} - 2h_{0r,r})_{,0m}\Big) + S^m\end{aligned} \tag{4.3.40}$$

S^ν being the non-linear terms with respect to $h_{\mu\nu}$ in the expressions of $\nabla_\mu G^{\mu\nu}$. In ignoring these terms expressions (4.3.39), (4.3.40) vanish due to the coordinate conditions. Hence, the right-hand members of the Bianchi identities are also to be annulled along the world lines of the bodies

$$\nabla_\mu G^{\mu\nu} = -\kappa \nabla_\mu (T^{\mu\nu} + T^{\mu\nu}). \tag{4.3.41}$$

Relation $\nabla_\mu T^{\mu\nu} = 0$ is satisfied identically (for the cosmological problem it reduces to (4.3.16)). As for the disturbing mass tensor, by putting

$$T^{\mu\nu} = \frac{\rho}{\sqrt{-g}} \frac{\mathrm{d}x^0}{\mathrm{d}s} \frac{\mathrm{d}x^\mu}{\mathrm{d}x^0} \frac{\mathrm{d}x^\nu}{\mathrm{d}x^0}$$

one has

$$T^{00} = \frac{\rho}{\sqrt{-g}} \frac{\mathrm{d}x^0}{\mathrm{d}s} \qquad T^{0m} = v^m T^{00} \qquad T^{mn} = v^m v^n T^{00} \tag{4.3.42}$$

and, consequently,

$$\nabla_\mu T^{0\mu} = T^{00}_{,0} + T^{0s}_{,s} + (2\Gamma^0_{00} + \Gamma^s_{0s})T^{00} + (3\Gamma^0_{0s} + \Gamma^r_{rs})T^{0s} + \Gamma^0_{rs} T^{rs} \tag{4.3.43}$$

$$\begin{aligned}\nabla_\mu T^{m\mu} = {} & T^{m0}_{,0} + T^{ms}_{,s} + \Gamma^m_{00} T^{00} + (\Gamma^0_{00} + \Gamma^s_{0s})T^{0m} + 2\Gamma^m_{0s} T^{0s} \\ & + \Gamma^m_{rs} T^{rs} + (\Gamma^0_{0s} + \Gamma^r_{rs})T^{ms}.\end{aligned} \tag{4.3.44}$$

Direct differentiation of (4.3.42) using the equation of continuity yields

$$\begin{aligned} T^{00}_{,0} + T^{0s}_{,s} = {} & -\tfrac{1}{2}T^{00}\{(AB^3 + B^3 h_{00} - AB^2 h_{ss})^{-1}[A_{,0}B^3 + 3AB^2 B_{,0} \\ & + A_{,s}B^3 v^s + 3AB^2 B_{,s} v^s + 3B^2(B_{,0} + B_{,s}v^s)h_{00} + B^3 h_{00,0} \\ & + B^3 v^s h_{00,s} - B^2(A_{,0} + A_{,s}v^s)h_{rr} \\ & - 2AB(B_{,0} + B_{,s}v^s)h_{rr} - AB^2 h_{rr,0} - AB^2 v^s h_{rr,s}] \\ & + (A - Bv^2 + h_{00} + 2h_{0s}v^s + h_{rs}v^r v^s)^{-1} \\ & \times (A_{,0} + A_{,s}v^s - B_{,0}v^2 - B_{,s}v^s v^2 - 2Bv^s \ddot{x}^s \\ & + h_{00,0} + h_{00,s}v^s + 2h_{0s,0}v^s + 2h_{0r,s}v^r v^s + 2h_{0s}\ddot{x}^s \\ & + h_{rs,0}v^r v^s + h_{kr,s}v^k v^r v^s + 2h_{rs}v^r \ddot{x}^s)\} \end{aligned} \tag{4.3.45}$$

$$\begin{aligned} T^{0m}_{,0} + T^{ms}_{,s} = {} & -\tfrac{1}{2}T^{00}\{-2\ddot{x}^m + (AB^3 + B^3 h_{00} - AB^2 h_{ss})^{-1} \\ & \times [A_{,0}B^3 + 3AB^2 B_{,0} + A_{,s}B^3 v^s \\ & + 3AB^2 B_{,s}v^s + 3B^2(B_{,0} + B_{,s}v^s)h_{00} \\ & + B^3 h_{00,0} + B^3 v^s h_{00,s} - B^2(A_{,0} + A_{,s}v^s)h_{rr} \\ & - 2AB(B_{,0} + B_{,s}v^s)h_{rr} - AB^2 h_{rr,0} - AB^2 v^s h_{rr,s}]v^m \\ & + (A - Bv^2 + h_{00} + 2h_{0s}v^s + h_{rs}v^r v^s)^{-1} \\ & \times (A_{,0} + A_{,s}v^s - B_{,0}v^2 - B_{,s}v^s v^2 - 2Bv^s\ddot{x}^s + h_{00,0} + h_{00,s}v^s \\ & + 2h_{0s,0}v^s + 2h_{0r,s}v^r v^s + 2h_{0s}\ddot{x}^s + h_{rs,0}v^r v^s \\ & + h_{kr,s}v^k v^r v^s + 2h_{rs}v^r \ddot{x}^s)v^m\}. \end{aligned} \tag{4.3.46}$$

Using the values of the Christoffel symbols one gets finally

$$\begin{aligned}
\nabla_\mu T^{0\mu} =& T^{00}(A - Bv^2 + h_{00} + 2h_{0s}v^s + h_{rs}v^r v^s)^{-1} \\
&\times \left\{ (Bv^s - h_{0s} - v^r h_{rs})\ddot{x}^s + \frac{1}{2}A_{,s}v^s + \left(B_{,0} - \frac{B}{2A}A_{,0}\right)v^2 \right. \\
&+ \left(\frac{1}{2}B_{,s} - \frac{B}{A}A_{,s}\right)v^s v^2 - \frac{B}{2A}B_{,0}v^4 - \frac{B}{2A}v^2 h_{00,0} \\
&+ \frac{1}{A}(\tfrac{1}{2}A - Bv^2)v^s h_{00,s} - v^s h_{0s,0} - \frac{B}{A}v^2 v^r v^s h_{0r,s} \\
&- \frac{1}{A}(A - \tfrac{1}{2}Bv^2)v^r v^s h_{rs,0} - \tfrac{1}{2}h_{kr,s}v^k v^r v^s \\
&+ \frac{B}{A^2}(\tfrac{1}{2}A_{,0} + A_{,s}v^s + \tfrac{1}{2}B_{,0}v^2)v^2 h_{00} + \left[-\frac{1}{2B}A_{,s} \right. \\
&+ \left(\frac{1}{A}A_{,0} - \frac{1}{B}B_{,0}\right)v^s + \frac{1}{2}\left(\frac{1}{A}A_{,s} + \frac{1}{B}B_{,s}\right)v^2 \\
&+ \left(\frac{2}{A}A_{,r} - \frac{1}{B}B_{,r}\right)v^r v^s + \frac{2}{A}B_{,0}v^2 v^s - \frac{1}{2A}B_{,s}v^4 \\
&\left. \left. + \frac{1}{A}B_{,r}v^r v^s v^2 \right] h_{0s} + \frac{1}{A}(\tfrac{1}{2}A_{,0} + A_{,s}v^s + \tfrac{1}{2}B_{,0}v^2)v^r v^s h_{rs} \right\}
\end{aligned} \tag{4.3.47}$$

$$\begin{aligned}
\nabla_\mu T^{m\mu} =& T^{00}(A - Bv^2 + h_{00} + 2h_{0s}v^s + h_{rs}v^r v^s)^{-1} \\
&\times \Big[(A - Bv^2 + h_{00} + 2v^s h_{0s} + v^r v^s h_{rs}) \\
&\times \left(\ddot{x}^m + \frac{1}{2B}A_{,m} + \frac{1}{B}B_{,0}v^m + \frac{1}{B}B_{,s}v^s v^m - \frac{1}{2B}B_{,m}v^2\right) \\
&- \tfrac{1}{2}(A_{,0} + A_{,s}v^s)v^m + \tfrac{1}{2}(B_{,0} + B_{,s}v^s)v^2 v^m \\
&+ (Bv^s - h_{0s} - v^r h_{rs})v^m \ddot{x}^s - \tfrac{1}{2}v^m(h_{00,0} + v^s h_{00,s}) \\
&- v^m v^s(h_{0s,0} + v^r h_{0s,r}) - \tfrac{1}{2}v^m v^r v^s(h_{rs,0} + v^k h_{rs,k}) \\
&+ \frac{1}{B}(A - Bv^2)[\tfrac{1}{2}h_{00,m} - h_{0m,0} + v^s h_{0s,m} - v^s h_{0m,s} - v^s h_{ms,0} \\
&+ \tfrac{1}{2}v^r v^s h_{rs,m} - v^r v^s h_{mr,s} + \frac{1}{A}(\tfrac{1}{2}A_{,0} + A_{,s}v^s + \tfrac{1}{2}B_{,0}v^2)h_{0m} \\
&+ \frac{1}{B}(\tfrac{1}{2}A_{,s} + B_{,0}v^s - \tfrac{1}{2}B_{,s}v^2 + B_{,r}v^r v^s)h_{ms}] \Big].
\end{aligned} \tag{4.3.48}$$

By equating expressions (4.3.47) and (4.3.48) to zero and combining them one obtains again equations (4.3.38).

The method used here needs to be further developed. First of all, this is concerned with solving the wave equation (4.3.36). Let us add that all these results may serve as an illustration of the facilities of the system GRATOS of tensor operations by computer (Tarasevich *et al* 1987).

4.4 GRAVITATIONAL RADIATION AND MOTION IN A BINARY SYSTEM

4.4.1 Quadrupole formula of the gravitational radiation

For a long time the study of the motion of compact objects and gravitational radiation has presented a severe problem in relativistic celestial mechanics. Until recently there existed only one solution, which was subject to criticism. According to this solution, based on the linearized field equations, the system of gravitating bodies loses energy by radiating gravitational waves. The loss of energy is determined by the expression

$$\frac{\mathrm{d}W}{\mathrm{d}t} = -\frac{G}{5c^5} B_{ik} B_{ik} \tag{4.4.1}$$

with

$$B_{ik} = \frac{\mathrm{d}^3}{\mathrm{d}t^3} D_{ik}(t) \tag{4.4.2}$$

D_{ik} being the quadrupole moments of the system. To be more specific, functions D_{ik} are the coefficients of the expansion of the gravitational potential U of the system of mass M at a large distance from the system

$$U = \frac{GM}{r} + \frac{1}{2} G D_{ik}(t) \left(\frac{1}{r}\right)_{,ik} + \dots. \tag{4.4.3}$$

For the system of point masses

$$U = \sum_A \frac{GM_A}{|\boldsymbol{x} - \boldsymbol{x}_A|} \tag{4.4.4}$$

and then

$$D_{ik}(t) = \sum_A M_A (x^i_A x^k_A - \tfrac{1}{3}\delta_{ik} x^m_A x^m_A). \tag{4.4.5}$$

The derivation of the quadrupole formula of gravitational radiation can be found in many textbooks (see, for instance, Fock 1955). This formula has been applied to calculate the loss of energy in the binary pulsar system PSR 1913+16. Assuming that the motion is performed in the plane $x^3 = 0$ one has for the coordinates of masses M_1 and M_2

$$x_1 = -\frac{M_2}{M} r \cos f \qquad x_2 = \frac{M_1}{M} r \cos f$$

$$y_1 = -\frac{M_2}{M} r \sin f \qquad y_2 = \frac{M_1}{M} r \sin f. \tag{4.4.6}$$

$M = M_1 + M_2$, r and f are the radius vector and the true anomaly of the relative motion. In accordance with (4.4.5)

$$D_{11} = \frac{M_1 M_2}{M} r^2 (\tfrac{1}{6} + \tfrac{1}{2}\cos 2f) \qquad D_{22} = \frac{M_1 M_2}{M} r^2 (\tfrac{1}{6} - \tfrac{1}{2}\cos 2f)$$

$$D_{33} = -\frac{1}{3}\frac{M_1 M_2}{M} r^2 \qquad D_{12} = \frac{1}{2}\frac{M_1 M_2}{M} r^2 \sin 2f. \tag{4.4.7}$$

Differentiating these expressions one finds

$$B_{11} = \frac{M_1 M_2}{M} \frac{n^3 a^4}{r^2(1-e^2)^{1/2}} (\tfrac{13}{6} e \sin f + 4 \sin 2f + \tfrac{3}{2} e \sin 3f)$$

$$B_{22} = -\frac{M_1 M_2}{M} \frac{n^3 a^4}{r^2(1-e^2)^{1/2}} (\tfrac{17}{6} e \sin f + 4 \sin 2f + \tfrac{3}{2} e \sin 3f)$$

$$B_{33} = \frac{M_1 M_2}{M} \frac{n^3 a^4}{r^2(1-e^2)^{1/2}} \frac{2}{3} e \sin f$$

$$B_{12} = -\frac{M_1 M_2}{M} \frac{n^3 a^4}{r^2(1-e^2)^{1/2}} (\tfrac{5}{6} e \cos f + 4 \cos 2f + \tfrac{3}{2} e \cos 3f) \tag{4.4.8}$$

a, e, n being the semi-major axis of the relative orbit, its eccentricity and mean motion, respectively, and $n^2 a^3 = GM$. Therefore,

$$\begin{aligned} \frac{\mathrm{d}W}{\mathrm{d}t} &= -\frac{G}{5c^5}(B_{11}^2 + B_{22}^2 + B_{33}^2 + 2B_{12}^2) \\ &= -\frac{G}{5c^5}\left(\frac{M_1 M_2}{M}\right)^2 \frac{n^6 a^8}{r^4(1-e^2)}(32 + \tfrac{52}{3}e^2 + 64 e \cos f + \tfrac{44}{3} e^2 \cos 2f). \end{aligned} \tag{4.4.9}$$

Using (1.1.20), the Hansen coefficients

$$X_0^{-4,0} = (1 + \tfrac{1}{2}e^2)(1-e^2)^{-5/2} \qquad X_0^{-4,1} = e(1-e^2)^{-5/2}$$
$$X_0^{-4,2} = \tfrac{1}{4}e^2(1-e^2)^{-5/2}$$

one obtains for the averaged value of the loss of energy in motion on the elliptic orbit the well-known formula

$$\left[\frac{\mathrm{d}W}{\mathrm{d}t}\right] = -\frac{32}{5}\frac{G}{c^5}\left(\frac{M_1 M_2}{M}\right)^2 \frac{n^6 a^4}{(1-e^2)^{7/2}}\left(1 + \frac{73}{24}e^2 + \frac{37}{96}e^4\right). \tag{4.4.10}$$

The total energy of the two-body problem being related with the semi-major axis by

$$W = -\frac{G M_1 M_2}{2a} \tag{4.4.11}$$

the gravitational radiation as follows from (4.4.10) leads to a decrease of the semi-major axis with the rate

$$\dot{a} = -\frac{64}{5}\frac{G^3 M_1 M_2 M}{c^5 a^3 (1-e^2)^{7/2}}\left(1 + \frac{73}{24}e^2 + \frac{37}{96}e^4\right). \tag{4.4.12}$$

The quadrupole formula causes doubt in at least two aspects:

(1) its validity in application to compact objects for which the ratio U/c^2 is not small, and

(2) the correctness of the conclusion (4.4.12) as derived from the quadrupole formula.

4.4.2 Equations of motion of compact bodies with consideration of gravitational radiation

To elucidate these questions it is necessary to derive the equations of motion of compact bodies taking into account the radiation terms and by solving them to examine the orbital evolution in the two-body problem by the methods of celestial mechanics. Many authors have advanced the solution of this problem. As already mentioned, culminating contributions to this advance were made by Damour (1983, 1984, 1987a,b), Grishchuk and Kopejkin (1983, 1986) and Kopejkin (1985). These papers involve different techniques and even different initial statements. In the papers by Damour compact objects were examined from the start. By contrast, as stated by Grishchuk and Kopejkin, the compactness parameter disappears from the equations of motion of the 'ordinary' macroscopic bodies making them applicable for the description of the motion of compact bodies, such as black holes. The derivation of the equations of motion taking account of the radiation terms demands too much space to be reproduced here. Considering that all the details may be found in the papers cited above this exposition is restricted to giving without derivation the equations of motion of the two-body problem in harmonic coordinates taking into account the radiation corrections. These equations are of the form

$$\begin{aligned} a_1^i = {} & F_0^i(\boldsymbol{x}_1 - \boldsymbol{x}_2) + c^{-2}F_2^i(\boldsymbol{x}_1 - \boldsymbol{x}_2, \boldsymbol{v}_1, \boldsymbol{v}_2, \boldsymbol{a}_1, \boldsymbol{a}_2) \\ & + c^{-4}F_4^i(\boldsymbol{x}_1 - \boldsymbol{x}_2, \boldsymbol{v}_1, \boldsymbol{v}_2, \boldsymbol{a}_1, \boldsymbol{a}_2, \dot{\boldsymbol{a}}_2, \ddot{\boldsymbol{a}}_2) \\ & + c^{-5}F_5^i(\boldsymbol{x}_1 - \boldsymbol{x}_2, \boldsymbol{v}_1 - \boldsymbol{v}_2, \boldsymbol{a}_1 - \boldsymbol{a}_2, \dot{\boldsymbol{a}}_1 - \ddot{\boldsymbol{a}}_2, \ddot{\boldsymbol{a}}_1 - \ddot{\boldsymbol{a}}_2, \dddot{\boldsymbol{a}}_1 - \dddot{\boldsymbol{a}}_2) + \ldots . \end{aligned} \tag{4.4.13}$$

$\boldsymbol{x}_i$, $\boldsymbol{v}_i$ and $\boldsymbol{a}_i$ are vectors of position, velocity and acceleration, respectively, for the body i ($i = 1, 2$). The components a_2^i of the acceleration of the second body satisfy analogous equations. The post-Newtonian equations considered in section 4.1 correspond to retaining in (4.4.13) only F_0^i

and F_2^i and replacing in F_2^i the accelerations $\boldsymbol{a}_1$, $\boldsymbol{a}_2$ by their Newtonian expressions. Including the post-post-Newtonian terms F_4^i does not change the conservative form of the equations and does not prevent their presentation in the Lagrange form. Just the terms F_5^i of dissipative character are responsible for the loss of energy of the system due to the gravitational radiation. Eliminating in the right-hand members of (4.4.13) the accelerations and their derivatives on the basis of Newtonian and post-Newtonian equations one obtains the reduced equations with the right-hand members dependent only on coordinates and velocities. These reduced equations have the form

$$\begin{aligned}\ddot{x}_1^i =& A_0^i(\boldsymbol{x}_1-\boldsymbol{x}_2)+c^{-2}A_2^i(\boldsymbol{x}_1-\boldsymbol{x}_2,\boldsymbol{v}_1,\boldsymbol{v}_2)+c^{-4}A_4^i(\boldsymbol{x}_1-\boldsymbol{x}_2,\boldsymbol{v}_1,\boldsymbol{v}_2)\\ &+c^{-5}A_5^i(\boldsymbol{x}_1-\boldsymbol{x}_2,\boldsymbol{v}_1-\boldsymbol{v}_2)+\dots\end{aligned} \tag{4.4.14}$$

and similarly for the second body. Denoting

$$r=[(\boldsymbol{x}_1-\boldsymbol{x}_2)^2]^{1/2} \qquad N^i=r^{-1}(x_1^i-x_2^i) \qquad v^i=v_1^i-v_2^i \tag{4.4.15}$$

one has for the functions occurring in the right-hand member (4.4.14)

$$A_0^i=-GM_2r^{-2}N^i \tag{4.4.16}$$

$$\begin{aligned}A_2^i =& GM_2r^{-2}\{N^i[-\boldsymbol{v}_1^2-2\boldsymbol{v}_2^2+4(\boldsymbol{v}_1\boldsymbol{v}_2)+\tfrac{3}{2}(\boldsymbol{N}\boldsymbol{v}_2)^2\\ &+(5GM_1+4GM_2)r^{-1}]+v^i(4\boldsymbol{N}\boldsymbol{v}_1-3\boldsymbol{N}\boldsymbol{v}_2)\}\end{aligned} \tag{4.4.17}$$

$$\begin{aligned}A_4^i =& GM_2r^{-2}\Bigg[N^i\Bigg(-2\boldsymbol{v}_2^4+4\boldsymbol{v}_2^2(\boldsymbol{v}_1\boldsymbol{v}_2)-2(\boldsymbol{v}_1\boldsymbol{v}_2)^2+\tfrac{3}{2}\boldsymbol{v}_1^2(\boldsymbol{N}\boldsymbol{v}_2)^2\\ &+\tfrac{9}{2}\boldsymbol{v}_2^2(\boldsymbol{N}\boldsymbol{v}_2)^2-6(\boldsymbol{v}_1\boldsymbol{v}_2)(\boldsymbol{N}\boldsymbol{v}_2)^2-\tfrac{15}{8}(\boldsymbol{N}\boldsymbol{v}_2)^4\\ &+\frac{GM_1}{r}[-\tfrac{15}{4}\boldsymbol{v}_1^2+\tfrac{5}{4}\boldsymbol{v}_2^2-\tfrac{5}{2}\boldsymbol{v}_1\boldsymbol{v}_2\\ &+\tfrac{39}{2}(\boldsymbol{N}\boldsymbol{v}_1)^2-39(\boldsymbol{N}\boldsymbol{v}_1)(\boldsymbol{N}\boldsymbol{v}_2)+\tfrac{17}{2}(\boldsymbol{N}\boldsymbol{v}_2)^2]\\ &+\frac{GM_2}{r}[4\boldsymbol{v}_2^2-8\boldsymbol{v}_1\boldsymbol{v}_2+2(\boldsymbol{N}\boldsymbol{v}_1)^2-4(\boldsymbol{N}\boldsymbol{v}_1)(\boldsymbol{N}\boldsymbol{v}_2)-6(\boldsymbol{N}\boldsymbol{v}_2)^2]\\ &+\frac{G^2}{r^2}(-\tfrac{57}{4}M_1^2-9M_2^2-\tfrac{69}{2}M_1M_2)\Bigg)\\ &+v^i\Bigg(\boldsymbol{v}_1^2(\boldsymbol{N}\boldsymbol{v}_2)+4\boldsymbol{v}_2^2(\boldsymbol{N}\boldsymbol{v}_1)-5\boldsymbol{v}_2^2(\boldsymbol{N}\boldsymbol{v}_2)-4(\boldsymbol{v}_1\boldsymbol{v}_2)(\boldsymbol{N}\boldsymbol{v}_1)\\ &+4(\boldsymbol{v}_1\boldsymbol{v}_2)(\boldsymbol{N}\boldsymbol{v}_2)-6(\boldsymbol{N}\boldsymbol{v}_1)(\boldsymbol{N}\boldsymbol{v}_2)^2+\tfrac{9}{2}(\boldsymbol{N}\boldsymbol{v}_2)^3\\ &+\frac{GM_1}{r}(-\tfrac{63}{4}\boldsymbol{N}\boldsymbol{v}_1+\tfrac{55}{4}\boldsymbol{N}\boldsymbol{v}_2)+\frac{GM_2}{r}(-2\boldsymbol{N}\boldsymbol{v}_1-2\boldsymbol{N}\boldsymbol{v}_2)\Bigg)\Bigg]\end{aligned} \tag{4.4.18}$$

$$A_5^i = \tfrac{4}{5}G^2 M_1 M_2 r^{-3}\left[N^i(\boldsymbol{Nv})\left(3v^2 - 6\frac{GM_1}{r} + \frac{52}{3}\frac{GM_2}{r}\right)\right.$$
$$\left. + v^i\left(-v^2 + 2\frac{GM_1}{r} - 8\frac{GM_2}{r}\right)\right]. \tag{4.4.19}$$

Thus, it remains to deal with the purely celestial mechanics problem of studying the motion of bodies in accordance with equations (4.4.14).

4.4.3 Relative motion in a binary system

As shown in the papers cited above and as may be verified by direct calculation, equations (4.4.14) of the two-body problem admit in the approximation under discussion the integral of the centre of mass motion of the form

$$C^i = \tilde{M}_1 x_1^i + \tilde{M}_2 x_2^i - \tfrac{7}{4}c^{-4}GM_1M_2(\boldsymbol{Nv}_1 + \boldsymbol{Nv}_2)v^i$$
$$- \frac{4}{5}\frac{GM_1M_2}{c^5 M}(M_1 - M_2)\left(v^2 - \frac{2GM}{r}\right)v^i \tag{4.4.20}$$

with

$$\tilde{M}_1 = M_1 + c^{-2}\left(\tfrac{1}{2}M_1 v_1^2 - \frac{GM_1M_2}{2r}\right) + c^{-4}\left[\tfrac{3}{8}M_1 v_1^4 + \frac{GM_1M_2}{r}\right.$$
$$\times\left(\tfrac{19}{8}v_1^2 - \tfrac{7}{8}v_2^2 - \tfrac{7}{4}v_1v_2 - \tfrac{1}{8}(\boldsymbol{Nv}_1)^2 + \tfrac{1}{8}(\boldsymbol{Nv}_2)^2\right.$$
$$\left.\left. - \tfrac{1}{4}(\boldsymbol{Nv}_1)(\boldsymbol{Nv}_2) - \frac{5}{4}\frac{GM_1}{r} + \frac{7}{4}\frac{GM_2}{r}\right)\right] \tag{4.4.21}$$

and similarly for $\tilde{M}_2$. C^i are the linear functions of time with coefficients representing the constants of the integrals of momentum and the centre of mass. Choosing the barycentric system implying $C^i = 0$ and defining $x^i = x_1^i - x_2^i$ one has

$$x_1^i = \frac{M_2}{M}x^i + c^{-2}(M_1 - M_2)\frac{M_1M_2}{2M^3}\left(v^2 - \frac{GM}{r}\right)x^i + \ldots \tag{4.4.22}$$

$$x_2^i = -\frac{M_1}{M}x^i + c^{-2}(M_1 - M_2)\frac{M_1M_2}{2M^3}\left(v^2 - \frac{GM}{r}\right)x^i + \ldots \tag{4.4.23}$$

$$v_1^i = \frac{M_2}{M}v^i + c^{-2}(M_1 - M_2)\frac{M_1M_2}{2M^3}\left[\left(v^2 - \frac{GM}{r}\right)v^i - \frac{GM}{r}(\boldsymbol{Nv})N^i\right] + \ldots \tag{4.4.24}$$

$$v_2^i = -\frac{M_1}{M}v^i + c^{-2}(M_1 - M_2)\frac{M_1 M_2}{2M^3} \times \left[\left(\boldsymbol{v}^2 - \frac{GM}{r}\right)v^i - \frac{GM}{r}(\boldsymbol{N}\boldsymbol{v})N^i\right] + \dots \tag{4.4.25}$$

Taking the difference of equations (4.4.14) and substituting (4.4.24), (4.4.25) into the right-hand members one obtains the equations of relative motion

$$\ddot{x}^i = B_0^i + c^{-2}B_2^i + c^{-4}B_4^i + c^{-5}B_5^i + \dots \tag{4.4.26}$$

with

$$B_0^i = -\frac{GM}{r^3}x^i \tag{4.4.27}$$

$$B_2^i = \frac{GM}{r^3}\left\{x^i\left[\left(4 + 2\frac{M_1M_2}{M^2}\right)\frac{GM}{r} + \frac{3}{2}\frac{M_1M_2}{M^2}(\boldsymbol{N}\boldsymbol{v})^2 - \left(1 + 3\frac{M_1M_2}{M^2}\right)\boldsymbol{v}^2\right] + \left(4 - 2\frac{M_1M_2}{M^2}\right)(x^kv^k)v^i\right\} \tag{4.4.28}$$

$$\begin{aligned}
B_4^i = \frac{GM}{r^3}\Bigg\{ & x^i\Bigg[\frac{M_1M_2}{M^2}\left(-3 + 4\frac{M_1M_2}{M^2}\right)\boldsymbol{v}^4 + \frac{15}{8}\frac{M_1M_2}{M^2} \\
& \times\left(-1 + 3\frac{M_1M_2}{M^2}\right)(\boldsymbol{N}\boldsymbol{v})^4 + \frac{M_1M_2}{M^2}\left(\frac{9}{2} - 6\frac{M_1M_2}{M^2}\right)\boldsymbol{v}^2(\boldsymbol{N}\boldsymbol{v})^2 \\
& + \frac{M_1M_2}{M^2}\left(\frac{13}{2} - 2\frac{M_1M_2}{M^2}\right)\frac{GM}{r}\boldsymbol{v}^2 \\
& + \left(2 + 25\frac{M_1M_2}{M^2} + 2\frac{M_1^2M_2^2}{M^4}\right)\frac{GM}{r}(\boldsymbol{N}\boldsymbol{v})^2 \\
& - \left(9 + \frac{87}{4}\frac{M_1M_2}{M^2}\right)\frac{G^2M^2}{r^2}\Bigg] \\
& + v^i(x^kv^k)\Bigg[\frac{M_1M_2}{M^2}\left(\frac{15}{2} + 2\frac{M_1M_2}{M^2}\right)\boldsymbol{v}^2 \\
& - \frac{M_1M_2}{M^2}\left(\frac{9}{2} + 3\frac{M_1M_2}{M^2}\right)(\boldsymbol{N}\boldsymbol{v})^2 \\
& - \left(2 + \frac{41}{2}\frac{M_1M_2}{M^2} + 4\frac{M_1^2M_2^2}{M^4}\right)\frac{GM}{r}\Bigg]\Bigg\}
\end{aligned} \tag{4.4.29}$$

$$B_5^i = \frac{8}{5}\frac{G^2M_1M_2}{r^3}\left[\frac{1}{r^2}\left(3\boldsymbol{v}^2 + \frac{17}{3}\frac{GM}{r}\right)(x^kv^k)x^i - \left(\boldsymbol{v}^2 + 3\frac{GM}{r}\right)v^i\right]. \tag{4.4.30}$$

The terms B_0^i represent the Newtonian part of the equations of the two-body problem. B_2^i are the post-Newtonian terms. Solution of (4.4.26) in the post-Newtonian approximation, i.e. taking into account only B_0^i and B_2^i, is obtained as the solution of equation (1.1.25) with the right-hand member (3.1.100) for the values

$$\alpha = \frac{1}{2}\frac{M_1M_2}{M^2} \qquad \sigma = 2 + \frac{M_1M_2}{M^2} \qquad 2\epsilon = 1 + 3\frac{M_1M_2}{M^2} \qquad \mu = 2 - \frac{M_1M_2}{M^2}. \tag{4.4.31}$$

This solution is given by (3.1.102)–(3.1.111). Similar solutions have been produced by many authors (for example, Damour and Deruelle 1985, Soffel *et al* 1987a). B_4^i are the post-post-Newtonian terms. The post-post-Newtonian solution caused by these terms and not differing essentially from the post-Newtonian solution has been examined in detail in Damour and Schäfer (1988). Post-post-Newtonian equations of motion of the two-body problem may still be presented in the Lagrange form. The appropriate Lagrangian depends not only on the coordinates and velocities but on the accelerations as well (Damour *et al* 1989). B_5^i are the radiation terms due to the gravitational radiation of the two-body problem.

4.4.4 Spin–orbital terms

Before examining the radiation terms B_5^i it is suitable to investigate the orbital evolution in the two-body problem accounting for the proper rotations of the bodies. Retaining in the post-Newtonian equations (4.1.24) the spin terms they give the following contributions to B_2^i:

$$\Delta B_2^i = \frac{GM}{r^3}\left(-\frac{1}{2}[(\boldsymbol{S} - 3(\boldsymbol{NS})\boldsymbol{N}) \times \boldsymbol{v}]^i - \frac{3}{r}(\boldsymbol{NS}_2)S_1^i - \frac{3}{r}(\boldsymbol{NS}_1)S_2^i \right.$$
$$\left. - \frac{3}{r^2}[\boldsymbol{S}_1\boldsymbol{S}_2 - 5(\boldsymbol{NS}_1)(\boldsymbol{NS}_2)]x^i\right). \tag{4.4.32}$$

$\boldsymbol{S}_1$ and $\boldsymbol{S}_2$ are the proper angular momenta of the bodies divided by the appropriate masses (the spin vectors)

$$\boldsymbol{S}_k = \frac{2}{M_k}I_k\boldsymbol{\omega}_k \qquad \boldsymbol{S} = \left(3 + \frac{M_1}{M}\right)\boldsymbol{S}_1 + \left(3 + \frac{M_2}{M}\right)\boldsymbol{S}_2. \tag{4.4.33}$$

$\boldsymbol{\omega}_k$ are the angular velocities of the bodies, I_k are their moments of inertia (in the sense of the definition (4.2.40)). The terms due to the spin interaction are retained here but the terms quadratic in each spin are omitted. The solution of the post-Newtonian two-body problem considering B_2^i and ΔB_2^i is given here within the framework of the secular perturbation theory.

In the post-Newtonian approximation the right-hand member of the equations of the perturbed two-body problem admits the disturbing function (1.1.34)

$$R = c^{-2}\left\{\frac{1}{8}\left(1 - 3\frac{M_1M_2}{M^2}\right)\boldsymbol{v}^4 + \frac{GM}{r}\left[\frac{1}{2}\left(3 + \frac{M_1M_2}{M^2}\right)\boldsymbol{v}^2 + \frac{1}{2}\frac{M_1M_2}{M^2}(\boldsymbol{Nv})^2 - \frac{GM}{2r^2}\right] + \frac{GM}{r^3}\{\tfrac{1}{2}\boldsymbol{S}(\boldsymbol{v}\times\boldsymbol{r}) + (\boldsymbol{S}_1\boldsymbol{S}_2)[1 - 3(\boldsymbol{NS}_1)(\boldsymbol{NS}_2)]\}\right\}. \quad (4.4.34)$$

The first-order secular perturbations of the osculating elements are determined by equaitons (1.1.44). Substituting into (4.4.34) the solution (1.1.9) of the two-body problem one obtains

$$R = GM\left[-\frac{M_1M_2}{M^2}\frac{GMp}{2r^3} + \left(3 + \frac{1}{2}\frac{M_1M_2}{M^2}\right)\frac{GM}{r^2} + \left(-2 + \frac{1}{2}\frac{M_1M_2}{M^2}\right)\frac{GM}{ra} + \left(1 - 3\frac{M_1M_2}{M^2}\right)\frac{GM}{8a^2} - (GMp)^{1/2}\frac{\boldsymbol{Sk}}{2r^3} - \frac{1}{2r^3}[\boldsymbol{S}_1\boldsymbol{S}_2 - 3(\boldsymbol{S}_1\boldsymbol{k})(\boldsymbol{S}_2\boldsymbol{k})] + \tfrac{3}{2}[(\boldsymbol{S}_1\boldsymbol{Q})(\boldsymbol{S}_2\boldsymbol{Q}) - (\boldsymbol{S}_1\boldsymbol{P})(\boldsymbol{S}_2\boldsymbol{P})]\frac{\cos 2f}{r^3} - \tfrac{3}{2}[(\boldsymbol{S}_1\boldsymbol{P})(\boldsymbol{S}_2\boldsymbol{Q}) + (\boldsymbol{S}_1\boldsymbol{Q})(\boldsymbol{S}_2\boldsymbol{P})]\frac{\sin 2f}{r^3}\right]. \quad (4.4.35)$$

With Hansen coefficients

$$X_0^{-1,0} = 1 \qquad X_0^{-2,0} = (1-e^2)^{-1/2} \qquad X_0^{-3,0} = (1-e^2)^{-3/2} \qquad X_0^{-3,2} = 0$$

there results

$$[R] = GM\left[\frac{3GM}{a^2(1-e^2)^{1/2}} + \left(-15 + \frac{M_1M_2}{M^2}\right)\frac{GM}{8a^2} - \frac{\sqrt{GM}\,\boldsymbol{Sk}}{2a^{5/2}(1-e^2)} - \frac{1}{2a^3(1-e^2)^{3/2}}[\boldsymbol{S}_1\boldsymbol{S}_2 - 3(\boldsymbol{S}_1\boldsymbol{k})(\boldsymbol{S}_2\boldsymbol{k})]\right]. \quad (4.4.36)$$

In addition, from (4.4.34) it follows that

$$\dot{\boldsymbol{r}}\frac{\partial R}{\partial\dot{\boldsymbol{r}}} = GM\left[\left(1 - 3\frac{M_1M_2}{M^2}\right)\frac{\dot{\boldsymbol{r}}^4}{2GM} + (3 + M_1M_2)\frac{\dot{\boldsymbol{r}}^2}{r} + \frac{M_1M_2}{M^2}\frac{(\boldsymbol{r}\dot{\boldsymbol{r}})^2}{r^3} - (GMp)^{1/2}\frac{\boldsymbol{Sk}}{2r^3}\right] \quad (4.4.37)$$

$$\left[\dot{r}\frac{\partial R}{\partial \dot{r}}\right] = GM\left[\left(8 - 3\frac{M_1 M_2}{M^2}\right)\frac{GM}{a^2(1-e^2)^{1/2}}\right.$$

$$\left. + \frac{1}{2}\left(-9 + 5\frac{M_1 M_2}{M^2}\right)\frac{GM}{a^2} - \frac{\sqrt{GM}\,\boldsymbol{S}\boldsymbol{k}}{2a^{5/2}(1-e^2)}\right]. \quad (4.4.38)$$

As seen from (4.4.36), $[R]$ depends only on the elements a, e, i and Ω. Therefore, equations (1.1.44) imply first of all

$$\frac{\mathrm{d}a}{\mathrm{d}t} = 0 \qquad \frac{\mathrm{d}e}{\mathrm{d}t} = 0. \quad (4.4.39)$$

The equations for the inclination and the longitude of node are reduced to the autonomous canonical system with one degree of freedom:

$$\frac{\mathrm{d}\cos i}{\mathrm{d}t} = \frac{1}{na^2(1-e^2)^{1/2}}\frac{\partial[R]}{\partial\Omega}$$

$$\frac{\mathrm{d}\Omega}{\mathrm{d}t} = -\frac{1}{na^2(1-e^2)^{1/2}}\frac{\partial[R]}{\partial\cos i} \quad (4.4.40)$$

or in more detail

$$\frac{\mathrm{d}i}{\mathrm{d}t} = GM\left(\frac{\boldsymbol{S}\boldsymbol{l}}{2a^3(1-e^2)^{3/2}} - \frac{3}{2}\frac{(\boldsymbol{S}_1\boldsymbol{k})(\boldsymbol{S}_2\boldsymbol{l}) + (\boldsymbol{S}_2\boldsymbol{k})(\boldsymbol{S}_1\boldsymbol{l})}{\sqrt{GM}a^{7/2}(1-e^2)^2}\right)$$

$$\sin i\frac{\mathrm{d}\Omega}{\mathrm{d}t} = GM\left(\frac{\boldsymbol{S}\boldsymbol{m}}{2a^3(1-e^2)^{3/2}} - \frac{3}{2}\frac{(\boldsymbol{S}_1\boldsymbol{k})(\boldsymbol{S}_2\boldsymbol{m}) + (\boldsymbol{S}_2\boldsymbol{k})(\boldsymbol{S}_1\boldsymbol{m})}{\sqrt{GM}a^{7/2}(1-e^2)^2}\right). \quad (4.4.41)$$

This system has the integral

$$[R] = \text{constant} \quad (4.4.42)$$

and may be solved by quadratures. From (4.4.42) it can be seen that $\cos i$ is a periodic function of Ω. In accordance with the general theory of solution of systems like (4.4.40) the longitude of node in its turn is represented by a trigonometric series in multiples of some angular variable linearly related to time. Depending on the coefficients of the function (4.4.36) this series may contain a term proportional to time. The cases of presence or absence of such a term correspond to the rotation or the libration of the line of node respectively.

Actual integration of the system (4.4.40) presents no difficulties. Choosing the equatorial plane of the body of mass M_2 as the reference plane one has $S_2^1 = S_2^2 = 0$, $S_2^3 = S_2$ (the lower index numerates the bodies, the upper index identifies the vector components). Considering that a and e are constant, from (4.4.42) it follows that

$$\sin i\left[\left(3+\frac{M_1}{M}\right)(GMp)^{1/2}-3S_2\cos i\right](-S_1^1\sin\Omega+S_1^2\cos\Omega)$$
$$=C+(GMp)^{1/2}S^3\cos i-3S_1^3S_2\cos^2 i \qquad (4.4.43)$$

C being an arbitrary constant. The relation

$$S_1^1\cos\Omega+S_1^2\sin\Omega=[(S_1^1)^2+(S_1^2)^2-(S_1^2\cos\Omega-S_1^1\sin\Omega)^2]^{1/2}$$

enables one to exclude Ω from the first of equations (4.4.40). The result is

$$\frac{\mathrm{d}\cos i}{\mathrm{d}t}=\pm\frac{\sqrt{GM}}{2c^2a^{7/2}(1-e^2)^2}(a_0\cos^4 i+4a_1\cos^3 i$$
$$+6a_2\cos^2 i+4a_3\cos i+a_4)^{1/2} \qquad (4.4.44)$$

with

$$a_0=-9(S_2)^2(\boldsymbol{S}_1)^2$$

$$4a_1=6(GMp)^{1/2}S_2\left[S^3S_1^3+\left(3+\frac{M_1}{M}\right)[(S_1^1)^2+(S_1^2)^2]\right]$$

$$6a_2=\left[9(S_2)^2-\left(3+\frac{M_1}{M}\right)^2 GMp\right][(S_1^1)^2+(S_1^2)^2]-GMp(S^3)^2+6CS_1^3S_2$$

$$4a_3=-2(GMp)^{1/2}\left[CS^3+3\left(3+\frac{M_1}{M}\right)S_2[(S_1^1)^2+(S_1^2)^2]\right]$$

$$a_4=\left(3+\frac{M_1}{M}\right)^2 GMp[(S_1^1)^2+(S_1^2)^2]-C^2.$$

Thus, $\cos i$ in the general case is expressed in the known manner in terms of the elliptic functions of Weierstrass. Then Ω is determined from the trigonometric equations (4.4.43). Integration of the two last equations of (1.1.44) is performed by simple quadratures:

$$\frac{\mathrm{d}\omega}{\mathrm{d}t}=-\cos i\frac{\mathrm{d}\Omega}{\mathrm{d}t}+GM\left(\frac{3\sqrt{GM}}{a^{5/2}(1-e^2)}-\frac{\boldsymbol{S}\boldsymbol{k}}{a^3(1-e^2)^{3/2}}\right.$$
$$\left.-\frac{3}{2\sqrt{GM}a^{7/2}(1-e^2)^2}[\boldsymbol{S}_1\boldsymbol{S}_2-3(\boldsymbol{S}_1\boldsymbol{k})(\boldsymbol{S}_2\boldsymbol{k})]\right) \qquad (4.4.45)$$

$$\frac{dl_0}{dt} = GM\left[\left(-15 + 9\frac{M_1M_2}{M^2}\right)\frac{\sqrt{GM}}{a^{5/2}(1-e^2)^{1/2}} + \left(6 - 7\frac{M_1M_2}{M^2}\right)\frac{\sqrt{GM}}{a^{5/2}}\right.$$
$$\left. - \frac{3}{2\sqrt{GM}a^{7/2}(1-e^2)^{3/2}}[\boldsymbol{S}_1\boldsymbol{S}_2 - 3(\boldsymbol{S}_1\boldsymbol{k})(\boldsymbol{S}_2\boldsymbol{k})]\right]. \tag{4.4.46}$$

In the particular case when a body of mass M_1 is a satellite with a negligibly small spin one obtains again the results of section 3.3.3.

Using equations (1.1.26), (1.1.27) and expressions (4.4.28), (4.4.32) for $\boldsymbol{F}$ it may be possible to derive equations (4.4.39), (4.4.40), (4.4.45) and (4.4.46) without using the disturbing function. For example, the use of (1.1.29) enables one to write the equations determining the secular variations of the area vector and the Laplace vector (Barker and O'Connell 1976)

$$\dot{\boldsymbol{c}} = \boldsymbol{\Omega} \times \boldsymbol{c} \qquad \dot{\boldsymbol{f}} = \boldsymbol{\Omega} \times \boldsymbol{f} \tag{4.4.47}$$

with

$$\boldsymbol{\Omega} = GM\left(\frac{3n}{a(1-e^2)}\boldsymbol{k} + \frac{1}{2a^3(1-e^2)^{3/2}}[\boldsymbol{S} - 3(\boldsymbol{S}\boldsymbol{k})\boldsymbol{k}]\right.$$
$$\left. - \frac{3}{2na^5(1-e^2)^2}[(\boldsymbol{S}_2\boldsymbol{k})\boldsymbol{S}_1 + (\boldsymbol{S}_1\boldsymbol{k})\boldsymbol{S}_2 + (\boldsymbol{S}_1\boldsymbol{S}_2 - 5(\boldsymbol{S}_1\boldsymbol{k})(\boldsymbol{S}_2\boldsymbol{k}))\boldsymbol{k}]\right). \tag{4.4.48}$$

This is equivalent to equations (4.4.39), (4.4.40) and (4.4.45). But the technique applied above gives immediately the canonical equations facilitating the solution of the problem.

4.4.5 Radiation terms

Solution of equation (4.4.26) in all details is rather time-consuming due to the necessity to treat not only terms linear in B_4^i but also terms that are quadratic in B_2^i. It might be suitable to take here as the intermediary the post-Newtonian solution with B_0^i and B_2^i and then by variation of arbitrary constants to take linearly into account B_4^i and B_5^i as suggested in Ashby (1986). But the post-post-Newtonian solution caused by B_4^i and the quadratic contribution by B_2^i does not differ essentially from the post-Newtonian solution (4.4.39), (4.4.40), (4.4.45) and (4.4.46). The influence of the terms B_5^i is of more interest. By (1.1.29) applied to $F^i = B_5^i$ one finds

$$\dot{\boldsymbol{c}} = -\frac{8}{5}\frac{G^2M_1M_2}{c^5r^3}\left(v^2 + 3\frac{GM}{r}\right)\boldsymbol{c} \tag{4.4.49}$$

$$\dot{\boldsymbol{f}} = \frac{8}{5}\frac{G^2 M_1 M_2}{c^5 r^3}\left\{\left[\left(3\boldsymbol{v}^2 + \frac{17}{3}\frac{GM}{r}\right)\frac{(\boldsymbol{r}\boldsymbol{v})^2}{r^2} - 2\left(\boldsymbol{v}^2 + 3\frac{GM}{r}\right)\boldsymbol{v}^2\right]\boldsymbol{r}\right.$$
$$\left. + \left(-\boldsymbol{v}^2 + \frac{1}{3}\frac{GM}{r}\right)(\boldsymbol{r}\boldsymbol{v})\boldsymbol{v}\right\}. \qquad (4.4.50)$$

Substitution of the Keplerian values for the coordinates and velocities results in

$$\dot{\boldsymbol{c}} = \frac{8}{5}\frac{G^3 M M_1 M_2}{c^5 a^4}\left(\frac{a}{r}\right)^3\left(5\frac{a}{r} - 1\right)\boldsymbol{c} \qquad (4.4.51)$$

$$\dot{\boldsymbol{f}} = \frac{8}{5}\frac{G^4 M^2 M_1 M_2}{c^5 a^4 (1-e^2)}\left(\frac{a}{r}\right)^3\left(\boldsymbol{P}\left\{e\left(\frac{5}{6} - \frac{1}{2}\frac{r}{a}\right)\right.\right.$$
$$+ \left[14 - \frac{133}{12}e^2 - 20\frac{p}{r} + \left(-2 + \frac{5}{4}e^2\right)\frac{r}{a}\right]\cos f$$
$$\left. + e\left(\frac{1}{2}\frac{r}{a} - \frac{5}{6}\right)\cos 2f + e^2\left(\frac{3}{4}\frac{r}{a} - \frac{35}{12}\right)\cos 3f\right\}$$
$$+ \boldsymbol{Q}\left\{\left[14 - \frac{83}{12}e^2 - 20\frac{p}{r} + \left(-2 + \frac{3}{4}e^2\right)\frac{r}{a}\right]\sin f\right.$$
$$\left.\left. + e\left(\frac{1}{2}\frac{r}{a} - \frac{5}{6}\right)\sin 2f + e^2\left(\frac{3}{4}\frac{r}{a} - \frac{35}{12}\right)\sin 3f\right\}\right). \qquad (4.4.52)$$

Performing averaging with the aid of the Hansen coefficients (1.1.20) one obtains the equations for the secular perturbations

$$\dot{\boldsymbol{c}} = -\frac{4}{5}\frac{G^3 M M_1 M_2}{c^5 a^4}\frac{8 + 7e^2}{(1-e^2)^{5/2}}\boldsymbol{c}. \qquad (4.4.53)$$

$$\dot{\boldsymbol{c}} = -\frac{1}{15}\frac{G^4 M^2 M_1 M_2}{c^5 a^4}\frac{(304 + 121e^2)e}{(1-e^2)^{5/2}}\boldsymbol{P}. \qquad (4.4.54)$$

Scalar products of (4.4.53) and $\boldsymbol{l}$, $\boldsymbol{m}$, $\boldsymbol{k}$ yield respectively

$$\frac{di}{dt} = 0 \qquad \frac{d\Omega}{dt} = 0 \qquad (4.4.55)$$

$$\frac{dp}{dt} = -\frac{8}{5}\frac{G^3 M M_1 M_2}{c^5 a^3}\frac{8 + 7e^2}{(1-e^2)^{3/2}}. \qquad (4.4.56)$$

Similarly, multiplying (4.4.54) by $\boldsymbol{Q}$ and $\boldsymbol{P}$ gives

$$\frac{d\omega}{dt} = 0 \qquad (4.4.57)$$

$$\frac{de}{dt} = -\frac{1}{15}\frac{G^3MM_1M_2}{c^5a^4}\frac{304+121e^2}{(1-e^2)^{5/2}}e \tag{4.4.58}$$

Combination of (4.4.56) and (4.4.58) leads again to (4.4.12). Thus, due to the gravitational radiation the expressions for semi-major axis and eccentricity contain secular terms governing the evolution of motion in the two-body problem. The expression for the mean longitude contains a term quadratic with respect to time and involving a secular decrease of the period of motion.

4.4.6 Motion and radiation

Consideration of the gravitational radiation affects significantly the evolution of motion. Along with this, celestial mechanics treatment of the gravitational radiation is of importance to elucidate the general structure of the GRT equations of motion. The unreduced equations (4.4.13) containing in their right-hand members the derivatives of order from 2 to 5 inclusively clearly demonstrate that their general structure remains to be clarified. Do the iteration methods of constructing the right-hand members of (4.4.13) converge? For any fixed order $i\,(i \geq 4)$ equations (4.4.13) are the differential equations with smallest parameter at the highest derivative. How rich is the variety of their solutions in comparison with the solutions of the reduced equations (4.4.14) obtained by removing the higher order derivatives with the aid of the lower order approximations? The founders of celestial mechanics believed that it would be possible to calculate the motion of all bodies in the Universe provided that their positions and velocities were given for the initial moment of time. This is not true for the unreduced equations (4.4.13). What is the correct statement of the Cauchy problem for these equations? Some of these questions are discussed in the papers by Damour but it is clear that in relativistic celestial mechanics there remain many interesting unsolved problems.

5

Equations of Motion of Solar System Bodies

5.1 EQUATIONS OF MOTION OF EARTH'S ARTIFICIAL SATELLITES

5.1.1 Barycentric equations

We now proceed to the formulation of specific equations of motion of the Solar System bodies. Let us start with the equations of motion of Earth's artificial satellites. The corresponding results are valid for any satellites of the planets. It is to be noted that the Newtonian parts of the equations may be given here only within the accuracy necessary for the derivation of the relativistic parts. The extension of the Newtonian parts to the level of modern accuracy is performed without difficulty. BRS satellite equations are given by (4.1.21). The BRS equations of motion of the Earth are given by (4.1.24). The difference of these equations yields the BRS satellite equations in relative coordinates. Expanding the regular parts of the potentials $\bar{U}_E$, $\bar{U}^i_E$, $\bar{W}_E$ as functions of $\boldsymbol{x}$ in series in powers of $\boldsymbol{r}_E = \boldsymbol{x} - \boldsymbol{x}_E$, one has

$$\ddot{r}^i_E = U_{E,i} - H^i_E + \bar{U}_{E,ik}(\boldsymbol{x}_E) r^k_E + \frac{1}{2}\bar{U}_{E,ikm}(\boldsymbol{x}_E) r^k_E r^m_E + \ldots + c^{-2}(G^i - \bar{G}^i) \tag{5.1.1}$$

with

$$\begin{aligned} G^i - \bar{G}^i = & - 4U_E U_{E,i} - 3\dot{U}_E \dot{x}^i - U_{E,k}\dot{x}^k\dot{x}^i + U_{E,i}\dot{x}^k\dot{x}^k \\ & + 4\dot{U}^i_E - 4U^k_{E,i}\dot{x}^k + W_{E,i} - 4U_E\bar{U}_{E,i}(\boldsymbol{x}_E) - 4U_{E,i}\bar{U}_E(\boldsymbol{x}_E) \\ & - 3c\bar{U}_{E,0}(\boldsymbol{x}_E)\dot{r}^i_E - 4\bar{U}_{E,k}(\boldsymbol{x}_E)(\dot{x}^k\dot{x}^i - v^k_E v^i_E) \\ & + \bar{U}_{E,i}(\boldsymbol{x}_E)(\dot{x}^k\dot{x}^k - v^2_E) + 4\bar{U}^i_{E,k}(\boldsymbol{x}_E)\dot{r}^k_E - 4\bar{U}^k_{E,i}(\boldsymbol{x}_E)\dot{r}^k_E \\ & + [-4U_E\bar{U}_{E,ik}(\boldsymbol{x}_E) - 4U_{E,i}\bar{U}_{E,k}(\boldsymbol{x}_E) - 4\bar{U}_E(\boldsymbol{x}_E)\bar{U}_{E,ik}(\boldsymbol{x}_E) \\ & - 4\bar{U}_{E,i}(\boldsymbol{x}_E)\bar{U}_{E,k}(\boldsymbol{x}_E) + \bar{W}_{E,ik}(\boldsymbol{x}_E) \\ & - 3c\bar{U}_{E,0k}(\boldsymbol{x}_E)\dot{x}^i - 4\bar{U}_{E,km}(\boldsymbol{x}_E)\dot{x}^m\dot{x}^i \end{aligned}$$

$$+ \bar{U}_{E,ik}(\boldsymbol{x}_E)\dot{x}^m\dot{x}^m + 4c\bar{U}^i_{E,0k}(\boldsymbol{x}_E) + 4\bar{U}^i_{E,km}(\boldsymbol{x}_E)\dot{x}^m$$
$$- 4\bar{U}^m_{E,ik}(\boldsymbol{x}_E)\dot{x}^m]r^k_E - 2U_{E,i}\bar{U}_{E,km}(\boldsymbol{x}_E)r^k_E r^m_E + \dots. \quad (5.1.2)$$

The potentials U_E, U^i_E, W_E are determined by (4.1.11)–(4.1.14) with the single value of the summation index $A = E$. Perturbing bodies $A, B, \dots$ (the Sun, the Moon, etc) are regarded as the material points. Therefore, the regular parts of the potentials have the form

$$\bar{U}_E = \sum_{A\neq E} \frac{GM_A}{r_A} + \dots \quad (5.1.3)$$

$$\bar{U}^i_E = \sum_{A\neq E} \frac{GM_A}{r_A} v^i_A + \dots \quad (5.1.4)$$

$$\bar{W}_E = \tfrac{3}{2}\sum_{A\neq E} \frac{GM_A}{r_A} v^2_A - \sum_{A\neq E}\sum_{B\neq A} \frac{G^2 M_A M_B}{r_A r_{AB}} + \tfrac{1}{2}\sum_{A\neq E} GM_A \frac{\partial^2 r_A}{\partial t^2} + \dots. \quad (5.1.5)$$

The most cumbersome operation in calculating (5.1.2) is to differentiate the functions W_E and $\bar{W}_E$. The appropriate derivatives are

$$W_{E,i} = \frac{1}{2}\frac{GM_E}{r_E}\left(-a^i_E + \frac{1}{r^2_E} a^k_E r^k_E r^i_E\right)$$
$$+ \frac{GM_E}{r^3_E}\left[\left(\bar{U}_E(\boldsymbol{x}_E) - 2v^2_E + \frac{3}{2r^2_E}(r^k_E v^k_E)^2\right)r^i_E - v^k_E r^k_E v^i_E\right]$$
$$+ 3\frac{G}{r^5_E} v^2_E \left(I^{mm}_E r^i_E + 2I^{im}_E r^m_E - \frac{5}{r^2_E} I^{mn}_E r^m_E r^n_E r^i_E\right)$$
$$+ \frac{3}{2}\frac{G}{r^5_E} v^i_E v^k_E \left(I^{mm}_E r^k_E + 2I^{km}_E r^m_E - \frac{5}{r^2_E} I^{mn}_E r^m_E r^n_E r^k_E\right)$$
$$+ \frac{3}{2}\frac{G}{r^5_E} v^k_E v^m_E \left(I^{km}_E r^i_E + 2I^{ik}_E r^m_E - \frac{10}{r^2_E} I^{kn}_E r^m_E r^n_E r^i_E\right)$$
$$- \frac{15}{4}\frac{G}{r^7_E}(r^k_E v^k_E)^2\left(I^{mm}_E r^i_E + 2I^{im}_E r^m_E - \frac{7}{r^2_E} I^{mn}_E r^m_E r^n_E r^i_E\right)$$
$$+ \frac{G}{r^3_E}\omega^j_E\left(4\epsilon_{kjn} v^k_E I^{in}_E + \epsilon_{ijn} v^m_E I^{mn}_E - \frac{3}{r^2_E}\epsilon_{kjn} I^{mn}_E (v^i_E r^k_E r^m_E\right.$$
$$+ v^m_E r^k_E r^i_E + 4v^k_E r^m_E r^i_E) - \frac{3}{r^2_E}\epsilon_{kjn} I^{in}_E r^k_E r^m_E v^m_E$$
$$\left. - \frac{3}{r^2_E}\epsilon_{ijn} I^{kn}_E r^k_E r^m_E v^m_E + \frac{15}{r^4_E}\epsilon_{kjn} I^{mn}_E r^k_E r^m_E r^i_E r^s_E v^s_E\right) \quad (5.1.6)$$

$$
\begin{aligned}
\bar{W}_{E,ik}(\boldsymbol{x}_E) = \sum_{A\neq E} \frac{GM_A}{r_{EA}^3} \Bigg[\Bigg(& \sum_{B\neq A} \frac{GM_B}{r_{AB}} - 2v_A^2 + \tfrac{1}{2} r_{EA}^m a_A^m \\
& + \frac{3}{2r_{EA}^2} (r_{EA}^m v_A^m)^2 \Bigg) \delta_{ik} + \tfrac{1}{2}(r_{EA}^k a_A^i + r_{EA}^i a_A^k) - v_A^i v_A^k \\
& + \frac{3}{r_{EA}^2} (r_{EA}^k v_A^i + r_{EA}^i v_A^k)(r_{EA}^m v_A^m) \\
& + \frac{3}{r_{EA}^2} r_{EA}^i r_{EA}^k \Bigg(2v_A^2 - \sum_{B\neq A} \frac{GM_B}{r_{AB}} - \tfrac{1}{2} r_{EA}^m a_A^m \\
& - \frac{5}{2r_{EA}^2} (r_{EA}^m v_A^m)^2 \Bigg) \Bigg] .
\end{aligned}
\tag{5.1.7}
$$

Substitution of all these values into (5.1.2) results in the following expressions of the relativistic right-hand members of equations (5.1.1):

$$
G^i - \bar{G}^i = \sum_{n=1}^{5} (\varphi_n^i + g_n^i) \tag{5.1.8}
$$

with

$$
\varphi_1^i = \frac{GM_E}{r_E^3} \left(4 \frac{GM_E}{r_E} r_E^i - \dot{r}_E^k \dot{r}_E^k r_E^i + 4 r_E^k \dot{r}_E^k \dot{r}_E^i \right) \tag{5.1.9}
$$

$$
g_1^i = \frac{GM_E}{r_E^3} \left[\left(2\dot{r}_E^k v_E^k + v_E^2 + \frac{3}{2r_E^2} (r_E^k v_E^k)^2 \right) r_E^i + r_E^k v_E^k \dot{r}_E^i \right] \tag{5.1.10}
$$

$$
\begin{aligned}
\varphi_2^i = & 4 \frac{G}{r_E^3} \omega_E^j I_E^{mn} \left[\epsilon_{ijn} \left(\delta_{km} - \frac{3}{r_E^2} r_E^k r_E^m \right) - \epsilon_{kjn} \left(\delta_{im} - \frac{3}{r_E^2} r_E^i r_E^m \right) \right] \dot{r}_E^k \\
& - 9 \frac{G}{r_E^5} \epsilon_{kjn} \omega_E^j \dot{r}_E^i I_E^{mn} r_E^k r_E^m
\end{aligned}
\tag{5.1.11}
$$

$$
\begin{aligned}
g_2^i = \frac{G}{r_E^3} \omega_E^j \Bigg[& \epsilon_{ijn} v_E^m I_E^{mn} - \frac{3}{r_E^2} \epsilon_{kjn} v_E^m I_E^{mn} r_E^k r_E^i \\
& + \frac{3}{r_E^2} r_E^s v_E^s \left(-\epsilon_{kjn} I_E^{in} r_E^k - \epsilon_{ijn} I_E^{kn} r_E^k + \frac{5}{r_E^2} \epsilon_{kjn} I_E^{mn} r_E^k r_E^m r_E^i \right) \Bigg]
\end{aligned}
\tag{5.1.12}
$$

$$
\begin{aligned}
\varphi_3^i = & 4 \frac{G^2 M_E}{r_E^6} \left(-2 I_E^{kk} r_E^i - 3 I_E^{ik} r_E^k + \frac{9}{r_E^2} I_E^{km} r_E^k r_E^m r_E^i \right) \\
& + \frac{3}{2} \frac{G}{r_E^5} \dot{r}_E^n \dot{r}_E^n \left(I_E^{kk} r_E^i + 2 I_E^{ik} r_E^k - \frac{5}{r_E^2} I_E^{km} r_E^k r_E^m \right) \\
& + 6 \frac{G}{r_E^5} \dot{r}_E^i \dot{r}_E^n \left(-I_E^{kk} r_E^n - 2 I_E^{kn} r_E^k + \frac{5}{r_E^2} I_E^{km} r_E^k r_E^m r_E^n \right)
\end{aligned}
\tag{5.1.13}
$$

$$
\begin{aligned}
g_3^i = & -3\frac{G}{r_E^5}(v_E^k \dot{r}_E^k)\left(I_E^{mm} r_E^i + 2I_E^{im} r_E^m - \frac{5}{r_E^2} I_E^{mn} r_E^m r_E^n r_E^i\right) \\
& - \frac{3}{2}\frac{G}{r_E^5} v_E^2 \left(I_E^{kk} r_E^i + 2I_E^{ik} r_E^k - \frac{5}{r_E^2} I_E^{mn} r_E^m r_E^n r_E^i\right) \\
& + \frac{3}{2}\frac{G}{r_E^5} \dot{r}_E^i v_E^k \left(-I_E^{mm} r_E^k - 2I_E^{km} r_E^m + \frac{5}{r_E^2} I_E^{mn} r_E^m r_E^n r_E^k\right) \\
& + \frac{3}{2}\frac{G}{r_E^5} v_E^k v_E^m \left(I_E^{km} r_E^i + 2I_E^{ik} r_E^m - \frac{10}{r_E^2} I_E^{kn} r_E^m r_E^n r_E^i\right) \\
& + \frac{15}{4}\frac{G}{r_E^7}(r_E^k v_E^k)^2 \left(-I_E^{mm} r_E^i - 2I_E^{im} r_E^m + \frac{7}{r_E^2} I_E^{mn} r_E^m r_E^n r_E^i\right)
\end{aligned}
\tag{5.1.14}
$$

$$
\varphi_4^i = 2\frac{GM_E}{r_E}\sum_{A\neq E}\frac{GM_A}{r_{EA}^3}\left(r_E^i - \frac{6}{r_{EA}^2} r_{EA}^i r_{EA}^k r_E^k + \frac{3}{r_E^2 r_{EA}^2}(r_E^k r_{EA}^k)^2 r_E^i\right)
\tag{5.1.15}
$$

$$
g_4^i = 5\frac{GM_E}{r_E^3} r_E^i \bar{U}_E(\boldsymbol{x}_E) + \frac{1}{2}\frac{GM_E}{r_E}\left(-a_E^i + \frac{9}{r_E^2} r_E^i r_E^k a_E^k\right)
\tag{5.1.16}
$$

$$
\begin{aligned}
\varphi_5^i = & [-4\bar{U}_{E,km}(\boldsymbol{x}_E)(\dot{r}_E^i + v_E^i)\dot{r}_E^m - \bar{U}_{E,km}(\boldsymbol{x}_E)(\dot{r}_E^i + v_E^i)v_E^m \\
& - 3\dot{a}_E^k(\dot{r}_E^i + v_E^i) + \bar{U}_{E,ik}(\boldsymbol{x}_E)(\dot{r}_E^m + v_E^m)(\dot{r}_E^m + v_E^m) + 4\dot{\bar{U}}^i_{E,k}(\boldsymbol{x}_E) \\
& - 4\bar{U}^m_{E,ik}(\boldsymbol{x}_E)(\dot{r}_E^m + v_E^m) + 4\bar{U}^i_{E,km}(\boldsymbol{x}_E)\dot{r}_E^m + \bar{W}_{E,ik}(\boldsymbol{x}_E) \\
& - 4\bar{U}_E(\boldsymbol{x}_E)\bar{U}_{E,ik}(\boldsymbol{x}_E) - 4a_E^i a_E^k] r_E^k + \ldots
\end{aligned}
\tag{5.1.17}
$$

$$
\begin{aligned}
g_5^i = & 4\bar{U}^i_{E,k}(\boldsymbol{x}_E)\dot{r}_E^k - 4\bar{U}^k_{E,i}(\boldsymbol{x}_E)\dot{r}_E^k + a_E^i(\dot{r}_E^k \dot{r}_E^k + 2\dot{r}_E^k v_E^k) \\
& - a_E^k(4\dot{r}_E^k \dot{r}_E^i + 4\dot{r}_E^k v_E^i + \dot{r}_E^i v_E^k) - 3\dot{\bar{U}}_E(\boldsymbol{x}_E)\dot{r}_E^i.
\end{aligned}
\tag{5.1.18}
$$

The relativistic terms (5.1.8) consist of ten groups. The first group of terms φ_1^i, dependent only on the Earth's mass M_E, represent the Schwarzschild terms (5.1.9). The orbital motion of the Earth in BRS involves the second group of the terms g_1^i (5.1.10) dependent on M_E and v_E^k. For close satellites of the Earth these terms are of the most importance but because of their kinematical origin they should disappear in converting to GRS. The terms φ_2^i (5.1.11) involving components ω_E^k represent the Lense–Thirring terms generated by the Earth's rotation. The orbital motion of the Earth leads to the spin–orbital terms g_2^i (5.1.12) dependent on ω_E^k and v_E^k. These terms should also disappear in converting to GRS. The second-order moments of inertia of the Earth are responsible for the quadrupole group of terms φ_3^i dependent on I_E^{mn} (5.1.13). Along with this the orbital motion of the Earth results in the large group of terms g_3^i (5.1.14) dependent on I_E^{mn} and v_E^k. These terms should vanish in transforming to GRS. Thus, all terms φ_n^i, n =1, 2, 3, represent the terms describing the one-body problem considered in Chapter 3. The terms g_n^i, n =1, 2, 3, are also related with the problem of one body moving in the reference system at hand (BRS). All further terms (n =4, 5) are due to the external masses. The terms φ_4^i(5.1.15), dependent on masses M_E and M_A, describe the non-linear coupling of the gravitational fields of the Earth and the external masses. In BRS this coupling gives also the terms g_4^i (5.1.16) dependent on the superposition of M_E and the potential $\bar{U}_E$ $(\boldsymbol{x}_E)$ or its first derivatives. Converting to GRS should annul these terms. The terms φ_5^i (5.1.17), proportional to the satellite coordinates r_E^k, describe the tidal perturbations due to the external masses. In converting to GRS these perturbations may change a little but their form should be retained. Of particular interest are the terms g_5^i (5.1.18), due again to the external masses but dependent only on the satellite relative velocity components $\dot{r}_E^k$ and not on its relative coordinates r_E^k. These terms should disappear in converting to GRS but they are specific for BRS. They may be rewritten explicitly as

$$g_5^i = \sum_{A\neq E} \frac{GM_A}{r_{EA}^3}[4(v_E^i - v_A^i)r_{EA}^k\dot{r}_E^k - 2r_{EA}^i(v_E^k - v_A^k)\dot{r}_E^k + 4r_{EA}^k(v_E^k - v_A^k)\dot{r}_E^i$$
$$+ 2r_{EA}^i v_A^k \dot{r}_E^k + r_{EA}^k v_A^k \dot{r}_E^i - r_{EA}^i \dot{r}_E^k \dot{r}_E^k + 4r_{EA}^k \dot{r}_E^k \dot{r}_E^i]. \quad (5.1.19)$$

When applied to the perturbations from the Sun, the third term in the square brackets is small, of the order of the eccentricity of the heliocentric orbit of the Earth, the fourth and the fifth terms are small being of the order of the barycentric velocity of the Sun and two last terms are quadratic relative to the satellite velocity and are small for this reason. The first two terms may be rewritten in vector form as

$$g_5^i = \sum_{A\neq E} \frac{GM_A}{r_{EA}^3}\{3[\dot{\boldsymbol{r}}_E \times (\dot{\boldsymbol{r}}_{EA} \times \boldsymbol{r}_{EA})]^i + (\boldsymbol{r}_{EA}\dot{\boldsymbol{r}}_E)\dot{r}_{EA}^i + (\dot{\boldsymbol{r}}_{EA}\dot{\boldsymbol{r}}_E)r_{EA}^i + \ldots\}. \quad (5.1.20)$$

The second and the third terms here determine the perturbations which depend on the orbital elements of the satellite. The first term in the form of the double vector product gives the Coriolis terms describing the effect of geodesic precession. Due to this precession the perigee and the node of any satellite of the Earth including the Moon move at a rate of 1.91" per century. The principal terms in (5.1.18) responsible for the geodesic precession are

$$g_5^i = 2a_E^i v_E^k \dot{r}_E^k - 4v_E^i a_E^k \dot{r}_E^k + \ldots.$$

The presence of these terms is characteristic of the BRS investigation of the motion of the Solar System bodies with respect to any body of this system (with respect to the Earth in the case under discussion).

Thus, in the expression (5.1.8) consisting of ten groups, the terms g_n^i (n =1, 2, 3, 4, 5) are of kinematic origin and should disappear in converting to GRS.

5.1.2 Transformation to the GRS equations of motion

Transformation from BRS to GRS is performed by (4.2.7) and (4.2.8) with (4.2.11), (4.2.17), (4.2.18) and (4.2.25). The functions F^{ik} responsible for the geodesic precession result, as stated below, in vanishing the Coriolis terms (5.1.20) in the GRS satellite equations. The derivation of the GRS satellite equations from the corresponding BRS equations involves three steps:

(1) converting the acceleration $\ddot{r}_E^i$ to the acceleration $\mathrm{d}^2 w^i/\mathrm{d}u^2$,
(2) performing transformation (4.2.8) in the Newtonian right-hand sides of the BRS equations, and
(3) re-defining dipole and quadrupole moments of the geopotential in GRS in accordance with (4.2.38) and (4.2.39).

The first step presents no difficulties. Differentiating twice expressions (4.2.7) and (4.2.8) with respect to t and substituting the results into the relation

$$\frac{\mathrm{d}^2 w^i}{\mathrm{d}u^2} = \frac{1}{\dot{u}}\frac{\mathrm{d}}{\mathrm{d}t}\left(\frac{\dot{w}^i}{\dot{u}}\right) = \frac{\ddot{w}^i}{\dot{u}^2} - \frac{\dot{w}^i}{\dot{u}^3}\ddot{u}$$

one obtains

$$\begin{aligned}\frac{\mathrm{d}^2 w^i}{\mathrm{d}u^2} &= \ddot{r}_E^i + c^{-2}[2(\dot{S} + v_E^k\dot{w}^k + a_E^k w^k)\ddot{w}^i + (\tfrac{1}{2}v_E^i v_E^k + F^{ik} + D^{ik} \\ &\quad + v_E^k \dot{w}^i + 2D^{ikm}w^m)\ddot{w}^k + (\ddot{S} + 2a_E^k\dot{w}^k + \dot{a}_E^k w^k)\dot{w}^i \\ &\quad + (v_E^k a_E^i + v_E^i a_E^k + 2\dot{F}^{ik} + 2\dot{D}^{ik} + 2D^{ikm}\dot{w}^m)\dot{w}^k + (\tfrac{1}{2}v_E^k \dot{a}_E^i + a_E^k a_E^i \\ &\quad + \tfrac{1}{2}v_E^i \dot{a}_E^k + \ddot{F}^{ik} + \ddot{D}^{ik} + 4\dot{D}^{ikm}\dot{w}^m)w^k + \ddot{D}^{ikm}w^k w^m]. \qquad (5.1.21)\end{aligned}$$

Now the transformation (4.2.8) is to be substituted into the Newtonian right-hand side of (5.1.1). Putting

$$U_E(\boldsymbol{w}) = \frac{GM_E}{\rho} + GI_E^k\frac{w^k}{\rho^3} + \frac{1}{2\rho^3}\left(-\delta_{km} + \frac{3}{\rho^2}w^k w^m\right)GI_E^{km} + \dots \qquad (5.1.22)$$

with the previous designation $\rho = (w^k w^k)^{1/2}$ one has

$$U_E = U_E(\boldsymbol{w}) - c^{-2}[(\tfrac{1}{2}v_E^k v_E^m + F^{km} + D^{km})w^m + D^{kmn}w^m w^n]\frac{\partial U_E(\boldsymbol{w})}{\partial w^k}. \qquad (5.1.23)$$

Differentiating

$$U_{E,i} = \frac{\partial U_E}{\partial w^k}\frac{\partial w^k}{\partial x^i} = \frac{\partial U_E(\boldsymbol{w})}{\partial w^i}$$
$$- c^{-2}[(\tfrac{1}{2}v_E^k v_E^m + F^{km} + D^{km})w^m + D^{kmn}w^m w^n]\frac{\partial^2 U_E(\boldsymbol{w})}{\partial w^k \partial w^i} \tag{5.1.24}$$

and substituting into (5.1.1) one gets

$$\ddot{r}_E^i = \frac{\partial U_E(\boldsymbol{w})}{\partial w^i} - H_E^i + \bar{U}_{E,ik}(\boldsymbol{x}_E)w^k + \tfrac{1}{2}\bar{U}_{E,ikm}(\boldsymbol{x}_E)w^k w^m + \dots$$
$$+ c^{-2}(G^i - \bar{G}^i) - c^{-2}[(\tfrac{1}{2}v_E^k v_E^m + F^{km} + D^{km})w^m + D^{kmn}w^m w^n]$$
$$\times \left(\frac{\partial^2 U_E(\boldsymbol{w})}{\partial w^i \partial w^k} + \bar{U}_{E,ik}(\boldsymbol{x}_E) + \bar{U}_{E,ijk}w^j + \dots\right). \tag{5.1.25}$$

The last step to derive the GRS satellite equations of motion is to transform the dipole and quadrupole moments occurring in (5.1.22). In the relativistic parts the products of the quadrupole moments with the external mass terms are here everywhere neglected. Therefore, the right-hand member of (4.2.38) is reduced to the first term alone. From this

$$U_E(\boldsymbol{w}) = \hat{U}_E + c^{-2}\frac{G}{\rho^3}I_E^{mn}\left[\frac{1}{2}\left(v_E^m - \frac{3}{\rho^2}v_E^k w^k w^m\right)v_E^n\right.$$
$$\left. + \epsilon_{kjn}\omega_E^j w^k\left(-v_E^m + \frac{3}{\rho^2}v_E^s w^s w^m\right)\right] \tag{5.1.26}$$

$\hat{U}_E$ being the GRS geopotential (4.2.5). One has further

$$\frac{\partial U_E(\boldsymbol{w})}{\partial w^i} = \hat{U}_{E,i} + c^{-2}\frac{G}{\rho^3}I_E^{mn}\left[\frac{3}{2\rho^2}\left(-v_E^m w^i - v_E^i w^m - v_E^k w^k \delta_{im}\right.\right.$$
$$\left. + \frac{5}{\rho^2}v_E^k w^k w^m w^i\right)v_E^n + \epsilon_{kjn}\omega_E^j\left(-\delta_{ik} + \frac{3}{\rho^2}w^i w^k\right)v_E^m$$
$$\left. + \frac{3}{\rho^2}\epsilon_{kjn}\omega_E^j v_E^s w^s\left(\delta_{ik}w^m + \delta_{im}w^k - \frac{5}{\rho^2}w^i w^k w^m\right)\right]. \tag{5.1.27}$$

The partial derivative of $\hat{U}_E$ with respect to w^i (under fixed u) is again denoted here by a comma followed by the appropriate index. In differentiating $\hat{U}_E$ there appears the term with the time u derivative of the form $-c^{-1}v_E^i\hat{U}_{E,0}$ due to (4.2.7). But this term cancels out with one of the terms resulting from the differentiation of the relativistic part of (5.1.26).

Combining the results (5.1.21), (5.1.25) and (5.1.27) one obtains the GRS satellite equations of motion in the form

$$\frac{d^2w^i}{du^2} = \hat{U}_{E,i} - H_E^i + \bar{U}_{E,ik}(\boldsymbol{x}_E)w^k + \tfrac{1}{2}\bar{U}_{E,ikm}(\boldsymbol{x}_E)w^k w^m + \ldots + c^{-2}\Phi^i \tag{5.1.28}$$

where the relativistic right-hand member Φ^i is determined by

$$\begin{aligned}\Phi^i =\, & G^i - \bar{G}^i + 2(\dot{S} + v_E^k \dot{w}^k + a_E^k w^k)(\hat{U}_{E,i} + \bar{U}_{E,im}(\boldsymbol{x}_E)w^m \\ & + \tfrac{1}{2}\bar{U}_{E,imn}(\boldsymbol{x}_E)w^m w^n + \ldots) \\ & + (\tfrac{1}{2}v_E^i v_E^k + F^{ik} + D^{ik} + v_E^k \dot{w}^i + 2D^{ijk}w^j) \\ & \times [\hat{U}_{E,k} + \bar{U}_{E,km}(\boldsymbol{x}_E)w^m + \ldots] + (\ddot{S} + 2a_E^k \dot{w}^k + \dot{a}_E^k w^k)\dot{w}^i \\ & + (v_E^k a_E^i + v_E^i a_E^k + 2\dot{F}^{ik} + 2\dot{D}^{ik} + 2D^{ikm}\dot{w}^m)\dot{w}^k \\ & + (\tfrac{1}{2}v_E^k \dot{a}_E^i + a_E^k a_E^i + \tfrac{1}{2}v_E^i \dot{a}_E^k + \ddot{F}^{ik} \\ & + \ddot{D}^{ik} + 4\dot{D}^{ikm}\dot{w}^m)w^k + \ddot{D}^{ikm}w^k w^m - [(\tfrac{1}{2}v_E^k v_E^m + F^{km} + D^{km})w^m \\ & + D^{kmn}w^m w^n][\hat{U}_{E,ik} + \bar{U}_{E,ik}(\boldsymbol{x}_E) + \ldots] \\ & + \frac{3}{2}\frac{G}{\rho^5}I_E^{mn}\left(-v_E^m w^i - v_E^i w^m - v_E^k w^k \delta_{im} + \frac{5}{\rho^2}v_E^k w^k w^m w^i\right)v_E^n \\ & + 3\frac{G}{\rho^5}\epsilon_{kjn}\omega_E^j v_E^s w^s I_E^{mn}\left(\delta_{ik}w^m + \delta_{im}w^k - \frac{5}{\rho^2}w^i w^k w^m\right) \\ & + \frac{G}{\rho^3}\epsilon_{kjn}\omega_E^j v_E^m I_E^{mn}\left(-\delta_{ik} + \frac{3}{\rho^2}w^i w^k\right).\end{aligned} \tag{5.1.29}$$

This expression enables one to see the contribution of each term of the transformations (4.2.7),(4.2.8) in forming the right-hand sides of the GRS equations. It is of importance that expressions (5.1.21) and (5.1.24) are quite rigorous with respect to w^i, which is the consequence of the closed form of (4.2.8) relative to w^i. The dots in (5.1.29) mean that the higher degrees in w^i are neglected in expanding the Newtonian right-hand member (5.1.1).

5.1.3 GRS equations of motion of a satellite

Substituting now (4.2.11), (4.2.17), (4.2.18) and (4.2.25) one finds

$$\Phi^i = G^i - \bar{G}^i - \sum_{n=1}^{5} g_n^i + \Delta\varphi_5^i \tag{5.1.30}$$

with

$$\Delta\varphi_5^i = [2\dot{a}_E^m \dot{w}^m \delta_{ik} + \ddot{\bar{U}}(\boldsymbol{x}_E)\delta_{ik} + 2v_E^i \dot{a}_E^k + a_E^i a_E^k - v_E^k \dot{a}_E^i + 3\dot{a}_E^k \dot{w}^i$$

$$
\begin{aligned}
&- 2\dot{a}^i_E \dot{w}^k - 2\dot{\bar{U}}^i_{E,k}(\boldsymbol{x}_E) + 2\dot{\bar{U}}^k_{E,i}(\boldsymbol{x}_E) \\
&- \tfrac{1}{2} v^k_E v^m_E \bar{U}_{E,im}(\boldsymbol{x}_E) + \tfrac{1}{2} v^i_E v^m_E \bar{U}_{E,km}(\boldsymbol{x}_E) \\
&+ v^2_E \bar{U}_{E,ik}(\boldsymbol{x}_E) + 2 v^m_E \dot{w}^m \bar{U}_{E,ik}(\boldsymbol{x}_E) + v^m_E \dot{w}^i \bar{U}_{E,km}(\boldsymbol{x}_E) \\
&+ F^{km} \bar{U}_{E,im}(\boldsymbol{x}_E) + F^{im} \bar{U}_{E,km}(\boldsymbol{x}_E) + 2\bar{U}_E(\boldsymbol{x}_E)\bar{U}_{E,ik}(\boldsymbol{x}_E)] w^k + \ldots .
\end{aligned}
\tag{5.1.31}
$$

Comparing with (5.1.8) it is seen that all the terms g^i_n (n =1, 2, 3, 4, 5) cancel out. As a result the right-hand side of the GRS satellite equations takes the simple form

$$
\Phi^i = \sum_{n=1}^{5} \varphi^i_n + \Delta\varphi^i_5 . \tag{5.1.32}
$$

φ^i_n (n =1, 2, 3, 4) are given by (5.1.9), (5.1.11), (5.1.13), (5.1.15) and

$$
\begin{aligned}
\varphi^i_5 + \Delta\varphi^i_5 = {} & [4\dot{w}^m v^m_E \bar{U}_{E,ik}(\boldsymbol{x}_E) - 4\dot{w}^m \bar{U}^m_{E,ik}(\boldsymbol{x}_E) - 4\dot{w}^m v^i_E \bar{U}_{E,km}(\boldsymbol{x}_E) \\
& + 4\dot{w}^m \bar{U}^i_{E,km}(\boldsymbol{x}_E) + 2\dot{a}^m_E \dot{w}^m \delta_{ik} - 2\dot{a}^i_E \dot{w}^k - 4\dot{w}^m \dot{w}^i \bar{U}_{E,km}(\boldsymbol{x}_E) \\
& + \bar{U}_{E,ik}(\boldsymbol{x}_E) \dot{w}^m \dot{w}^m + F^{im} \bar{U}_{E,km}(\boldsymbol{x}_E) \\
& + F^{km} \bar{U}_{E,im}(\boldsymbol{x}_E) + \ddot{\bar{U}}_E(\boldsymbol{x}_E)\delta_{ik} - 3 a^i_E a^k_E - 2\bar{U}_E(\boldsymbol{x}_E)\bar{U}_{E,ik}(\boldsymbol{x}_E) \\
& + \bar{W}_{E,ik}(\boldsymbol{x}_E) + 2\dot{\bar{U}}^i_{E,k}(\boldsymbol{x}_E) + 2\dot{\bar{U}}^k_{E,i}(\boldsymbol{x}_E) \\
& - v^k_E \dot{a}^i_E - v^i_E \dot{a}^k_E - 4 v^m_E \bar{U}^m_{E,ik}(\boldsymbol{x}_E) - \tfrac{1}{2} v^i_E v^m_E \bar{U}_{E,km}(\boldsymbol{x}_E) \\
& - \tfrac{1}{2} v^k_E v^m_E \bar{U}_{E,im}(\boldsymbol{x}_E) + 2 v^2_E \bar{U}_{E,ik}(\boldsymbol{x}_E)] w^k + \ldots .
\end{aligned}
\tag{5.1.33}
$$

Finally, the GRS satellite equations of motion are of the form

$$
\frac{d^2 w^i}{du^2} = F^i_0 + F^i_1 + F^i_2 + F^i_3 + \ldots + c^{-2} \sum_{n=1}^{6} \Phi^i_n \tag{5.1.34}
$$

with

$$
F^i_0 = -\frac{G\hat{M}^E}{\rho^3} w^i \tag{5.1.35}
$$

$$
F^i_1 = \frac{3}{2}\frac{G}{\rho^5}\left(\hat{I}^{kk}_E w^i + 2\hat{I}^{ik}_E w^k - \frac{5}{\rho^2} \hat{I}^{km}_E w^k w^m w^i \right) \tag{5.1.36}
$$

$$
F^i_2 = \bar{U}_{E,ik}(\boldsymbol{x}_E) w^k + \tfrac{1}{2} \bar{U}_{E,ikm}(\boldsymbol{x}_E) w^k w^m + \ldots \tag{5.1.37}
$$

$$
F^i_3 = -H^i_E = -\tfrac{1}{2} \hat{M}_E^{-1} \hat{I}^{km}_E \bar{U}_{E,ikm}(\boldsymbol{x}_E) \tag{5.1.38}
$$

$$
\Phi^i_1 = \varphi^i_1 = \frac{GM_E}{\rho^3}\left[\left(4\frac{GM_E}{\rho} - \dot{w}^k \dot{w}^k \right) w^i + 4 w^k \dot{w}^k \dot{w}^i \right] \tag{5.1.39}
$$

$$\Phi_2^i = \varphi_2^i = \frac{4}{\rho^3} G\omega_E^j I_E^{mn} \left[\epsilon_{ijn} \left(\delta_{km} - \frac{3}{\rho^2} w^k w^m \right) - \epsilon_{kjn} \left(\delta_{im} - \frac{3}{\rho^2} w^i w^m \right) \right] \dot{w}^k - \frac{9G}{\rho^5} \epsilon_{kjn} \omega_E^j I_E^{mn} w^k w^m \dot{w}^i \tag{5.1.40}$$

$$\begin{aligned} \Phi_3^i = \varphi_3^i = 4 \frac{G^2 M_E}{\rho^6} & \left(-2 I_E^{kk} w^i - 3 I_E^{ik} w^k + \frac{9}{\rho^2} I_E^{km} w^k w^m w^i \right) \\ & + \frac{3}{2} \frac{G}{\rho^5} \dot{w}^n \dot{w}^n \left(I_E^{kk} w^i + 2 I_E^{ik} w^k - \frac{5}{\rho^2} I_E^{km} w^k w^m w^i \right) \\ & + 6 \frac{G}{\rho^5} \dot{w}^i \dot{w}^n \left(-I_E^{kk} w^n - 2 I_E^{kn} w^k + \frac{5}{\rho^2} I_E^{km} w^k w^m w^n \right) \end{aligned} \tag{5.1.41}$$

$$\Phi_4^i = \varphi_4^i = 2 \frac{GM_E}{\rho} \sum_{A \neq E} \frac{GM_A}{r_{EA}^3} \times \left(w^i - \frac{6}{r_{EA}^2} r_{EA}^i r_{EA}^k w^k + \frac{3}{\rho^2 r_{EA}^2} (w^k r_{EA}^k)^2 w^i \right) \tag{5.1.42}$$

$$\begin{aligned} \Phi_5^i = \sum_{A \neq E} \frac{GM_A}{r_{EA}^3} w^k \Bigg(& -6 \dot{w}^m (v_E^m - v_A^m) \delta_{ik} + \frac{6}{r_{EA}^2} \dot{w}^m r_{EA}^m r_{EA}^n (v_E^n - v_A^n) \delta_{ik} \\ & + 6 \dot{w}^k (v_E^i - v_A^i) - \frac{6}{r_{EA}^2} \dot{w}^k r_{EA}^i r_{EA}^m (v_E^m - v_A^m) \\ & + \frac{12}{r_{EA}^2} r_{EA}^i r_{EA}^k (v_E^m - v_A^m) \dot{w}^m - \frac{12}{r_{EA}^2} r_{EA}^k r_{EA}^m (v_E^i - v_A^i) \dot{w}^m - \dot{w}^m \dot{w}^m \delta_{ik} \\ & + 4 \dot{w}^i \dot{w}^k - \frac{12}{r_{EA}^2} r_{EA}^k r_{EA}^m \dot{w}^m \dot{w}^i + \frac{3}{r_{EA}^2} r_{EA}^i r_{EA}^k \dot{w}^m \dot{w}^m \Bigg) \end{aligned} \tag{5.1.43}$$

$$\begin{aligned} \Phi_6^i = \sum_{A \neq E} \frac{GM_A}{r_{EA}^3} w^k \Bigg\{ \Bigg[& -r_{EA}^m a_E^m + \tfrac{3}{2} r_{EA}^m a_A^m - 3(v_E^m - v_A^m)(v_E^m - v_A^m) \\ & + \frac{3}{r_{EA}^2} (r_{EA}^m v_E^m)^2 - \frac{6}{r_{EA}^2} (r_{EA}^m v_E^m)(r_{EA}^n v_A^n) + \frac{9}{2 r_{EA}^2} (r_{EA}^m v_A^m)^2 \\ & + \frac{G}{r_{EA}} (M_E + 2M_A) + \sum_{B \neq A, E} GM_B \left(\frac{1}{r_{AB}} + \frac{2}{r_{EB}} \right) \Bigg] \delta_{ik} \end{aligned}$$

$$+\frac{3}{r_{EA}^2} r_{EA}^m (r_{EA}^k F^{im} + r_{EA}^i F^{km}) + 3(v_E^i - v_A^i)(v_E^k - v_A^k)$$
$$+\frac{3}{r_{EA}^2} r_{EA}^i r_{EA}^m (-\frac{3}{2} v_E^k v_E^m + v_E^k v_A^m + 2 v_E^m v_A^k - v_A^k v_A^m)$$
$$+\frac{6}{r_{EA}^2} r_{EA}^i r_{EA}^k (v_E^m - v_A^m)(v_E^m - v_A^m)$$
$$+\frac{3}{r_{EA}^2} r_{EA}^k r_{EA}^m (-\tfrac{3}{2} v_E^i v_E^m + v_E^i v_A^m + 2 v_E^m v_A^i - v_A^i v_A^m)$$
$$-\frac{15}{2r_{EA}^4} r_{EA}^i r_{EA}^k (r_{EA}^m v_A^m)^2 - \tfrac{3}{2} r_{EA}^k a_A^i - \tfrac{3}{2} r_{EA}^i a_A^k$$
$$-\frac{3}{2r_{EA}^2} r_{EA}^i r_{EA}^k r_{EA}^m a_A^m - \frac{3}{r_{EA}^3} G(M_E + 3M_A) r_{EA}^i r_{EA}^k$$
$$-3r_{EA}^i \sum_{B \neq A,E} GM_B \left(\frac{r_{EA}^k}{r_{EA}^2 r_{AB}} + \frac{2r_{EA}^k}{r_{EA}^2 r_{EB}} + \frac{r_{EB}^k}{r_{EB}^3} \right) \Bigg\}. \tag{5.1.44}$$

Let us note once again the physical meaning of the acceleration terms. F_0^i are the Keplerian terms. F_1^i are the Newtonian terms due to the Earth's non-sphericity. F_2^i are the Newtonian perturbing accelerations due to the external masses. F_3^i are the Newtonian terms caused by the non-geodesic motion of the Earth (coupling of the external mass action and the Earth quadrupole moment effects). Next, one has the relativistic perturbing accelerations. Φ_1^i are the Schwarzschild terms. Φ_2^i are the Lense–Thirring terms due to the Earth rotation. Φ_3^i are the relativistic quadrupole terms. For the Earth approximated by an oblate spheroid in rotation with constant angular velocity ω around the polar axis one has, by choosing the equatorial reference system,

$$\hat{I}_E^{11} = \hat{I}_E^{22} = \tfrac{1}{2}C \qquad \hat{I}_E^{33} = A - \tfrac{1}{2}C \qquad Q = G(A - C)$$

$$\omega_E^1 = \omega_E^2 = 0 \qquad \omega_E^3 = \omega$$

with the principal moments of inertia A, C and the quadrupole moment Q. Using $\boldsymbol{s} = (0, 0, 1)$ as the unit vector along the polar axis one has in vector form

$$\boldsymbol{F}_1 = \frac{3Q}{\rho^5} \left[\frac{1}{2} \left(1 - 5\frac{z^2}{\rho^2} \right) \boldsymbol{w} + z\boldsymbol{s} \right] \tag{5.1.45}$$

$$\boldsymbol{F}_3 = -\tfrac{1}{2}(G\hat{M}_E)^{-1} Q \,\mathrm{grad}\bar{U}_{E,33}(\boldsymbol{x}_E) \tag{5.1.46}$$

$$\boldsymbol{\Phi}_2 = \frac{2G}{\rho^3} C\omega \left(\dot{\boldsymbol{w}} \times \boldsymbol{s} + 3\frac{z}{\rho^2} (\boldsymbol{w} \times \dot{\boldsymbol{w}}) \right) \tag{5.1.47}$$

$$\Phi_3 = \frac{Q}{\rho^5}\left\{\left[4\left(-2+9\frac{z^2}{\rho^2}\right)\frac{GM_E}{\rho} + \frac{3}{2}\left(1-5\frac{z^2}{\rho^2}\right)\dot{w}^2\right]w \right.$$
$$\left. + 3\left(-4\frac{GM_E}{\rho} + \dot{w}^2\right) zs - 6\left[\left(1-5\frac{z^2}{\rho^2}\right)(w\dot{w}) + 2z\dot{z}\right]\dot{w}\right\} \tag{5.1.48}$$

coinciding with (3.3.27). Proceeding further, Φ_4^i are the relativistic terms due to the interaction of the Earth and the external masses. Φ_5^i and Φ_6^i give in sum (5.1.33). The terms Φ_5^i may be interpreted as the relativistic tidal perturbations from the external masses. The terms Φ_6^i representing the relativistic corrections to the Newtonian force F_2^i contain contributions due to the geodesic precession F^{km}. This is due to the fact that the Newtonian external mass perturbations F_2^i are expressed in terms of the BRS space coordinates of the external masses and the BRS time coordinate t. Considering that the motion of the Solar System bodies is given in ephemeris astronomy just as in BRS it is not suitable to perform the transformations (4.2.7) and (4.2.8) in F_2^i.

5.1.4 Estimation of the relativistic terms

Let us examine now the magnitude of the relativistic perturbing accelerations from the external masses considering two perturbing bodies, the Sun S and the Moon L, and retaining in (5.1.42)–(5.1.44) only the terms with the summation indices $A = S$ and $A = L$. All three terms in Φ_4^i have order $G^2 M_E M_A / r_{EA}^3$. In Φ_5^i the first six terms are of order $GM_A \rho v_{EA} \dot{w}/r_{EA}^3$ whereas the last four terms are of order $GM_A \rho \dot{w}^2/r_{EA}^3$, v_{EA} being the characteristic heliocentric velocity of the Earth ($A = S$) or the geocentric velocity of the Moon ($A = L$) and $\dot{w}$ being the characteristic geocentric velocity of the satellite. For not too eccentric satellite orbits these latter terms are of the same order as the terms Φ_4^i and the first six terms differ from them by the ratio $v_{EA}/\dot{w}$. Finally, most of the terms in Φ_6^i are of order $(GM_A)^2 \rho / r_{EA}^4$. In addition, the Schwarzschild terms Φ_1^i are of order $(GM_E)^2/\rho^3$, the Lense–Thirring terms Φ_2^i are of order $GM_E A_E^2 \omega_E \dot{w}/\rho^3$ (A_E being the Earth's radius) and the quadrupole terms Φ_3^i have order $(GM_E)^2 \alpha_E A_E^2/\rho^5$ (where α_E is the oblateness of the Earth).

It is of interest to give the estimations of the relativistic perturbations characteristic for BRS. g_1^i and g_2^i may be estimated as $G^2 M_E M_A/(\rho^2 r_{EA})$ and $GM_E A_E^2 \omega_E v_{EA}/\rho^3$, respectively. The terms g_3^i may be of order $G^2 M_E M_A \alpha_E A_E^2/(\rho^4 r_{EA})$ or $GM_E \alpha_E A_E^2 v_{EA} \dot{w}/\rho^4$. The order of g_4^i is $G^2 M_E M_A/(\rho r_{EA}^2)$. Finally, the terms g_5^i are of order $GM_A v_{EA} \dot{w}/r_{EA}^2$ or $G^2 M_E M_A/(\rho r_{EA}^2)$.

For comparison, the Newtonian perturbations may be estimated as

$$F_0^i \simeq GM_E/\rho^2 \qquad F_1^i \simeq GM_E A_E^2 \alpha_E/\rho^4$$
$$F_2^i \simeq GM_A \rho/r_{EA}^3 \qquad F_3^i \simeq GM_A A_E^2 \alpha_E/r_{EA}^4.$$

Replacing in all these estimations $v_{ES} \simeq (GM_S/r_{ES})^{1/2}$, $v_{EL} \simeq (GM_E/r_{EL})^{1/2}$, $\dot{w} \simeq (GM_E/\rho)^{1/2}$ and using numerical values $GM_E = 3.99\times10^5$ km^3 s^{-2}, $GM_S = 1.33\times10^{11}$ km^3 s^{-2}, $GM_L = 4.91\times10^3$ km^3 s^{-2}, $c = 3\times10^5$ km s^{-1}, $m_E = 0.44$ cm, $m_S = 1.5$ km, $A_E = 6.4\times10^3$ km, $\alpha_E = 3.4\times10^{-3}$, $\omega_E = 7.3\times10^{-5}$ s^{-1}, $r_{ES} = 1.5\times10^8$ km, $r_{EL} = 3.84\times10^5$ km, $v_{ES} = 29.8$ km s^{-1}, $v_{EL} = 1.02$ km s^{-1} one gets numerical estimates of the accelerations for different types of Earth satellites (table 5.1). This table demonstrates the 'compression' of the GRS treatment of the satellite motion as compared with the BRS treatment.

5.1.5 Historical remarks

Until recently in the relativistic treatment of Earth satellite motion one took into account only the Schwarzschild and Lense–Thirring perturbations. These perturbations are presented for a variety of satellite orbits in, for example, Cugusi and Proverbio (1978). At present, consideration is more often given to the refined relativistic effects due to the Earth's oblateness (Soffel *et al* 1988, Soffel 1989, Heimberger *et al* 1990) and to the influence of the Sun and the Moon. The solar–lunar perturbations were examined initially in BRS (Martin *et al* 1985, Bordovitsyna *et al* 1985, Vincent 1986). Ashby and Bertotti (1984) were the first to suggest the investigation of Earth satellite motion in a geocentric reference system. The fact that dynamical perturbations in the geocentric system give immediately the correct order of the magnitude of the measurable relativistic effects has been explicitly demonstrated for the Moon (Soffel *et al* 1986) and for Earth satellites (Zhu *et al* 1988). In a subsequent paper Ashby and Bertotti (1986) succeeded in constructing explicitly a geocentric system by means of the technique of the generalized Fermi normal coordinates, thus enabling them to derive the Earth satellite equations of motion by direct application of the geodesic principle. The method developed here is given in Brumberg and Kopejkin (1989b). The same equations of motion were also obtained in this paper by applying the geodesic principle to the GRS metric. In an analogous way the equations of satellite motion have been derived by Voinov (1990). Investigations of the Earth satellite motion in the barycentric and geocentric reference systems have also been performed by Krivov (1988), Nordtvedt (1988), Ries *et al* (1988) and Yao *et al* (1988).

Table 5.1 Perturbing accelerations in Earth satellite motion (expressed in cm s^{-2}).

	GEOS-1	LAGEOS	NAVSTAR	Geosynchr.	Moon
ρ (km)	8000	12 000	26 000	42 000	384 000
F_0	6.2×10^{2}	2.8×10^{2}	5.9×10^{1}	2.3×10^{1}	2.7×10^{-1}
F_1	1.4×10^{0}	2.7×10^{-1}	1.2×10^{-2}	1.8×10^{-3}	2.6×10^{-7}
$F_2(L)$	6.9×10^{-5}	1.0×10^{-4}	2.3×10^{-4}	3.6×10^{-4}	—
$F_2(S)$	3.2×10^{-5}	4.7×10^{-5}	1.0×10^{-4}	1.7×10^{-4}	1.5×10^{-3}
$F_3(L)$	3.1×10^{-9}	3.1×10^{-9}	3.1×10^{-9}	3.1×10^{-9}	—
$F_3(S)$	3.7×10^{-12}	3.7×10^{-12}	3.7×10^{-12}	3.7×10^{-12}	3.7×10^{-12}
$c^{-2}g_1(L)$	8.8×10^{-11}	3.3×10^{-11}	8.3×10^{-12}	3.2×10^{-12}	—
$c^{-2}g_1(S)$	6.1×10^{-6}	2.7×10^{-6}	5.8×10^{-7}	2.2×10^{-7}	2.7×10^{-9}
$c^{-2}g_2(L)$	2.6×10^{-9}	7.8×10^{-10}	7.6×10^{-11}	1.8×10^{-11}	—
$c^{-2}g_2(S)$	7.7×10^{-8}	2.3×10^{-8}	2.2×10^{-9}	5.3×10^{-10}	7.0×10^{-13}
$c^{-2}g_3(L)$	1.1×10^{-10}	1.8×10^{-11}	5.4×10^{-13}	6.2×10^{-14}	—
$c^{-2}g_3(S)$	1.3×10^{-8}	2.6×10^{-9}	1.2×10^{-10}	1.8×10^{-11}	2.5×10^{-15}
$c^{-2}g_4(L)$	1.8×10^{-12}	1.2×10^{-12}	5.6×10^{-13}	3.5×10^{-13}	—
$c^{-2}g_4(S)$	3.3×10^{-10}	2.2×10^{-10}	1.0×10^{-10}	6.2×10^{-11}	6.8×10^{-12}
$c^{-2}g_5(L)$	2.7×10^{-13}	2.2×10^{-13}	1.5×10^{-13}	1.2×10^{-13}	—
$c^{-2}g_5(S)$	1.4×10^{-9}	1.1×10^{-9}	7.7×10^{-10}	6.0×10^{-10}	2.0×10^{-10}
$c^{-2}\Phi_1$	3.5×10^{-7}	1.0×10^{-7}	1.0×10^{-8}	2.4×10^{-9}	3.1×10^{-12}
$c^{-2}\Phi_2$	1.8×10^{-8}	4.4×10^{-9}	3.0×10^{-10}	5.5×10^{-11}	2.4×10^{-14}
$c^{-2}\Phi_3$	7.5×10^{-10}	9.9×10^{-11}	2.1×10^{-12}	1.9×10^{-13}	3.0×10^{-18}
$c^{-2}\Phi_4(L)$	3.8×10^{-14}	3.8×10^{-14}	3.8×10^{-14}	3.8×10^{-14}	—
$c^{-2}\Phi_4(S)$	1.7×10^{-14}	1.7×10^{-14}	1.7×10^{-14}	1.7×10^{-14}	1.7×10^{-14}
$c^{-2}\Phi_5(L)$	3.8×10^{-14}	3.8×10^{-14}	3.8×10^{-14}	3.8×10^{-14}	—
$c^{-2}\Phi_5(S)$	7.4×10^{-14}	9.0×10^{-14}	1.3×10^{-13}	1.7×10^{-13}	5.1×10^{-13}
$c^{-2}\Phi_6(L)$	9.9×10^{-18}	1.5×10^{-17}	3.2×10^{-17}	5.2×10^{-17}	—
$c^{-2}\Phi_6(S)$	3.1×10^{-13}	4.7×10^{-13}	1.0×10^{-12}	1.6×10^{-12}	1.5×10^{-11}

5.2 MOTION OF THE MAJOR PLANETS

5.2.1 Barycentric metric and barycentric equations

The BRS metric and the BRS equations of motion are given in section 4.1 taking into account quadrupole and spin terms. Keeping in mind practical applications, these results are simplified here, on the one hand, by considering only non-rotating point masses. On the other hand, they are generalized by introducing the coordinate parameters α, ν (Brumberg 1972) and the main parameters β, γ of the PPN formalism (Will and Nordtvedt 1972).

Then the field metric of N non-rotating point masses is described in the form

$$\begin{aligned}
ds^2 = \Bigg[& 1 - 2\sum_i \frac{m_i}{r_i} + 2(\beta-\alpha)\left(\sum_i \frac{m_i}{r_i}\right)^2 + (4\beta-2)\sum_i \frac{m_i}{r_i}\sum_{j\neq i}\frac{m_j}{r_{ij}} \\
& + 2\alpha\sum_i \frac{m_i}{r_i^3}\sum_{j\neq i} m_j\left(\frac{1}{r_{ij}} - \frac{1}{r_j}\right)(\boldsymbol{r}_i\boldsymbol{r}_{ij}) - c^{-2}(2\gamma+1)\sum_i \frac{m_i}{r_i}\dot{\boldsymbol{x}}_i^2 \\
& + c^{-2}(\nu-1)\sum_i \frac{m_i}{r_i}\left(\dot{\boldsymbol{x}}_i^2 - \boldsymbol{r}_i\ddot{\boldsymbol{x}}_i - \frac{1}{r_i^2}(\boldsymbol{r}_i\dot{\boldsymbol{x}}_i)^2\right)\Bigg] c^2\,dt^2 \\
& + 2c^{-1}\sum_i \frac{m_i}{r_i}\left((2\gamma+2-\alpha-\tfrac{1}{2}\nu)\dot{\boldsymbol{x}}_i + (\alpha+\tfrac{1}{2}\nu)\frac{1}{r_i^2}(\boldsymbol{r}_i\dot{\boldsymbol{x}}_i)\boldsymbol{r}_i\right) d\boldsymbol{x}\,c\,dt \\
& - \left(1 + 2(\gamma-\alpha)\sum_i \frac{m_i}{r_i}\right)(d\boldsymbol{x})^2 - 2\alpha\sum_i \frac{m_i}{r_i^3}(\boldsymbol{r}_i d\boldsymbol{x})^2
\end{aligned} \tag{5.2.1}$$

with $\boldsymbol{r}_i = \boldsymbol{x} - \boldsymbol{x}_i$, $\boldsymbol{r}_{ij} = \boldsymbol{x}_i - \boldsymbol{x}_j$, $m_i = GM_i/c^2$, M_i being the masses of the bodies. Denoting the harmonic coordinates corresponding to $\alpha = \nu = 0$ by a tilde one has

$$\tilde{t} = t - \frac{\nu}{2c^2}\sum_i \frac{m_i}{r_i}(\boldsymbol{r}_i\dot{\boldsymbol{x}}_i) \qquad \tilde{\boldsymbol{x}} = \boldsymbol{x} - \alpha\sum_i \frac{m_i}{r_i}\boldsymbol{r}_i \tag{5.2.2}$$

and

$$\tilde{\boldsymbol{x}}_i = \boldsymbol{x}_i - \alpha\sum_{j\neq i}\frac{m_i}{r_{ij}}\boldsymbol{r}_{ij} \tag{5.2.3}$$

$$\tilde{r}_{ij} = r_{ij} - \alpha(m_i + m_j) + \alpha\frac{\boldsymbol{r}_{ij}}{r_{ij}}\sum_{k\neq i,j} m_k\left(\frac{\boldsymbol{r}_{jk}}{r_{jk}} - \frac{\boldsymbol{r}_{ik}}{r_{ik}}\right). \tag{5.2.4}$$

The last formula determines the mutual distance between bodies in different quasi-Galilean coordinate systems.

The equations of motion of the N-point mass problem associated with (5.2.1) may be presented in the form

$$\ddot{\boldsymbol{x}}_i = -\sum_{j\neq i}\frac{GM_j}{r_{ij}^3}\boldsymbol{r}_{ij} + \sum_{j\neq i} m_j(A_{ij}\boldsymbol{r}_{ij} + B_{ij}\dot{\boldsymbol{r}}_{ij}) \tag{5.2.5}$$

with

$$\begin{aligned}
A_{ij} = & \frac{\dot{\boldsymbol{x}}_i^2}{r_{ij}^3} - (\gamma+1+\alpha)\frac{\dot{\boldsymbol{r}}_{ij}^2}{r_{ij}^3} + \frac{3}{2r_{ij}^5}(\boldsymbol{r}_{ij}\dot{\boldsymbol{x}}_j)^2 + \frac{3\alpha}{r_{ij}^5}(\boldsymbol{r}_{ij}\dot{\boldsymbol{r}}_{ij})^2 \\
& + G[(2\gamma+2\beta+1-2\alpha)M_i + (2\gamma+2\beta-2\alpha)M_j]\frac{1}{r_{ij}^4}
\end{aligned}$$

$$+\sum_{k\neq i,j} GM_k \left[(2\gamma+2\beta-\alpha)\frac{1}{r_{ij}^3 r_{ik}} + (2\beta-1-2\alpha)\frac{1}{r_{ij}^3 r_{jk}} + \frac{\alpha}{r_{ij} r_{ik}^3}\right.$$

$$+\frac{2(\gamma+1)}{r_{ij} r_{jk}^3} + (-2\gamma-\tfrac{3}{2}+\alpha)\frac{1}{r_{ik} r_{jk}^3} - \frac{\alpha}{r_{ij}^3 r_{jk}}$$

$$\left. + \left(\frac{\alpha}{r_{ik}^3} - \frac{\alpha+1/2}{r_{jk}^3} - \frac{3\alpha}{r_{ij}^2 r_{ik}} + \frac{3\alpha}{r_{ij}^2 r_{jk}}\right)\frac{1}{r_{ij}^3}(\boldsymbol{r}_{ij}\boldsymbol{r}_{ik})\right] \tag{5.2.6}$$

$$B_{ij} = \frac{1}{r_{ij}^3}[(2\gamma+2-2\alpha)(\boldsymbol{r}_{ij}\dot{\boldsymbol{r}}_{ij}) + \boldsymbol{r}_{ij}\dot{\boldsymbol{x}}_j]. \tag{5.2.7}$$

These equations have the standard form of the equations of the perturbed motion in celestial mechanics and are convenient for numerical integration. They differ from the equations of the PPN formalism (Estabrook 1969, Anderson 1974) employed in particular at JPL for the actual calculation of the ephemerides of the major planets (Standish *et al* 1976) by the fact that the right-hand members of (5.2.5) are resolved only into vectors $\boldsymbol{r}_{ij}$, $\dot{\boldsymbol{r}}_{ij}$ and do not contain second derivatives.

If metric (5.2.1) and equations (5.2.5) are referred to the barycentric system then

$$\sum_i M_i\left(1+\frac{1}{2c^2}\dot{\boldsymbol{x}}_i^2 - \tfrac{1}{2}\sum_{j\neq i}\frac{m_j}{r_{ij}}\right)\boldsymbol{x}_i = 0. \tag{5.2.8}$$

For analytical or qualitative examination it is suitable to put equations (5.2.5) in the Lagrange form with Lagrangian

$$L = \tfrac{1}{2}\sum_i GM_i\dot{\boldsymbol{x}}_i^2 + \tfrac{1}{2}\sum_i\sum_{j\neq i}\frac{G^2M_iM_j}{r_{ij}} + c^{-2}\left[\tfrac{1}{8}\sum_i GM_i(\dot{\boldsymbol{x}}_i^2)^2\right.$$

$$+\sum_i\sum_{j\neq i}\frac{G^2M_iM_j}{r_{ij}}\left((\gamma+\tfrac{1}{2}-\alpha)\dot{\boldsymbol{x}}_i^2 + (-\gamma-\tfrac{3}{4}+\alpha)\dot{\boldsymbol{x}}_i\dot{\boldsymbol{x}}_j\right.$$

$$\left. -(\tfrac{1}{4}+\alpha)\frac{1}{r_{ij}^2}(\boldsymbol{r}_{ij}\dot{\boldsymbol{x}}_i)(\boldsymbol{r}_{ij}\dot{\boldsymbol{x}}_j) + \frac{\alpha}{r_{ij}^2}(\boldsymbol{r}_{ij}\dot{\boldsymbol{x}}_i)^2\right)$$

$$+(-\beta+\tfrac{1}{2}+\alpha)\sum_i GM_i\left(\sum_{j\neq i}\frac{GM_j}{r_{ij}}\right)^2$$

$$\left. +\alpha\sum_i\sum_{j\neq i}\sum_{k\neq i,j}\frac{G^3M_iM_jM_k}{r_{ij}^3 r_{ik}}\boldsymbol{r}_{ij}\boldsymbol{r}_{jk}\right]. \tag{5.2.9}$$

5.2.2 Heliocentric equations

The barycentric equations (5.2.5) may be easily converted to heliocentric equations. Referring the zero index to the Sun and introducing instead of $\boldsymbol{x}_i$

($i = 0, 1, 2, \ldots$) the heliocentric position vectors of the planets $\boldsymbol{R}_i = \boldsymbol{x}_i - \boldsymbol{x}_0$ and the vector

$$\boldsymbol{R} = \left(\sum_i M_i \boldsymbol{x}_i\right)\left(\sum_i M_i\right)^{-1} \tag{5.2.10}$$

of the Newtonian centre of mass, one obtains

$$\begin{aligned}\ddot{\boldsymbol{R}}_i = & - G(M_0 + M_i)\frac{\boldsymbol{R}_i}{R_i^3} + \sum_{j\neq i} GM_j \left(\frac{\boldsymbol{R}_j - \boldsymbol{R}_i}{r_{ij}^3} - \frac{\boldsymbol{R}_j}{R_j^3}\right) \\ & + (m_0 A_{i0} + m_i A_{0i})\boldsymbol{R}_i + (m_0 B_{i0} + m_i B_{0i})\dot{\boldsymbol{R}}_i \\ & + \sum_{j\neq i} m_j [A_{ij}(\boldsymbol{R}_i - \boldsymbol{R}_j) + A_{0j}\boldsymbol{R}_j + B_{ij}(\dot{\boldsymbol{R}}_i - \dot{\boldsymbol{R}}_j) + B_{0j}\dot{\boldsymbol{R}}_j].\end{aligned} \tag{5.2.11}$$

The summation is performed over $j = 1, 2, \ldots$ ($j \neq i$). The vector $\boldsymbol{R}$ may be determined from (5.2.8). Equations (5.2.11), of course, have nothing to do with the heliocentric reference system in a dynamical sense. One may construct the heliocentric reference system and the relevant equations of motion by the same technique as was used for GRS and the GRS satellite equations. The independent argument should therefore be the heliocentric time, i.e. the coordinate time of the heliocentric reference system. In distinction from the GRS satellite equations of motion one cannot expand here the right-hand members in powers of the heliocentric coordinates of the planets. However, for the representation of planetary motion such a procedure is not necessary at present, considering that the direct relativistic mutual perturbations of the planets are quite negligible. In fact, one often ignores in the right-hand members (5.2.11) all relativistic terms containing planetary masses M_i ($i = 1, 2, \ldots$) as factors. Then the coefficients of the solar relativistic terms contain only the Schwarzschild terms

$$A_{i0} = -(\gamma + \alpha)\frac{\dot{\boldsymbol{R}}_i^2}{R_i^3} + \frac{3\alpha}{R_i^5}(\boldsymbol{R}_i\dot{\boldsymbol{R}}_i)^2 + (2\beta + 2\gamma - 2\alpha)\frac{GM_0}{R_i^4} \tag{5.2.12}$$

$$B_{i0} = (2\gamma + 2 - 2\alpha)(\boldsymbol{R}_i\dot{\boldsymbol{R}}_i)\frac{1}{R_i^3}. \tag{5.2.13}$$

At present, in calculating the relativistic effects in the motion of the major planets it is sufficient to consider the Schwarzschild problem only. Therefore, in integrating numerically one may omit in (5.2.5) or (5.2.11) all relativistic coefficients except for (5.2.12) and (5.2.13). In analytical consideration of the relativistic perturbations one may in virtue of the smallness of eccentricities and inclinations use expressions (3.1.81), (3.1.96), (3.1.97) or (3.1.90), (3.1.98), (3.1.99) for the spherical coordinates of the planets.

5.2.3 Numerical results

Numerical results for the major planets of the Solar System have been obtained on the basis of the equations involving coefficients (5.2.12) and (5.2.13) by Lestrade and Bretagnon (1982) as a supplement to the analytical theory of motion of the major planets VSOP-82 produced in Bureau des Longitudes (Bretagnon 1982). In using these results one should keep in mind two facts. First of all, the solution of the Schwarzschild problem is presented in this paper by expanding the perturbations of the osculating elements into trigonometric series with respect to one trigonometric variable, i.e. the mean longitude of the planet. This solution may be obtained from the closed form solution (3.1.102)–(3.1.105) with the aid of the trigonometric expansions in mean longitude λ. Secondly, this paper gives not only the Schwarzschild terms proportional to m_0 but also the terms proportional to $m_0 M_j$ caused by interaction of the Newtonian planetary perturbations and the Schwarzschild perturbations (these terms may be called the indirect planetary relativistic perturbations). The Newtonian theory of motion of the major planets VSOP-82 contains both secular and mixed terms, retaining only the mean longitudes as the trigonometric arguments. Therefore, the relativistic terms under discussion are of the same structure. As for the order of smallness of these terms, it is to be noted that this order is the same as the order of the terms in the right-hand members (5.2.11) omitted in the paper considered (the direct planetary relativistic perturbations). Hence, the terms proportional to $m_0 M_j$ obtained by Lestrade and Bretagnon include only a portion of the terms of this order, which are inadequate to estimate the real magnitude of such terms. The Schwarzschild terms calculated by Lestrade and Bretagnon are reproduced briefly in the book by Soffel (1989). Only the Schwarzschild advances of the perihelia of the orbits of the inner planets are given here. In accordance with (3.1.66) the advance for one century is determined by

$$\Delta\pi = \frac{3m_0 n}{a(1-e^2)}. \qquad (5.2.14)$$

The relativistic gravitational parameter of the Sun m_0 and the semi-major axis a of the orbit of the planet should be expressed in the same units of length. n is the mean motion of the planet for one century. With the currently adopted values $GM_0 = 132\,712 \times 10^{15}$ m^3 s^{-2}, $c = 299\,792.5 \times 10^3$ m s^{-1}, 1 AU= $149\,598 \times 10^6$ m, $m_0 = 1476.6$ m, one obtains the following results:

	Mercury	Venus	Earth	Mars
a (AU)	0.387 10	0.723 33	1.000 00	1.523 69
e	0.205 60	0.006 84	0.016 77	0.093 27
n (″/year)	5381 016	2106 641	1295 977	689 050
$\Delta\pi$ (″/century)	42.98	8.62	3.84	1.35
$e\Delta\pi$ (″/century)	8.837	0.059	0.064	0.126

Some details concerning the calculation of the Schwarzschild advance of the perihelion of Mercury are given in Nobili and Will (1986).

5.3 MOTION OF THE MOON

5.3.1 BRS and GRS treatment

The problem of the motion of the Moon has always been of particular interest in celestial mechanics. Almost all the famous specialists in celestial mechanics have to a greater or lesser extent contributed to this domain. To the end of the nineteenth and the beginning of the twentieth centuries numerous efforts resulted in a variety of very efficient theories, the most accurate of them being the Hill–Brown theory. The second half of the twentieth century saw the elaboration of new, even more refined theories. This became necessary in relation to the investigation of motion of the Moon by astrodynamic tools and the development of new observation techniques, primarily lunar laser ranging (LLR). At present, the most accurate theories are numerical theory LE200 produced at JPL and analytical theory ELP-2000 from the Bureau des Longitudes (Chapront and Chapront-Touzé 1981, Chapront-Touzé and Chapront 1983). Relativistic effects are taken into account in LE200 by simultaneously integrating the post-Newtonian equations (5.2.5) of the major planets and the Moon. In ELP-2000 the relativistic perturbations are added separately (Lestrade and Chapront-Touzé 1982). In both cases one deals with the BRS relativistic theories of the motion of the Moon. In the case of the Moon the advantages of constructing the GRS theories are not as evident as in the case of Earth artificial satellites. Certainly, as explicitly stated in Soffel *et al* (1986) and reflected in table 5.1 the magnitude of the relativistic perturbations in the motion of the Moon is much less in GRS than in BRS. But the Moon is more often considered together with the major planets and to use different reference systems (including timescales) for the major planets and the Moon may be not convenient. Perhaps both BRS and GRS theories of the motion of the

Moon are needed. So far, an extensive GRS theory of the motion of the Moon has not been constructed although the main perturbations in the motion of the Moon in Fermi coordinates have been derived (Soffel 1989). The main BRS perturbations in the motion of the Moon are given below as determined by Brumberg and Ivanova (1985).

5.3.2 Physical considerations

Before applying (5.2.5) to the problem at hand some remarks concerning the physical foundation of the point mass model are needed. In many papers on the PPN formalism the equations of motion of celestial bodies are used in a perfect fluid model (Will 1974). These equations are based on a greater number of physical assumptions than the equations of motion of point masses. In particular, the equations for massive fluid bodies include the Nordtvedt effect, i.e. the violation of the principle of equivalence for the massive bodies. In considering only the main PPN formalism parameters the expression describing this violation contains the factor $4\beta - \gamma - 3$ (Will 1971). Thus, in the equations for point masses the parameters β and γ act as characteristics of the external gravitational field (5.2.1). In addition, in the equations for massive bodies they enter as characteristics of the internal structure model as well. In the equations of motion of the Moon derived within the framework of the PPN formalism for massive bodies the Nordtvedt effect turns out to be the most significant post-Newtonian effect. But discussion of LLR observations demonstrates the absence of such an effect (Williams *et al* 1976, Shapiro *et al* 1976), resulting in the conclusion that $4\beta - \gamma - 3 = 0$. But such a conclusion uses implicitly physical assumptions of the adopted massive body model. Also, one may note the paper (Kreinovich 1975) questioning the correctness of common considerations of the Nordtvedt effect and concluding that it has a negligibly small effect on the motion of the Moon (independent of values for β and γ). It seems that the point mass model has not lost its role in the PPN formalism and consideration of the motion of the Moon on the basis of this model is meaningful.

5.3.3 Equations of relative motion

Returning to (5.2.5) let the indices 1, 2 and 3 refer to the Earth, the Sun and the Moon, respectively. Introduce instead of $\boldsymbol{x}_1$, $\boldsymbol{x}_2$ and $\boldsymbol{x}_3$ the vector $\boldsymbol{r}_0$ of the Newtonian centre of mass of three bodies, the geocentric position vector $\boldsymbol{r}$ of the Moon and the heliocentric position vector $\boldsymbol{R}$ of the Newtonian centre of mass of the Earth–Moon system. Then

$$\boldsymbol{x}_1 = \boldsymbol{r}_0 + \frac{M_2}{M}\boldsymbol{R} - \frac{M_3}{M_c}\boldsymbol{r} \qquad \boldsymbol{x}_2 = \boldsymbol{r}_0 - \frac{M_c}{M}\boldsymbol{R} \qquad \boldsymbol{x}_3 = \boldsymbol{r}_0 + \frac{M_2}{M}\boldsymbol{R} + \frac{M_1}{M_c}\boldsymbol{r} \tag{5.3.1}$$

with $M_c = M_1 + M_3$, $M = M_2 + M_c$. From (5.3.1) there results

$$\boldsymbol{r}_{12} = \boldsymbol{R} - \frac{M_3}{M_c}\boldsymbol{r} \qquad \boldsymbol{r}_{32} = \boldsymbol{R} + \frac{M_1}{M_c}\boldsymbol{r} \qquad \boldsymbol{r}_{31} = \boldsymbol{r}. \tag{5.3.2}$$

The equation determining $\boldsymbol{r}_0$ follows immediately from (5.2.8)

$$\begin{aligned}\boldsymbol{r}_0 = &-\tfrac{1}{2}c^{-2}\left(\frac{M_2}{M}\right)^2 \boldsymbol{R}\left\{\frac{M_c}{M}\left(1-\frac{M_c}{M_2}\right)\left[\dot{\boldsymbol{R}}^2 - \frac{GM}{M_c}\left(\frac{M_1}{r_{12}}+\frac{M_3}{r_{32}}\right)\right]\right.\\ &\left.+\frac{M_1M_3}{M_2M_c}\left(\dot{\boldsymbol{r}}^2 - 2\frac{GM_c}{r}\right)\right\} - \frac{1}{2}c^{-2}\frac{M_1M_3}{MM_c}\boldsymbol{r}\left[GM_2\left(\frac{1}{r_{12}}-\frac{1}{r_{32}}\right)\right.\\ &\left.+\frac{M_1-M_3}{M_c}\left(\dot{\boldsymbol{r}}^2-\frac{GM_c}{r}\right)+2\frac{M_2}{M}\dot{\boldsymbol{R}}\dot{\boldsymbol{r}}\right].\end{aligned} \tag{5.3.3}$$

Differential equations for $\boldsymbol{R}$ and $\boldsymbol{r}$ follow from (5.2.5) and (5.3.1)

$$\ddot{\boldsymbol{r}} = -\frac{GM_c}{r^3}\boldsymbol{r} + GM_2\left(\frac{\boldsymbol{r}_{12}}{r_{12}^3} - \frac{\boldsymbol{r}_{32}}{r_{32}^3}\right) + m_2(A\boldsymbol{R} + B\dot{\boldsymbol{R}} + C\boldsymbol{r} + D\dot{\boldsymbol{r}}) \tag{5.3.4}$$

$$\ddot{\boldsymbol{R}} = -\frac{M}{M_c}\left(\frac{GM_1}{r_{12}^3}\boldsymbol{r}_{12} + \frac{GM_3}{r_{32}^3}\boldsymbol{r}_{32}\right) + m_2(A'\boldsymbol{R} + B'\dot{\boldsymbol{R}} + C'\boldsymbol{r} + D'\dot{\boldsymbol{r}}) \tag{5.3.5}$$

with

$$\begin{aligned} A &= A_{32} - A_{12} \qquad B = B_{32} - B_{12}\\ C &= \frac{M_1}{M_c}A_{32} + \frac{M_3}{M_c}A_{12} + \frac{M_1}{M_2}A_{31} + \frac{M_3}{M_2}A_{13}\\ D &= \frac{M_1}{M_c}B_{32} + \frac{M_3}{M_c}B_{12} + \frac{M_1}{M_2}B_{31} + \frac{M_3}{M_2}B_{13}\end{aligned} \tag{5.3.6}$$

and

$$\begin{aligned} A' &= \frac{M_1}{M_c}A_{12} + \frac{M_3}{M_c}A_{32} + \frac{M_1}{M_2}A_{21} + \frac{M_3}{M_2}A_{23}\\ B' &= \frac{M_1}{M_c}B_{12} + \frac{M_3}{M_c}B_{32} + \frac{M_1}{M_2}B_{21} + \frac{M_3}{M_2}B_{23}\\ C' &= \frac{M_1M_3}{M_c^2}(A_{32} - A_{12}) + \frac{M_1M_3}{M_2M_c}(A_{23} - A_{21} + A_{31} - A_{13})\\ D' &= \frac{M_1M_3}{M_c^2}(B_{32} - B_{12}) + \frac{M_1M_3}{M_2M_c}(B_{23} - B_{21} + B_{31} - B_{13}).\end{aligned} \tag{5.3.7}$$

Within the post-Newtonian approximation equations (5.3.4) and (5.3.5) are quite rigorous. Under actual practical requirements one may put $C' =$

$D' = 0$ in the right-hand side of (5.3.5) and retain in A' and B' only the Schwarzschild terms

$$A' = -(\gamma+\alpha)\frac{\dot{\boldsymbol{R}}^2}{R^3} + \frac{3\alpha}{R^5}(\boldsymbol{R}\dot{\boldsymbol{R}})^2 + (2\beta+2\gamma-2\alpha)\frac{GM_2}{R^4}$$
$$B' = (2\gamma+2-2\alpha)(\boldsymbol{R}\dot{\boldsymbol{R}})\frac{1}{R^3}. \qquad (5.3.8)$$

Then equation (5.3.5) is related to (5.3.4) only by means of the coordinates $\boldsymbol{r}$ of the Moon entering into the Newtonian terms of the right-hand member (5.3.5). In simultaneous numerical integration of the equations for the major planets and the Moon it is suitable to replace the equations for the Moon and the Earth by equations (5.3.4) and (5.3.5), adding the appropriate terms for the planetary perturbations.

If one neglects in (5.3.6) the Schwarzschild terms, the terms due to the motion of the Earth and the terms of the Earth–Sun coupling then the most important solar terms are

$$A = 8(-\beta-\gamma+\alpha)\frac{GM_2}{R^6}(\boldsymbol{R}\boldsymbol{r}) + 3(\gamma+\alpha)\frac{\dot{\boldsymbol{R}}^2}{R^5}(\boldsymbol{R}\boldsymbol{r}) - 2(\gamma+\alpha)\frac{\dot{\boldsymbol{R}}\dot{\boldsymbol{r}}}{R^3}$$
$$+ \frac{3\alpha}{R^5}(\boldsymbol{R}\dot{\boldsymbol{R}})\left(2\boldsymbol{R}\dot{\boldsymbol{r}} + 2\boldsymbol{r}\dot{\boldsymbol{R}} - \frac{5}{R^2}(\boldsymbol{R}\dot{\boldsymbol{R}})(\boldsymbol{R}\boldsymbol{r})\right)$$
$$B = 2(\gamma+1-\alpha)\frac{1}{R^3}\left(\boldsymbol{R}\dot{\boldsymbol{r}} + \boldsymbol{r}\dot{\boldsymbol{R}} - \frac{3}{R^2}(\boldsymbol{R}\dot{\boldsymbol{R}})(\boldsymbol{R}\boldsymbol{r})\right) \qquad (5.3.9)$$
$$C = 2(\beta+\gamma-\alpha)\frac{GM_2}{R^4} - (\gamma+\alpha)\frac{\dot{\boldsymbol{R}}^2}{R^3} + \frac{3\alpha}{R^5}(\boldsymbol{R}\dot{\boldsymbol{R}})^2$$
$$D = 2(\gamma+1-\alpha)(\boldsymbol{R}\dot{\boldsymbol{R}})\frac{1}{R^3}.$$

These coefficients take an even simpler form when substituting the circular motion for the Earth–Moon centre of mass:

$$\dot{\boldsymbol{R}}^2 = GM_2/R \qquad \boldsymbol{R}\dot{\boldsymbol{R}} = 0.$$

Then

$$A = (-8\beta-5\gamma+11\alpha)\frac{GM_2}{R^6}(\boldsymbol{R}\boldsymbol{r}) - 2(\gamma+\alpha)\frac{\dot{\boldsymbol{R}}\dot{\boldsymbol{r}}}{R^3}$$
$$B = 2(\gamma+1-\alpha)(\boldsymbol{R}\dot{\boldsymbol{r}} + \boldsymbol{r}\dot{\boldsymbol{R}})\frac{1}{R^3} \qquad (5.3.10)$$
$$C = (2\beta+\gamma-3\alpha)\frac{GM_2}{R^4} \qquad D = 0.$$

In particular, these terms describe the effect of geodesic precession. In fact, in the limit $(r/R) \to 0$ in the relativistic right-hand side of equation (5.3.4) under (5.3.10) there remains only

$$(\ddot{\boldsymbol{r}})_{rel} = \frac{m_2}{R^3}(2\gamma+1)[(\boldsymbol{R}\times\dot{\boldsymbol{R}})\times\dot{\boldsymbol{r}}] + \frac{m_2}{R^3}(1-2\alpha)[(\dot{\boldsymbol{R}}\dot{\boldsymbol{r}})\boldsymbol{R} + (\boldsymbol{R}\dot{\boldsymbol{r}})\dot{\boldsymbol{R}}]. \quad (5.3.11)$$

The first term describes the geodesic precession with rate

$$(2\gamma+1)\frac{m_2}{R}N \tag{5.3.12}$$

N being the mean motion of the Earth–Moon barycentre. Thus, one comes again to the terms (5.1.20). Such terms are absent in the GRS theory of the Moon's motion.

5.3.4 Equations in Lagrange form

When applying the Lagrangian (5.2.9) to the equations of motion of the Earth–Moon–Sun system one has to substitute (5.1.1) in (5.2.9). Firstly, one separates the terms determining the motion of the Newtonian centre of mass

$$\begin{aligned}
L_0 = {} & \tfrac{1}{2}GM\dot{\boldsymbol{r}}_0^2 + \tfrac{1}{2}c^{-2}(\boldsymbol{R}\dot{\boldsymbol{r}}_0)\frac{M_2}{M}\left[G(M_2-M_c)\left(\frac{M_c}{M}\dot{\boldsymbol{R}}^2 - \frac{GM_1}{r_{12}} - \frac{GM_3}{r_{32}}\right)\right.\\
& \left. + \frac{GM_1M_3}{M_c}\left(\dot{\boldsymbol{r}}^2 - 2\frac{GM_c}{r}\right)\right] + \tfrac{1}{2}c^{-2}\frac{GM_1M_3}{M_c}(\dot{\boldsymbol{r}}\dot{\boldsymbol{r}}_0)\\
& \times\left[\frac{M_2}{M}\left(2\dot{\boldsymbol{R}}\dot{\boldsymbol{r}} + \frac{GM}{r_{12}} - \frac{GM}{r_{32}}\right) + \frac{M_1-M_3}{M_c}\left(\dot{\boldsymbol{r}}^2 - \frac{GM_c}{r}\right)\right]\\
& - \tfrac{1}{2}c^{-2}\frac{G^2M_2M_1}{r_{12}^3}(\boldsymbol{r}_{12}\dot{\boldsymbol{r}}_0)\left(\frac{M_2-M_c}{M}(\boldsymbol{r}_{12}\dot{\boldsymbol{R}}) - \frac{M_3}{M_c}(\boldsymbol{r}_{12}\dot{\boldsymbol{r}})\right)\\
& - \tfrac{1}{2}c^{-2}\frac{G^2M_2M_3}{r_{32}^3}(\boldsymbol{r}_{32}\dot{\boldsymbol{r}}_0)\left(\frac{M_2-M_c}{M}(\boldsymbol{r}_{32}\dot{\boldsymbol{R}}) + \frac{M_1}{M_c}(\boldsymbol{r}_{32}\dot{\boldsymbol{r}})\right)\\
& - \tfrac{1}{2}c^{-2}\frac{G^2M_1M_3}{r^3}(\boldsymbol{r}\dot{\boldsymbol{r}}_0)\left(\frac{M_1-M_3}{M_c}(\boldsymbol{r}\dot{\boldsymbol{r}}) + 2\frac{M_2}{M}(\boldsymbol{R}\boldsymbol{r})\right).
\end{aligned}\tag{5.3.13}$$

The motion of the Newtonian centre of mass is determined by the equation $\partial L_0/\partial\dot{\boldsymbol{r}}_0 = 0$ and its integration again leads to (5.3.3). Within the post-Newtonian approximation the heliocentric motion of the Newtonian Earth–Moon barycentre and the geocentric motion of the Moon $\boldsymbol{r}$ are separated from the motion of the Newtonian centre of mass $\boldsymbol{r}_0$ of the whole Earth–Moon–Sun system. Hence, L_0 may be omitted and the Lagrangian of (5.3.4) and (5.3.5) takes the form

$$\begin{aligned}
L = {} & \frac{1}{2}\frac{GM_2M_c}{M}\dot{\boldsymbol{R}}^2 + \frac{1}{2}\frac{GM_1M_3}{M_c}\dot{\boldsymbol{r}}^2 + \frac{G^2M_2M_1}{r_{12}} + \frac{G^2M_2M_3}{r_{32}} + \frac{G^2M_1M_3}{r}\\
& + m_2\left[\frac{1}{8}\frac{M_c}{M}\left(1 - 3\frac{M_2M_c}{M^2}\right)\dot{\boldsymbol{R}}^4 + \left(-\alpha+\gamma+\frac{1}{2}+\frac{1}{2}\frac{M_2M_c}{M^2}\right)\right.\\
& \times\left(\frac{GM_1}{r_{12}} + \frac{GM_3}{r_{32}}\right)\dot{\boldsymbol{R}}^2 + \left(\alpha + \frac{1}{2}\frac{M_2M_c}{M^2}\right)\left(\frac{GM_1}{r_{12}^3} + \frac{GM_3}{r_{32}^3}\right)(\boldsymbol{R}\dot{\boldsymbol{R}})^2
\end{aligned}$$

$$
\begin{aligned}
&+(\alpha-\beta+\tfrac{1}{2})G^2\left((M_2+M_1)\frac{M_1}{r_{12}^3}+(M_2+M_3)\frac{M_3}{r_{32}^3}\right)\Bigg]\\
&+m_2\frac{GM_1M_3}{M_c}\Bigg\{\left(\frac{1}{r_{12}}-\frac{1}{r_{32}}\right)\\
&\times\left[\left(2\alpha-2\gamma-1-\frac{1}{2}\frac{M_c}{M}\right)(\boldsymbol{R}\dot{\boldsymbol{r}})-\alpha\frac{GM_c}{r^3}(\boldsymbol{R}\boldsymbol{r})\right]\\
&-\left(\frac{1}{r_{12}^3}-\frac{1}{r_{32}^3}\right)(\boldsymbol{R}\dot{\boldsymbol{R}})\left[\left(2\alpha+\frac{M_2M_c}{M^2}\right)(\dot{\boldsymbol{R}}\boldsymbol{r})+\left(2\alpha+\frac{1}{2}\frac{M_c}{M}\right)(\boldsymbol{R}\dot{\boldsymbol{r}})\right]\\
&+\frac{GM_2}{M^2}\left(\frac{1}{2}(\boldsymbol{R}\dot{\boldsymbol{r}})^2+\frac{1}{4}\dot{\boldsymbol{R}}^2\dot{\boldsymbol{r}}^2\right)-\frac{1}{2}\frac{M_2M_c}{M^2}\frac{1}{r}\left(\dot{\boldsymbol{R}}^2+\frac{1}{r^2}(\dot{\boldsymbol{R}}\boldsymbol{r})^2\right)\\
&+\frac{1}{M_c}(-\alpha+\gamma+\tfrac{1}{2})\left(\frac{M_3}{r_{12}}+\frac{M_1}{r_{32}}\right)\dot{\boldsymbol{r}}^2+\frac{1}{M_c}\left(\frac{M_3}{r_{12}^3}+\frac{M_1}{r_{32}^3}\right)\\
&\times\Bigg[\left(\alpha+\frac{1}{2}\frac{M_2M_c}{M^2}\right)(\dot{\boldsymbol{R}}\boldsymbol{r})^2+\alpha(\boldsymbol{R}\dot{\boldsymbol{r}})^2+\left(2\alpha+\frac{1}{2}\frac{M_c}{M}\right)(\boldsymbol{R}\dot{\boldsymbol{r}})(\dot{\boldsymbol{R}}\boldsymbol{r})\\
&+\left(2\alpha+\frac{1}{2}\frac{M_c}{M}\right)(\boldsymbol{R}\dot{\boldsymbol{R}})(\boldsymbol{r}\dot{\boldsymbol{r}})\Bigg]+(2\alpha-2\beta+1)\frac{GM_c}{r}\left(\frac{1}{r_{12}}+\frac{1}{r_{32}}\right)\\
&-\frac{\alpha}{r}\left(\frac{GM_1}{r_{12}}+\frac{GM_3}{r_{32}}\right)-\alpha\left(\frac{1}{r_{12}^3}+\frac{1}{r_{32}^3}\right)\\
&\times\left(\frac{GM_c}{r}R^2+G\frac{M_1-M_3}{r}(\boldsymbol{R}\boldsymbol{r})-\frac{GM_1M_3}{M_c}r\right)\\
&+\frac{1}{2}\frac{M_1-M_3}{GMM_c}(\boldsymbol{R}\dot{\boldsymbol{r}})\left(\dot{\boldsymbol{r}}^2-\frac{GM_c}{r}\right)-\frac{1}{2}\frac{M_1-M_3}{M}\frac{1}{r^3}(\boldsymbol{r}\dot{\boldsymbol{r}})(\dot{\boldsymbol{R}}\boldsymbol{r})\\
&-\frac{1}{M_c^2}\left(\frac{M_3^2}{r_{12}^3}-\frac{M_1^2}{r_{32}^3}\right)(\boldsymbol{r}\dot{\boldsymbol{r}})\left[\left(2\alpha+\frac{1}{2}\frac{M_c}{M}\right)(\dot{\boldsymbol{R}}\boldsymbol{r})+2\alpha(\boldsymbol{R}\dot{\boldsymbol{r}})\right]\\
&+(2\alpha-2\beta+1)\frac{GM_c}{r_{12}r_{32}}+\frac{1}{8GM_2}\left(1-3\frac{M_1M_3}{M_c^2}\right)\dot{\boldsymbol{r}}^4\\
&+\frac{\alpha}{M_c^3}\left(\frac{M_3^3}{r_{12}^3}+\frac{M_1^3}{r_{32}^3}\right)(\boldsymbol{r}\dot{\boldsymbol{r}})^2+\frac{M_c}{M_2}\left(-\alpha+\gamma+\frac{1}{2}+\frac{1}{2}\frac{M_1M_3}{M_c^2}\right)\frac{1}{r}\dot{\boldsymbol{r}}^2\\
&+\frac{M_c}{M_2}\left(\alpha+\frac{1}{2}\frac{M_1M_3}{M_c^2}\right)\frac{1}{r_3}(\boldsymbol{r}\dot{\boldsymbol{r}})^2+\frac{GM_c^2}{M_2}(\alpha-\beta+\tfrac{1}{2})\frac{1}{r^2}\\
&+\alpha\frac{GM_c}{r_{12}r_{32}}\left(\frac{1}{r_{12}^2}-\frac{1}{r_{32}^2}\right)(\boldsymbol{R}\boldsymbol{r})-\frac{\alpha}{r_{12}r_{32}}\left(\frac{GM_3}{r_{12}^2}+\frac{GM_1}{r_{32}^2}\right)r^2\Bigg\}. \quad (5.3.14)
\end{aligned}
$$

The Lagrangian (5.3.14) is quite rigorous and corresponds completely to equations (5.3.4) and (5.3.5).

5.3.5 Principal inequalities in the motion of the Moon

To provide the solution in analytical form the Lagrangian (5.3.14) is expanded in powers of the ratio r/R using relations (5.3.2). To proceed, one separates in (5.3.14) the terms dependent only on the heliocentric motion of the Newtonian Earth–Moon barycentre

$$L_c = \frac{GM_2M_c}{M}\left\{\tfrac{1}{2}\dot{\boldsymbol{R}}^2 + \frac{GM}{r} + c^{-2}\left[\frac{1}{8}\left(1 - 3\frac{M_2M_c}{M^2}\right)\dot{\boldsymbol{R}}^4 \right.\right.$$
$$+ \left(-\alpha + \gamma + \frac{1}{2} + \frac{1}{2}\frac{M_2M_c}{M^2}\right)\frac{GM}{r}\dot{\boldsymbol{R}}^2 + \left(\alpha + \frac{1}{2}\frac{M_2M_c}{M^2}\right)\frac{GM}{R^3}(\boldsymbol{R}\dot{\boldsymbol{R}})^2$$
$$\left.\left. + (\alpha - \beta + \tfrac{1}{2})\left(1 - 2\frac{M_1M_3}{MM_c}\right)\frac{G^2M^2}{R^2}\right]\right\}. \tag{5.3.15}$$

This enables us to consider the heliocentric motion of the Newtonian Earth–Moon barycentre as known and described by the Schwarzschild problem solution (3.1.90), (3.1.91) and (3.1.94). Therefore, choosing as the basic reference plane the orbital plane of the Newtonian Earth–Moon barycentre one obtains for the components of the vector $\boldsymbol{R} = (X', Y', Z')$

$$\frac{1}{A}(X' + \mathrm{i}Y') = \exp \mathrm{i}\lambda' + e'\{[-\tfrac{3}{2} + \tfrac{1}{2}\sigma(\alpha - \beta + \gamma + 1)]\exp \mathrm{i}\pi'$$
$$+ [\tfrac{1}{2} + \tfrac{1}{2}\sigma(\alpha - \beta + \gamma + 1)]\exp \mathrm{i}(2\lambda' - \pi')\} + \ldots \tag{5.3.16}$$

$$Z' = 0 \tag{5.3.17}$$

with

$$\dot{\lambda}' = N \qquad \dot{\pi}' = \sigma(-\beta + 2\gamma + 2)N \qquad [1 + \sigma(-3\alpha + 2\beta + \gamma)]N^2A^3 = GM \tag{5.3.18}$$

$\sigma = N^2A^2/c^2$ being a dimensionless relativistic small parameter ($\approx 10^{-8}$). The coordinates (5.3.16) and (5.3.17) are substituted into the Lagrangian and in addition to the expansion in the solar parallax A^{-1} one performs an expansion in the eccentricity e'. The actual solution of the equations of motion of the Moon is constructed in Brumberg and Ivanova (1985) in the pure analytical form of the Hill–Brown series but in contrast to the classic technique of undetermined coefficients an iteration method is applied which is more suitable for analytical manipulation on series by computer. The final expansions for the sidereal spherical coordinates of the Moon

$$\boldsymbol{r} = r\begin{pmatrix}\cos\varphi\cos v\\ \cos\varphi\sin v\\ \sin\varphi\end{pmatrix} \tag{5.3.19}$$

have the form

$$\frac{r}{a_0} = R_0 + eR_e + e'R_{e'} + \frac{a_0}{A_0}R_\alpha + ee'R_{ee'} + \ldots \qquad (5.3.20)$$

$$v - \lambda = V_0 + eV_e + e'V_{e'} + \frac{a_0}{A_0}V_\alpha + ee'V_{ee'} + \ldots \qquad (5.3.21)$$

$$\varphi = k(\Phi_k + e'\Phi_{ke'} + \ldots). \qquad (5.3.22)$$

λ, e and k are the mean longitude, eccentricity and inclination of the orbit of the Moon, respectively, with $\dot{\lambda} = n$. a_0 and A_0 are the semi-major axes determined from the measurable mean motions n, N and the masses M_c, M by the following equations:

$$N^2A_0^3 = GM \qquad n^2a_0^3 = GM_c. \qquad (5.3.23)$$

The coefficients of expansions (5.3.20)–(5.3.22) represent series with power variable

$$m = \frac{N}{n-N} \qquad (5.3.24)$$

and trigonometric variables

$$D = \lambda - \lambda_\odot = \lambda - \lambda' + 180^\circ \qquad l' = fD + l'_0 \qquad l = cD + l_0 \qquad F = gD + F_0. \qquad (5.3.25)$$

l' is the mean anomaly of the orbit of the Earth–Moon barycentre, and l and F are the mean anomaly and the argument of latitude of the orbit of the Moon, respectively. Denote the relativistic small parameters by

$$\epsilon = \sigma(-\tfrac{2}{3}\beta + \tfrac{2}{3}\gamma + 1) \qquad \epsilon_1 = \sigma(2\gamma + 1) \qquad \epsilon_2 = \sigma \qquad \epsilon_3 = \alpha\sigma. \qquad (5.3.26)$$

The initial terms of coefficients (5.3.20)–(5.3.22) are

$$\begin{aligned} R_0 = {} & 1 + 2\epsilon - \epsilon_1 - \frac{5}{4}\epsilon_2 + \frac{1}{2}\epsilon_3 + \frac{1}{3}\epsilon_1 m + \left(-\frac{1}{6} - \frac{5}{6}\epsilon - \frac{25}{96}\epsilon_2 + \frac{41}{48}\epsilon_3\right)m^2 \\ & + \left(\frac{1}{3} + \frac{5}{3}\epsilon - \frac{5}{9}\epsilon_1 - \frac{53}{48}\epsilon_2 + \frac{37}{24}\epsilon_3\right)m^3 \\ & + \left[\frac{1}{4}\epsilon_2 - \frac{1}{2}\epsilon_3 + \left(-1 - 3\epsilon + 2\epsilon_1 + \frac{29}{24}\epsilon_2 - \frac{5}{12}\epsilon_3\right)m^2\right. \\ & \left. + \left(-\frac{7}{6} - \frac{7}{2}\epsilon + \frac{31}{12}\epsilon_1 + \frac{37}{24}\epsilon_2 - \frac{3}{4}\epsilon_3\right)m^3\right]\cos 2D \\ & + \left[\left(\frac{7}{32}\epsilon_2 - \frac{7}{16}\epsilon_3\right)m^2 + \left(\frac{19}{48}\epsilon_2 - \frac{19}{24}\epsilon_3\right)m^3\right]\cos 4D + \ldots \qquad (5.3.27) \end{aligned}$$

$$R_e = \left[-\frac{1}{2} - \epsilon + \frac{1}{2}\epsilon_1 + \frac{5}{8}\epsilon_2 - \frac{1}{4}\epsilon_3 + \left(-\frac{1}{6}\epsilon_1 - \frac{75}{128}\epsilon_2 + \frac{75}{64}\epsilon_3\right) m \right.$$
$$\left. + \left(\frac{11}{24} + \frac{55}{24}\epsilon - \frac{3}{4}\epsilon_1 - \frac{3847}{1536}\epsilon_2 + \frac{3143}{768}\epsilon_3\right) m^2\right] \cos l$$
$$+ \left[-\frac{5}{16}\epsilon_2 + \frac{5}{8}\epsilon_3 + \left(-\frac{15}{16} - \frac{45}{16}\epsilon + \frac{45}{32}\epsilon_1 + \frac{75}{64}\epsilon_2 - \frac{15}{32}\epsilon_3\right) m \right.$$
$$\left. + \left(-\frac{95}{64} - \frac{105}{16}\epsilon + \frac{85}{32}\epsilon_1 + \frac{1501}{768}\epsilon_2 - \frac{361}{384}\epsilon_3\right) m^2\right] \cos(2D - l)$$
$$+ \left[\frac{3}{16}\epsilon_2 - \frac{3}{8}\epsilon_3 + \left(-\frac{17}{32} - \frac{51}{32}\epsilon + \frac{17}{16}\epsilon_1 \right.\right.$$
$$\left.\left. + \frac{87}{128}\epsilon_2 - \frac{19}{64}\epsilon_3\right) m^2\right] \cos(2D + l)$$
$$+ \left[\left(\frac{45}{128}\epsilon_2 - \frac{45}{64}\epsilon_3\right) m + \left(\frac{561}{512}\epsilon_2 - \frac{561}{256}\epsilon_3\right) m^2\right] \cos(4D - l)$$
$$+ \left(\frac{85}{256}\epsilon_2 - \frac{85}{128}\epsilon_3\right) m^2 \cos(4D + l) + \ldots \qquad (5.3.28)$$

$$R_{e'} = \left[-6\epsilon + \epsilon_1 + \frac{3}{2}\epsilon_2 + \frac{1}{2}\epsilon_3 + \left(\frac{3}{2} - \frac{35}{4}\epsilon \right.\right.$$
$$\left.\left. + \frac{1}{3}\epsilon_1 + \frac{73}{32}\epsilon_2 + \frac{53}{12}\epsilon_3\right) m^2\right] \cos l'$$
$$+ \left[\frac{1}{2}\epsilon_2 - \frac{5}{4}\epsilon_3 + \left(12\epsilon - 3\epsilon_1 - \frac{17}{4}\epsilon_2 - \frac{3}{2}\epsilon_3\right) m \right.$$
$$\left. + \left(-\frac{7}{2} + \frac{123}{4}\epsilon - \frac{1}{6}\epsilon_1 - \frac{701}{48}\epsilon_2 - \frac{2}{3}\epsilon_3\right) m^2\right] \cos(2D - l')$$
$$+ \left[\frac{3}{4}\epsilon_3 + \left(-12\epsilon + 3\epsilon_1 + \frac{17}{4}\epsilon_2 + \frac{3}{2}\epsilon_3\right) m \right.$$
$$\left. + \left(\frac{1}{2} - \frac{37}{4}\epsilon + \frac{3}{2}\epsilon_1 + \frac{367}{48}\epsilon_2 - 2\epsilon_3\right) m^2\right] \cos(2D + l')$$
$$+ \left(\frac{77}{64}\epsilon_2 - \frac{21}{8}\epsilon_3\right) m^2 \cos(4D - l')$$
$$+ \left(-\frac{7}{64}\epsilon_2 + \frac{7}{8}\epsilon_3\right) m^2 \cos(4D + l') + \ldots \qquad (5.3.29)$$

$$R_\alpha = \left[\left(3\epsilon - \frac{3}{4}\epsilon_1 - \frac{3}{4}\epsilon_2 \right) m^{-1} + 21\epsilon - 6\epsilon_1 - \frac{13}{2}\epsilon_2 + \frac{3}{8}\epsilon_3 \right.$$
$$\left. + \left(\frac{15}{16} + \frac{1781}{16}\epsilon - \frac{2029}{64}\epsilon_1 - \frac{4457}{128}\epsilon_2 - \frac{45}{64}\epsilon_3 \right) m \right] \cos D$$
$$+ \left[-\frac{3}{8}\epsilon_3 + \left(\frac{51}{16}\epsilon - \frac{51}{64}\epsilon_1 - \frac{275}{128}\epsilon_2 + \frac{45}{64}\epsilon_3 \right) m \right] \cos 3D + \dots \tag{5.3.30}$$

$$R_{ee'} = \left[\left(3\epsilon - \frac{3}{4}\epsilon_1 - \frac{5}{4}\epsilon_2 \right) m^{-1} + \frac{9}{2}\epsilon - \epsilon_1 - \frac{13}{8}\epsilon_2 - \frac{1}{8}\epsilon_3 \right.$$
$$\left. + \left(-\frac{21}{16} + \frac{25}{16}\epsilon + \frac{5}{32}\epsilon_1 - \frac{1063}{384}\epsilon_2 + \frac{141}{128}\epsilon_3 \right) m \right] \cos(l - l')$$
$$+ \left[\left(-3\epsilon + \frac{3}{4}\epsilon_1 + \frac{5}{4}\epsilon_2 \right) m^{-1} - \frac{3}{2}\epsilon + \frac{1}{2}\epsilon_1 + \frac{7}{8}\epsilon_2 - \frac{1}{8}\epsilon_3 \right.$$
$$\left. + \left(\frac{21}{16} - \frac{25}{16}\epsilon - \frac{5}{32}\epsilon_1 + \frac{313}{384}\epsilon_2 + \frac{209}{128}\epsilon_3 \right) m \right] \cos(l + l')$$
$$+ \left[\frac{45}{8}\epsilon - \frac{45}{32}\epsilon_1 - \frac{95}{32}\epsilon_2 + \frac{25}{16}\epsilon_3 \right.$$
$$\left. + \left(-\frac{35}{16} + \frac{435}{32}\epsilon + \frac{5}{384}\epsilon_1 - \frac{435}{64}\epsilon_2 - \frac{35}{32}\epsilon_3 \right) m \right] \cos(2D - l - l')$$
$$+ \left[-\frac{45}{8}\epsilon + \frac{45}{32}\epsilon_1 + \frac{75}{32}\epsilon_2 - \frac{15}{16}\epsilon_3 \right.$$
$$\left. + \left(\frac{15}{16} + \frac{45}{32}\epsilon - \frac{255}{128}\epsilon_1 + \frac{245}{64}\epsilon_2 \right) m \right] \cos(2D - l + l')$$
$$+ \left[\frac{3}{8}\epsilon_2 - \frac{15}{16}\epsilon_3 + \left(\frac{153}{16}\epsilon - \frac{153}{64}\epsilon_1 \right. \right.$$
$$\left. \left. - \frac{375}{128}\epsilon_2 - \frac{135}{64}\epsilon_3 \right) m \right] \cos(2D + l - l')$$
$$+ \left[\frac{9}{16}\epsilon_3 + \left(-\frac{153}{16}\epsilon + \frac{153}{64}\epsilon_1 \right. \right.$$
$$\left. \left. + \frac{375}{128}\epsilon_2 + \frac{135}{64}\epsilon_3 \right) m \right] \cos(2D + l + l')$$
$$+ \left(\frac{195}{128}\epsilon_2 - \frac{435}{128}\epsilon_3 \right) m \cos(4D - l - l')$$

$$+\left(-\frac{45}{128}\epsilon_2+\frac{225}{128}\epsilon_3\right)m\cos(4D-l+l')+\dots \tag{5.3.31}$$

$$\begin{aligned}V_0=&\left[-\frac{1}{4}\epsilon_2+\frac{1}{2}\epsilon_3+\left(\frac{11}{8}+\frac{11}{8}\epsilon-\frac{11}{8}\epsilon_1\right)m^2\right.\\&\left.+\left(\frac{13}{6}+\frac{13}{6}\epsilon-\frac{13}{4}\epsilon_1\right)m^3\right]\sin 2D+\left[\left(-\frac{11}{32}\epsilon_2+\frac{11}{16}\epsilon_3\right)m^2\right.\\&\left.+\left(-\frac{13}{24}\epsilon_2+\frac{13}{12}\epsilon_3\right)m^3\right]\sin 4D+\dots\end{aligned} \tag{5.3.32}$$

$$\begin{aligned}V_e=&\left[1+\left(\frac{15}{32}\epsilon_2-\frac{15}{16}\epsilon_3\right)m+\left(\frac{257}{128}\epsilon_2-\frac{257}{64}\epsilon_3\right)m^2\right]\sin l\\&+\left[\frac{1}{4}\epsilon_2-\frac{1}{2}\epsilon_3+\left(\frac{15}{8}+\frac{15}{8}\epsilon-\frac{15}{16}\epsilon_1\right)m\right.\\&\left.+\left(\frac{203}{32}+\frac{169}{16}\epsilon-\frac{203}{32}\epsilon_1\right)m^2\right]\sin(2D-l)\\&+\left[-\frac{1}{4}\epsilon_2+\frac{1}{2}\epsilon_3+\left(\frac{17}{16}+\frac{17}{16}\epsilon-\frac{17}{16}\epsilon_1\right)m^2\right]\sin(2D+l)\\&+\left[\left(-\frac{15}{32}\epsilon_2+\frac{15}{16}\epsilon_3\right)m+\left(-\frac{159}{128}\epsilon_2+\frac{159}{64}\epsilon_3\right)m^2\right]\sin(4D-l)\\&+\left(-\frac{39}{64}\epsilon_2+\frac{39}{32}\epsilon_3\right)m^2\sin(4D+l)+\dots\end{aligned} \tag{5.3.33}$$

$$\begin{aligned}V_{e'}=&\left[(12\epsilon-3\epsilon_1-5\epsilon_2)m^{-1}+12\epsilon-\tfrac{3}{2}\epsilon_1-5\epsilon_2\right.\\&+\left(-3+21\epsilon-\frac{11}{2}\epsilon_1-\frac{27}{2}\epsilon_2\right)m\\&\left.+\left(3-21\epsilon+4\epsilon_1+\frac{227}{16}\epsilon_2\right)m^2\right]\sin l'\\&+\left[-\frac{1}{2}\epsilon_2+\frac{5}{4}\epsilon_3+\left(-\frac{33}{2}\epsilon+\frac{33}{8}\epsilon_1+\frac{49}{8}\epsilon_2+\frac{3}{2}\epsilon_3\right)m\right.\\&\left.+\left(\frac{77}{16}-\frac{1745}{32}\epsilon+\frac{145}{24}\epsilon_1+\frac{2663}{96}\epsilon_2-\frac{3}{2}\epsilon_3\right)m^2\right]\sin(2D-l')\\&+\left[-\frac{3}{4}\epsilon_3+\left(\frac{33}{2}\epsilon-\frac{33}{8}\epsilon_1-\frac{49}{8}\epsilon_2-\frac{3}{2}\epsilon_3\right)m\right.\\&\left.+\left(-\frac{11}{16}+\frac{799}{32}\epsilon-\frac{41}{8}\epsilon_1-\frac{1409}{96}\epsilon_2+\frac{3}{2}\epsilon_3\right)m^2\right]\sin(2D+l')\end{aligned}$$

$$+\left(-\frac{121}{64}\epsilon_2+\frac{33}{8}\epsilon_3\right)m^2\sin(4D-l')$$
$$+\left(\frac{11}{64}\epsilon_2-\frac{11}{8}\epsilon_3\right)m^2\sin(4D+l')+\dots \tag{5.3.34}$$

$$\begin{aligned}V_\alpha=\Bigg[&\left(-6\epsilon+\frac{3}{2}\epsilon_1+\frac{5}{2}\epsilon_2\right)m^{-1}-45\epsilon+12\epsilon_1+\frac{31}{2}\epsilon_2-\frac{1}{8}\epsilon_3\\&+\left(-\frac{15}{8}-\frac{2029}{8}\epsilon+\frac{2279}{32}\epsilon_1+\frac{244}{3}\epsilon_2+\frac{15}{16}\epsilon_3\right)m\Bigg]\sin D\\&+\left[\frac{3}{8}\epsilon_3+\left(-\frac{51}{8}\epsilon+\frac{51}{32}\epsilon_1+\frac{29}{8}\epsilon_2-\frac{15}{16}\epsilon_3\right)m\right]\sin 3D+\dots\end{aligned} \tag{5.3.35}$$

$$\begin{aligned}V_{ee'}=\Bigg[&\left(-6\epsilon+\frac{3}{2}\epsilon_1+\frac{5}{2}\epsilon_2\right)m^{-1}-6\epsilon+\frac{3}{2}\epsilon_1+\frac{5}{2}\epsilon_2\\&+\left(\frac{21}{8}-\frac{111}{8}\epsilon+\frac{59}{16}\epsilon_1+\frac{333}{32}\epsilon_2-\frac{45}{32}\epsilon_3\right)m\Bigg]\sin(l-l')\\&+\Bigg[\left(6\epsilon-\frac{3}{2}\epsilon_1-\frac{5}{2}\epsilon_2\right)m^{-1}+6\epsilon-\frac{3}{2}\epsilon_1-\frac{5}{2}\epsilon_2\\&+\left(-\frac{21}{8}+\frac{111}{8}\epsilon-\frac{59}{16}\epsilon_1-\frac{283}{32}\epsilon_2-\frac{25}{32}\epsilon_3\right)m\Bigg]\sin(l+l')\\&+\Bigg[-\frac{45}{4}\epsilon+\frac{45}{16}\epsilon_1+\frac{83}{16}\epsilon_2-\frac{5}{4}\epsilon_3\\&+\left(\frac{35}{8}-\frac{809}{16}\epsilon+\frac{1627}{192}\epsilon_1+\frac{1661}{64}\epsilon_2-\frac{3}{16}\epsilon_3\right)m\Bigg]\sin(2D-l-l')\\&+\Bigg[\frac{45}{4}\epsilon-\frac{45}{16}\epsilon_1-\frac{75}{16}\epsilon_2+\frac{3}{4}\epsilon_3\\&+\left(-\frac{15}{8}+\frac{429}{16}\epsilon-\frac{249}{64}\epsilon_1-\frac{1261}{64}\epsilon_2+\frac{3}{16}\epsilon_3\right)m\Bigg]\sin(2D-l+l')\\&+\Bigg[-\frac{1}{2}\epsilon_2+\frac{5}{4}\epsilon_3+\Bigg(-\frac{153}{8}\epsilon+\frac{153}{32}\epsilon_1\\&+\frac{105}{16}\epsilon_2+\frac{45}{16}\epsilon_3\Bigg)m\Bigg]\sin(2D+l-l')\\&+\left[-\frac{3}{4}\epsilon_3+\left(\frac{153}{8}\epsilon-\frac{153}{32}\epsilon_1-\frac{105}{16}\epsilon_2-\frac{45}{16}\epsilon_3\right)m\right]\sin(2D+l+l')\\&+\left(-\frac{65}{32}\epsilon_2+\frac{145}{32}\epsilon_3\right)m\sin(4D-l-l')\\&+\left(\frac{15}{32}\epsilon_2-\frac{75}{32}\epsilon_3\right)m\sin(4D-l+l')+\dots\end{aligned} \tag{5.3.36}$$

$$
\begin{aligned}
\Phi_k = {} & \left[1+\frac{1}{4}\epsilon_2+\frac{1}{2}\epsilon_3+\left(\frac{3}{64}\epsilon_2-\frac{3}{32}\epsilon_3\right)m+\left(\frac{79}{256}\epsilon_2-\frac{79}{128}\epsilon_3\right)m^2\right]\sin F \\
& +\left[\frac{1}{8}\epsilon_2-\frac{1}{4}\epsilon_3+\left(\frac{3}{8}+\frac{3}{8}\epsilon-\frac{3}{16}\epsilon_1+\frac{3}{32}\epsilon_2+\frac{3}{16}\epsilon_3\right)m\right. \\
& \left.+\left(\frac{13}{32}+\frac{1}{8}\epsilon-\frac{19}{32}\epsilon_1+\frac{13}{128}\epsilon_2+\frac{13}{64}\epsilon_3\right)m^2\right]\sin(2D-F) \\
& +\left[-\frac{1}{8}\epsilon_2+\frac{1}{4}\epsilon_3+\left(\frac{11}{16}+\frac{11}{16}\epsilon-\frac{11}{16}\epsilon_1\right.\right. \\
& \left.\left.+\frac{11}{64}\epsilon_2+\frac{11}{32}\epsilon_3\right)m^2\right]\sin(2D+F) \\
& +\left[\left(-\frac{3}{64}\epsilon_2+\frac{3}{32}\epsilon_3\right)m+\left(\frac{31}{256}\epsilon_2-\frac{31}{128}\epsilon_3\right)m^2\right]\sin(4D-F) \\
& +\left(-\frac{33}{128}\epsilon_2+\frac{33}{64}\epsilon_3\right)m^2\sin(4D+F)+\ldots
\end{aligned}
\tag{5.3.37}
$$

$$
\begin{aligned}
\Phi_{ke'} = {} & \left[\left(-6\epsilon+\frac{3}{2}\epsilon_1+\frac{5}{2}\epsilon_2\right)m^{-1}-6\epsilon+\frac{3}{2}\epsilon_1+\frac{11}{4}\epsilon_2+\frac{1}{4}\epsilon_3\right. \\
& \left.+\left(\frac{3}{8}-9\epsilon+\frac{17}{8}\epsilon_1+\frac{309}{64}\epsilon_2+\frac{3}{64}\epsilon_3\right)m\right]\sin(F-l') \\
& +\left[\left(6\epsilon-\frac{3}{2}\epsilon_1-\frac{5}{2}\epsilon_2\right)m^{-1}+6\epsilon-\frac{3}{2}\epsilon_1-\frac{9}{4}\epsilon_2+\frac{1}{4}\epsilon_3\right. \\
& \left.+\left(-\frac{3}{8}+9\epsilon-\frac{17}{8}\epsilon_1-\frac{299}{64}\epsilon_2-\frac{17}{64}\epsilon_3\right)m\right]\sin(F+l') \\
& +\left[-\frac{9}{4}\epsilon+\frac{9}{16}\epsilon_1+\frac{19}{16}\epsilon_2-\frac{5}{8}\epsilon_3\right. \\
& \left.+\left(\frac{7}{8}-\frac{93}{16}\epsilon+\frac{217}{192}\epsilon_1+\frac{89}{32}\epsilon_2-\frac{1}{8}\epsilon_3\right)m\right]\sin(2D-F-l') \\
& +\left[\frac{9}{4}\epsilon-\frac{9}{16}\epsilon_1-\frac{15}{16}\epsilon_2+\frac{3}{8}\epsilon_3\right. \\
& \left.+\left(-\frac{3}{8}+\frac{9}{16}\epsilon+\frac{13}{64}\epsilon_1-\frac{55}{32}\epsilon_2+\frac{9}{16}\epsilon_3\right)m\right]\sin(2D-F+l') \\
& +\left[-\frac{1}{4}\epsilon_2+\frac{5}{8}\epsilon_3+\left(-\frac{99}{8}\epsilon+\frac{99}{32}\epsilon_1\right.\right. \\
& \left.\left.+\frac{303}{64}\epsilon_2+\frac{27}{32}\epsilon_3\right)m\right]\sin(2D+F-l') \\
& +\left[-\frac{3}{8}\epsilon_3+\left(\frac{99}{8}\epsilon-\frac{99}{32}\epsilon_1-\frac{303}{64}\epsilon_2-\frac{27}{32}\epsilon_3\right)m\right]\sin(2D+F+l') \\
& +\left(-\frac{13}{64}\epsilon_2+\frac{29}{64}\epsilon_3\right)m\sin(4D-F-l')
\end{aligned}
$$

$$+\left(\frac{3}{64}\epsilon_2-\frac{15}{64}\epsilon_3\right)m\sin(4D-F+l')+\ldots. \qquad (5.3.38)$$

As for the frequencies of the trigonometric arguments (5.3.25) one has

$$f=(1-\tfrac{3}{2}\epsilon-\tfrac{1}{2}\epsilon_1)m \qquad (5.3.39)$$

and c and g may be replaced by the rates of advances of the perigee and the node

$$\frac{\dot{\pi}}{n}=1-\frac{c}{1+m} \qquad \frac{\dot{\Omega}}{n}=1-\frac{g}{1+m}. \qquad (5.3.40)$$

Then

$$\begin{aligned}\frac{\dot{\pi}}{n}=&\tfrac{1}{2}\epsilon_1 m+\left(\frac{3}{4}+\frac{9}{4}\epsilon-\frac{5}{4}\epsilon_1\right)m^2+\left(\frac{177}{32}+\frac{153}{16}\epsilon-\frac{523}{64}\epsilon_1\right)m^3\\&+\left(\frac{1659}{128}+\frac{5775}{128}\epsilon-\frac{461}{16}\epsilon_1\right)m^4+\left(\frac{85\,205}{2048}+\frac{82\,165}{512}\epsilon-\frac{440\,729}{4096}\epsilon_1\right)m^5\\&+\left(\frac{3073\,531}{24\,576}+\frac{16\,498\,169}{24\,576}\epsilon-\frac{3214\,533}{8192}\epsilon_1\right)m^6\\&+\left(\frac{258\,767\,293}{589\,824}+\frac{862\,144\,879}{294\,912}\epsilon-\frac{1864\,831\,507}{1179\,648}\epsilon_1\right)m^7\\&+\left(\frac{12\,001\,004\,273}{7077\,888}+\frac{95\,700\,888\,409}{7077\,888}\epsilon-\frac{24\,617\,929\,057}{3538\,944}\epsilon_1\right)m^8+\ldots\end{aligned} \qquad (5.3.41)$$

$$\begin{aligned}\frac{\dot{\Omega}}{n}=&\tfrac{1}{2}\epsilon_1 m+\left(-\frac{3}{4}-\frac{3}{4}\epsilon-\frac{1}{4}\epsilon_1\right)m^2+\left(\frac{57}{32}+\frac{33}{16}\epsilon-\frac{51}{64}\epsilon_1\right)m^3\\&+\left(-\frac{123}{128}+\frac{15}{128}\epsilon-\frac{71}{64}\epsilon_1\right)m^4+\left(\frac{1925}{2048}-\frac{839}{512}\epsilon-\frac{9}{4096}\epsilon_1\right)m^5\\&+\left(-\frac{25\,667}{24\,576}-\frac{27\,835}{24\,576}\epsilon+\frac{5179}{8192}\epsilon_1\right)m^6\\&+\left(\frac{268\,309}{589\,824}+\frac{78\,319}{294\,912}\epsilon+\frac{494\,453}{1179\,648}\epsilon_1\right)m^7\\&+\left(-\frac{9662\,017}{7077\,888}-\frac{18\,270\,287}{7077\,888}\epsilon+\frac{310\,057}{442\,368}\epsilon_1\right)m^8+\ldots.\end{aligned} \qquad (5.3.42)$$

The solution (5.3.27)–(5.3.38), (5.3.41) and (5.3.42) is presented in the form of the series in powers of m. The convergence of the series in powers of m is rather slow and such power expansions are avoided if possible in modern theories of motion. But for calculating relativistic perturbations the power expansion technique turns out to be quite adequate. Expressions (5.3.27)–(5.3.38) enable one to reveal some details of interest such as, for example, the occurrence of negative powers of m in some coefficients. It may be noted in addition that the relativistic perturbations enter into the series for the coordinates of the Moon also in an implicit manner as relativistic contributions into the frequencies of the arguments l, l' and F in Newtonian terms. Such perturbations might be presented, of course, in explicit form.

5.3.6 Numerical perturbations and constants of the theory

For numerical estimations of the relativistic perturbations one takes $n = 17\,325\,593''\ \mathrm{yr}^{-1}$, $N = 1295\,977''\ \mathrm{yr}^{-1}$, $A/c = 499.005$ s, $m = 0.080\,85$, $\sigma = 9.8711 \times 10^{-9}$. R_e and V_e should be multiplied by the constant of eccentricity of the Moon, $e = 2E$. $R_{e'}$ and $V_{e'}$ are multiplied by the solar eccentricity e'. R_α and V_α enter with the factor equal to the ratio of the semi-major axes a_0/A_0. finally, Φ_k should be multiplied by the constant of inclination of the Moon, $k = 2\Gamma$. Numerical estimates of these constants are

$$E = 0.054\,90 \qquad e' = 0.016\,71 \qquad \Gamma = 0.044\,89$$

and

$$\frac{a_0}{A_0} = \left(\frac{M_c}{M}\frac{N^2}{n^2}\right)^{1/3} \qquad \frac{M_c}{M} = 3.0404 \times 10^{-6}.$$

Substituting these values into series (5.3.20)–(5.3.22) and returning by (5.3.26) to the initial parameters α, β and γ one obtains the trigonometric series with numerical coefficients given in Brumberg and Ivanova (1985). There is no need to reproduce these series here and it is sufficient to give the relativistic secular rates of motion of the perigee and the node of the Moon:

$$\Delta\dot{\pi} = 0.8328'' - 0.2568''\beta + 1.1520''\gamma \tag{5.3.43}$$

$$\Delta\dot{\Omega} = 0.5902'' + 0.0435''\beta + 1.2673''\gamma. \tag{5.3.44}$$

For GRT the relativistic secular rates (5.3.43) and (5.3.44) yield 1.7280″ and 1.9010″, respectively. The secular motion of the lunar perigee should be augmented by the Schwarzschild advance of 0.06″ per century. The values 1.79″ and 1.90″ for the relativistic motions of the perigee and the node of the Moon are quite consistent with the values of ELP2000 theory determined by a completely different technique (Chapront-Touzé and Chapront 1983).

The problem of comparison with observations will be discussed in the next chapter. Let us note once again that the results discussed here hold true only for the BRS relativistic perturbations and cannot be identified with physically measurable relativistic effects. The most significant relativistic term in the radius vector r has argument $2D$ and amplitude of order 100 cm. The physically measurable effects are two orders of magnitude lower, as is to be expected from the estimations of table 5.1. There are no discrepancies now between the theoretical and observational data concerning the secular advances of the perigee and the node (Chapront and Chapront-Touzé 1981). Moreover, the effect of geodesic precession was explicitly confirmed recently (Bertotti *et al* 1987, Shapiro *et al* 1988).

It remains to interpret the constants occurring in the series for the co-ordinates of the Moon. The mean motion of the Moon n and the Sun N, the masses of all three bodies and the light velocity c may be regarded as directly measurable quantities. Based on them one may calculate the values a_0 and A_0 of the semi-major axes. Eccentricity and inclination constants of the orbit of the Moon $2E$ and 2Γ are defined in Newtonian theory as coefficients in the principal terms $\sin l$ and $\sin F$ in the longitude and latitude of the Moon, respectively, and are regarded to be known from observations. In the relativistic theory such a definition becomes, generally speaking, coordinate dependent. Indeed, the coefficients in $\sin l$ and $\sin F$ in (5.3.33) and (5.3.37) contain the coordinate parameter α and, hence, cannot be considered as directly measurable quantities. If it is desirable to have coordinate-independent definitions of the parameters of orbit one should give other definitions for the eccentricity and inclination constants. In the Schwarzschild problem such a definition is possible because one is confined usually to a class of the coordinate systems distinguished from one another only by radial coordinate. In the problem of the motion of the Moon the corresponding class of the coordinate systems is determined by (5.2.3) and (5.2.4) or

$$\tilde{\boldsymbol{r}} = \boldsymbol{r} - \alpha\frac{m_c}{r}\boldsymbol{r} + \alpha m_2\left(\frac{\boldsymbol{r}_{12}}{r_{12}} - \frac{\boldsymbol{r}_{32}}{r_{32}}\right). \tag{5.3.45}$$

Expanding in powers of r/R one has

$$\begin{aligned}\tilde{\boldsymbol{r}} &= \boldsymbol{r} - \alpha\frac{m_c}{r}\boldsymbol{r} + \alpha\frac{m_2}{R}\left(-\boldsymbol{r} + \frac{1}{R^2}(\boldsymbol{R}\boldsymbol{r})\boldsymbol{R}\right)\\ &\quad + \alpha\frac{m_2}{R^3}\left[(\boldsymbol{R}\boldsymbol{r})\boldsymbol{r} + \frac{1}{2}\left(r^2 - \frac{3}{R^2}(\boldsymbol{R}\boldsymbol{r})^2\right)\boldsymbol{R}\right].\end{aligned} \tag{5.3.46}$$

Expressed in terms of the spherical coordinates we obtain

$$\tilde{r} = r\left[1 - \alpha\frac{m_c}{r} - \alpha\frac{m_2}{r}\left(1 - \frac{3}{2}\frac{(\boldsymbol{R}\boldsymbol{r})}{R^2}\right)\frac{(\boldsymbol{R}\times\boldsymbol{r})^2}{R^2r^2}\right] \tag{5.3.47}$$

$$\tilde{v} - v = \alpha\frac{m_2}{R}\left[\frac{\boldsymbol{R}\boldsymbol{r}}{Rr} + \frac{1}{2}\frac{r}{R}\left(1 - 3\frac{(\boldsymbol{R}\boldsymbol{r})^2}{R^2r^2}\right)\right]\sin(v - v_\odot) \tag{5.3.48}$$

$$\tilde{\varphi} = \varphi \left(1 - \alpha \frac{m_2}{R} \frac{(\boldsymbol{R}\boldsymbol{r})^2}{R^2 r^2} \right) . \tag{5.3.49}$$

In addition, from (5.3.46) and (5.3.47)

$$\frac{\tilde{\boldsymbol{r}}}{\tilde{r}} = \frac{\boldsymbol{r}}{r} + \alpha \frac{m_2}{R} \left[\frac{\boldsymbol{R}\boldsymbol{r}}{Rr} + \frac{1}{2} \left(1 - 3 \frac{(\boldsymbol{R}\boldsymbol{r})^2}{R^2 r^2} \right) \frac{r}{R} \right] \frac{\boldsymbol{r} \times (\boldsymbol{R} \times \boldsymbol{r})}{R r^2} . \tag{5.3.50}$$

The tilde again denotes harmonic coordinates. Performing the transformation (5.3.47)–(5.3.49) to harmonic coordinates one obtains expressions (5.3.20)–(5.3.22) with the condition $\epsilon_3 = 0$ in the coefficients.

6

Relativistic Reduction of Astrometric Measurements

6.1 GENERAL PRINCIPLES OF REDUCTION

6.1.1 Reduction of measurable quantities

The information given by astronomical observations characterizes not only the object of observation but the observer as well. This information depends on the position of the observer, its velocity and the value of the gravitational potential at the point of observation. In order to use information obtained by different observers or even by one and the same observer but at different moments of time it is necessary to perform a reduction of the observations, i.e. to refer them to some conventional point at some adopted moment of time. Depending on the problem at hand such a point may be the geocentre, the Solar System barycentre, a point infinitely far from the Solar System (under the assumption of the isolated existence of the Solar System) and so on.

In classical astronomy the main types of reduction are related to the position and velocity of the observer. For example, annual aberration and annual parallax represent reductions to the Solar System barycentre caused by the barycentric velocity of motion of the Earth and the difference of the observer's position from the Solar System barycentre. Diurnal aberration and diurnal parallax are reductions to the geocentre due to the diurnal rotation of the observer on the surface of the Earth and the difference of its location from the geocentre. These types of reduction enable one to reduce the results of different observers on the surface of the Earth to one point and thus to use them in a uniform manner as actual observations. Relativistic reduction introduces its own corrections to these classical types of reduction and in addition makes its specific contribution due to the influence of the gravitational field on the results of measurement.

The reduction described here may be called the reduction of measurable quantities. We recall that by measurable quantities we mean quantities independent of mathematical constructions such as reference systems. Such quantities may be both directly measurable quantities (angular distances between two light sources measured by the observer, intervals of the observer's proper time between two events occurring at the point of observation, ratio of the frequencies of emitted and received signals, etc) and quantities that are measurable in principle (for example, angular distances, time intervals and frequency ratios for an imagined observer in the geocentre or in the Solar System barycentre or infinitely far from the Solar System).

6.1.2 Reduction of coordinates

The second type of reduction specific to GRT is related to the reduction of coordinates, i.e. the transformation of coordinate-dependent quantities into measurable quantities. In Newtonian theory one deals, as a rule, with the physically privileged inertial coordinates which may be considered as measurable quantities. Such coordinates do not exist in GRT. Solution of any dynamical problem is influenced by the choice of a particular coordinate system. The aim of this second type of reduction is to exclude this influence and to present the results of dynamical solution in terms of measurable quantities. In principle, such a problem has a place in Newtonian theory by using non-inertial coordinates and is easily solved by transformation to inertial coordinates. To solve this problem in GRT one has to describe the measurement procedure in the same coordinates as used in treating the dynamical problem. Eventually, this reduces to the description of light propagation within the space-time metric of the dynamical problem. As a result, the solution of the dynamical problem will be expressed in measurable quantities which are, possibly, not directly measurable. For example, if the LLR problem has been solved in BRS then it gives the time for the round-trip light propagation for an infinitely far observer (the BRS coordinate time). This is a measurable quantity but, obviously, to obtain the directly measurable quantity, i.e. the time interval for the clock of a ground observer, one still has to perform a reduction of the first kind, related to the transformation of measurable quantities.

It should be noted that the terminology used here is not universally accepted and completely opposite statements can sometimes be met. For instance, some authors do not make any distinction at all between calculated and measurable quantities, incorporating calculation by means of definite algorithms within the measurement procedure. But it is of importance to realize what is kept in mind when formulating one or other statement. The terminology used above seems to be quite clear and will be used below.

Let us emphasize again that reduction of coordinates is aimed to eliminate coordinate-dependent quantities (coordinate parameters in particular) and to obtain expressions for quantities that are, in principle, measurable. The aim of measurement reduction is to transform from one set of measurable quantities to another set.

6.1.3 Basic relations between coordinate and measurable quantities

The basic relations for the problem of measurement in GRT have already been given in section 2.3. Here these relations are specified in application to astronomical problems.

Let some dynamical problem be solved in a reference system defined by the metric

$$ds^2 = g_{\mu\nu} dx^\mu dx^\nu \qquad x^0 = ct \tag{6.1.1}$$

$$g_{\mu\nu} = \eta_{\mu\nu} + h_{\mu\nu} \qquad |h_{\mu\nu}| \ll 1 \tag{6.1.2}$$

$$\eta_{00} = 1 \qquad \eta_{0i} = 0 \qquad \eta_{ij} = -\delta_{ij}. \tag{6.1.3}$$

The first problem of reduction to measurable quantities is the transformation to the observer's proper time τ. This is performed by integrating along the observer's world line the equation

$$d\tau = (g_{00} + 2c^{-1} g_{0i}\dot{x}^i + c^{-2} g_{ik}\dot{x}^i\dot{x}^k)^{1/2}\, dt \tag{6.1.4}$$

or in the post-Newtonian approximation

$$d\tau = (1 - \tfrac{1}{2}c^{-2}v^2 + \tfrac{1}{2}h_{00} + \underline{c^{-1}h_{0i}\dot{x}^i} + \ldots)\, dt \tag{6.1.5}$$

Here and below h_{00} and h_{ik} are regarded as second-order quantities with respect to the ratio v/c, $v = (x^k x^k)^{1/2}$ being the magnitude of the observer's coordinate velocity. As for h_{0i}, these components have third order for non-rotating reference systems and first order for rotating systems. Therefore, in (6.1.5) and all formulae below underlined terms should be omitted when treating non-rotating systems.

The second problem is to investigate light propagation in system (6.1.1). This is performed with the aid of equations (2.2.61) where the components h_{0i} are treated as third-order quantities. It is sufficient to examine light propagation in non-rotating systems. If necessary, the corresponding relations for rotating systems may be obtained by coordinate transformation. For non-rotating systems one may choose as initial values for the light particle its position at the initial moment of time $\boldsymbol{x}_0 = \boldsymbol{x}(t_0)$ and the unit vector $\boldsymbol{\sigma} = c^{-1}\dot{\boldsymbol{x}}(-\infty)$, $\boldsymbol{\sigma}^2 = 1$, characterizing the light direction at a remote past. For the problem with boundary conditions $\boldsymbol{x}(t_0) = \boldsymbol{x}_0$, $\boldsymbol{x}(t) = \boldsymbol{x}$ one has to express $\boldsymbol{\sigma}$ and the interval $t - t_0$ in terms of these boundary values. These formulae are adequate for discussing radio range and radiointerferometric measurements.

Measurement of the angular distance between two light emitters may be regarded as the basic idealized type of astrometric observation. Therefore, the third typical problem of a relativistic reduction is to calculate an expression for the angle between two light rays at the point of observation. For this purpose one performs the local $3+1$ splitting (2.3.11)

$$\mathrm{d}s^2 = c^2\mathrm{d}\tau^2 - \mathrm{d}\ell^2 \tag{6.1.6}$$

with

$$c\mathrm{d}\tau = \frac{1}{\sqrt{g_{00}}} g_{0\alpha}\,\mathrm{d}x^\alpha \tag{6.1.7}$$

and

$$\mathrm{d}\ell^2 = \gamma_{ik}\mathrm{d}x^i\mathrm{d}x^k \qquad \gamma_{ik} = \frac{1}{g_{00}} g_{0i}g_{0k} - g_{ik}. \tag{6.1.8}$$

In the post-Newtonian approximation

$$c\mathrm{d}\tau = (1 + \tfrac{1}{2}h_{00} - \tfrac{1}{8}h_{00}^2)c\,\mathrm{d}t + (h_{0i} - \underline{\tfrac{1}{2}h_{00}h_{0i})}\,\mathrm{d}x^i \tag{6.1.9}$$

$$\gamma_{ik} = \delta_{ik} - h_{ik} + \underline{h_{0i}h_{0k}}. \tag{6.1.10}$$

The three-dimensional form $\mathrm{d}\ell^2$ describes the local spatial relations at the point of observation. The scalar product and length of arbitrary three-dimensional vectors applied at the point of observation are defined by the formulae

$$(\boldsymbol{PQ})_{rel} = \gamma_{ik}P^iQ^k \qquad P_{rel} = (\gamma_{ik}P^iP^k)^{1/2} \tag{6.1.11}$$

thus yielding the following expression for the angle between these vectors:

$$\cos\psi = \frac{(\boldsymbol{PQ})_{rel}}{P_{rel}Q_{rel}}. \tag{6.1.12}$$

If $\boldsymbol{P}$ and $\boldsymbol{Q}$ are identified with the directions of two light rays at the point of observation then $\boldsymbol{P} = c^{-1}\dot{\boldsymbol{r}}_1(t)$, $\boldsymbol{Q} = c^{-1}\dot{\boldsymbol{r}}_2(t)$ and relation (6.1.12) represents the required expression for the measurable angular distance between two light emitters.

Along with mutual angular distances astrometric practice often requires the derivation of the direction towards a star or a planet in the reference frame of an observer. This is the fourth problem of relativistic reduction. The reference frame of an observer corresponding to the splitting (6.1.6) may be constructed by introducing the tetrad $\lambda^\nu_{(\mu)}$ with

$$\lambda^0_{(0)} = 1 \qquad \lambda^i_{(0)} = 0 \qquad \lambda^0_{(j)} = 0 \qquad \lambda^i_{(j)} = \delta_{ij} + \tfrac{1}{2}h_{ij} - \underline{\tfrac{1}{2}h_{0i}h_{0j}} \tag{6.1.13}$$

or in covariant components

$$\lambda_0^{(0)} = 1 \qquad \lambda_i^{(0)} = 0 \qquad \lambda_0^{(j)} = 0 \qquad \lambda_i^{(j)} = \delta_{ij} - \tfrac{1}{2}h_{ij} + \underline{\tfrac{1}{2}h_{0i}h_{0j}}. \tag{6.1.14}$$

The measurable space intervals are

$$\mathrm{d}x^{(i)} = \lambda_j^{(i)}\mathrm{d}x^j = \mathrm{d}x^i + \tfrac{1}{2}(-h_{ik} + \underline{h_{0i}h_{0k}})\,\mathrm{d}x^k \tag{6.1.15}$$

so that

$$\mathrm{d}\ell^2 = \delta_{ik}\mathrm{d}x^{(i)}\mathrm{d}x^{(k)}. \tag{6.1.16}$$

The 4-impulse of a photon has the following components in the metric (6.1.6):

$$p^0 = h\nu \qquad p^i = \frac{h\nu}{c}\frac{\mathrm{d}x^i}{\mathrm{d}\tau} \tag{6.1.17}$$

(ν is the frequency, h is Planck's constant). The components p^i may be calculated on the basis of the equations of light propagation followed by the change (6.1.9) from t to τ. Then the tetrad components enable one to calculate the invariant components of the 3-vector $\boldsymbol{p} = (p^{(1)}, p^{(2)}, p^{(3)})$

$$p^{(i)} = p^{\mu}\lambda_{\mu}^{(i)}. \tag{6.1.18}$$

With the normalization $h\nu = 1$, the vector $\boldsymbol{p}$ represents the unit vector determining the observed direction from which the light ray comes to the point of observation. Instead of (6.1.18) one may also use the relation

$$p^{(i)} = c^{-1}\frac{\mathrm{d}x^{(i)}}{\mathrm{d}\tau} = \left[1 - \underline{c^{-1}h_{0k}\frac{\mathrm{d}x^k}{\mathrm{d}t}} + \underline{\left(c^{-1}h_{0k}\frac{\mathrm{d}x^k}{\mathrm{d}t}\right)^2} - \tfrac{1}{2}h_{00}\right] c^{-1}\frac{\mathrm{d}x^i}{\mathrm{d}t}$$
$$+ \tfrac{1}{2}(-h_{ik} + \underline{h_{0i}h_{0k}})c^{-1}\frac{\mathrm{d}x^k}{\mathrm{d}t} \tag{6.1.19}$$

$\mathrm{d}x^i/\mathrm{d}t$ being the coordinate velocity of light. Expressions (6.1.18) or (6.1.19) require a relation of the form

$$\boldsymbol{\sigma} = \boldsymbol{p} + \Delta\boldsymbol{p}. \tag{6.1.20}$$

The correction $\Delta\boldsymbol{p}$ represents the relativistic term which must be added to the observed value of $\boldsymbol{p}$ in order to obtain the gravitationally unperturbed direction $\boldsymbol{\sigma}$ at infinity. It is easy to express this reduction in spherical coordinates. For example, if the direction $\boldsymbol{p}$ is given by spherical angles α, δ

$$\boldsymbol{p} = -(\cos\alpha\cos\delta, \sin\alpha\cos\delta, \sin\delta) \tag{6.1.21}$$

then, accurate to quadratic corrections, one has

$$\cos\delta\Delta\alpha = \sin\alpha\Delta p^1 - \cos\alpha\Delta p^2 + \sin\delta\Delta\alpha\Delta\delta \tag{6.1.22}$$

$$\Delta\delta = \sin\delta(\cos\alpha\Delta p^1 + \sin\alpha\Delta p^2) - \cos\delta\Delta p^3 - \tfrac{1}{2}\sin\delta\cos\delta(\Delta\alpha)^2. \quad (6.1.23)$$

The above formula (6.1.12) for the angle between two light rays with directions $\boldsymbol{p}_1$, $\boldsymbol{p}_2$ reduces now to the Euclidean form

$$\cos\psi = \boldsymbol{p}_1\boldsymbol{p}_2. \qquad (6.1.24)$$

The fifth problem of relativistic reduction is to take into account the velocity of the observer's motion. In the post-Newtonian approximation this problem may be solved independently of other problems by applying the formulae of relativistic aberration obtained from the Lorentz transformations. One may introduce also a reference system related to the moving observer and perform for this system the splitting (6.1.6), forming ψ and $\boldsymbol{p}$. Then these expressions will involve the observer's velocity in the initial system (6.1.1).

Finally, the main problem of relativistic reduction is the general theory of relativistic reference frames and their interrelations. This problem includes all the above problems as particular cases. Therefore, instead of going into the details of these particular problems elucidated, for instance, in Brumberg (1981, 1986), we shall address directly the general problem of relativistic reference frames (Brumberg 1987b, Kopejkin 1988, Brumberg and Kopejkin 1989a).

6.2 RELATIVISTIC THEORY OF ASTRONOMICAL REFERENCE SYSTEMS

6.2.1 General principles

The concept of reference frame is often used in a different sense in physics and astronomy, sometimes leading to misunderstanding. For astronomical applications, to avoid any confusion it is suitable to follow the operational definition given by Kovalevsky and Mueller (1981) and detailed by Kovalevsky (1985). In accordance with this definition the reference (coordinate) system is the primary mathematical construction to be given in GRT by a metric form. Instead of such a laconic description the reference system may be defined by formulating conditions that underlie the corresponding metric form. These conditions involve the following: (1) coordinate conditions characterizing the choice of a particular set of quasi-Galilean space-time coordinates; (2) the type of solution of the GRT field equations; (3) the choice of the world line of the origin of the reference system; and (4) the choice of the angular rotation velocity of the spatial axes.

The reference frame results from the matching of the reference system to some reference astronomical objects ('materialization' of a reference system). Such materialization is not necessary in solving questions relating, for instance, to timescales or units of measurement. Until recently, the most widespread approach in the relativistic theory of astronomical reference frames was to construct the proper reference frame of a fictitious or actual observer with the aid of the Fermi normal coordinates (the time axis is the world line of an observer, the three space axes are the spacelike geodesics orthogonal to the world line of the observer). Such an approach is treated in a number of papers, starting from Mast and Strathdee (1959) and ending with Ni and Zimmermann (1978). Many other aspects of constructing reference frames for massless observers have also been developed (Synge 1960, Møller 1972, Vladimirov 1982).

When applied to the geocentric frame this approach involves difficulties because one cannot consider the Earth as a massless body and neglect its own gravitational field. To overcome these difficulties the metric tensor of the whole Solar System is separated into a 'local' part due to the gravitational field of the Earth and an 'external' one caused by external bodies (and their interaction with the Earth). One constructs first the proper reference frame for the fictitious Earth moving in the 'external' field. Then the corresponding coordinate transformation is substituted into the full metric incorporating completely the gravitational influence of the Earth. This approach was developed in several papers at IAU Symposium no 114 (Bertotti 1986, Boucher 1986, Fujimoto and Grafarend 1986, Fukushima *et al* 1986a) with a final derivation by Ashby and Bertotti (1986), Fukushima (1988) and Soffel (1989). This technique of generalizing the Fermi normal coordinates is not unique since the separation of the metric tensor into 'local' and 'external' parts cannot be performed in a unique manner (Thorne and Hartle 1985). Moreover, the generalized Fermi coordinates have no physical privileges, in contrast to the Fermi coordinates of a massless observer.

Among other methods of constructing a geocentric system one may note the technique of Pavlov (1984a,b, 1985) based on linear transformations of space-time coordinates.

To date, it has been sufficient in astronomical practice to use the local $3+1$ splitting of the space-time at the point of observation and to introduce the local inertial reference system in the infinitely small vicinity of the point of observation. Reduction of observations is therefore performed just for this system (Murray 1983, Hellings 1986, Brumberg 1986). Such an approach may be realized by using the equations of section 2.3.4.

6.2.2 Harmonic reference systems

It is obvious that one may use any coordinates in constructing a reference system. But if a coordinate system is not dynamically adequate to the class of problems under consideration then both the solution of the dynamical problems (the subject of relativistic celestial mechanics) and the transformation to the observational data (the subject of relativistic astrometry) will contain a number of extra terms caused only by the inadequate choice of reference frame. These terms cancel out in the expressions for the measurable quantities (time intervals, angular distances, frequency ratios, etc) and the resulting relativistic effects turn out to be much smaller than the relativistic perturbations in the coordinate solution of the dynamical problems. On the other hand, if the coordinate system is dynamically adequate, then the coordinate solution of the dynamical problems will not contain any large terms of non-dynamical origin and will hardly change in converting to measurable quantities.

Reasoning from these considerations and using the technique of Kopejkin (1987) one may construct a hierarchy of relativistic reference systems (Brumberg and Kopejkin 1989a). This hierarchy includes a barycentric reference system (BRS), geocentric reference system (GRS), topocentric reference system (TRS) and satellite reference system (SRS). In constructing all these reference systems four conditions are satisfied: (1) all these systems have been built in harmonic coordinates; (2) the corresponding metric tensors represent dynamically adequate solutions of the GRT field equations for the relevant problems; (3) the origins of these systems are, respectively the Solar System barycentre (BRS), the geocentre (GRS), a ground station (TRS), and an Earth satellite (SRS); and (4) all these systems are dynamically non-rotating. Let us make some comments in this respect.

The choice of harmonic coordinates does not mean in any way that they have any physical privileges. But they are mathematically convenient, being defined by the explicit mathematical equations (2.1.16). Moreover, many problems of relativistic celestial mechanics and ephemeris astronomy has been solved in these coordinates (for instance, theories DE200/LE200 for the motion of the planets and the Moon). Needless to say, one can use any coordinates appropriate to a specific problem. But to apply the relativistic theories in ephemeris astronomy it is useful to have an agreement to use a certain type of GRT coordinate conditions. Such an agreement will facilitate the comparison of various results and may help to avoid ambiguities in dealing with coordinate-dependent quantities. In principle, one may suggest three possible ways of overcoming difficulties caused by the intrusion of coordinate-dependent quantities into ephemeris astronomy (Brumberg 1986): (1) constructing theories only in terms of measurable quantities; (2) using arbitrary coordinates and developing unambiguous procedures to compare measurable and calculated quantities; and (3) using the same type of coordinate conditions. For ephemeris astronomy, and especially for problems related to reference frames, timescales, units of measurement, etc, the third way seems to be the most appropriate.

The dynamical adequacy of solutions of field equations means their conformity with the principle of equivalence. In application to GRS, TRS and SRS this implies that the influence of the external masses manifests in these systems only as the tidal terms.

The origin of the reference system is fixed by quite definite mathematical conditions. Thus, the Solar System barycentre taken as the BRS origin is defined by removing the constants of the relativistic integrals of the centre of mass and the linear momentum. The geocentre serving as the GRS origin is determined by the BRS equations of motion of the Earth and setting the dipole term in the GRS expansion of the geopotential equal to zero. A ground station serving as the TRS origin moves in GRS due to the rotation of the Earth and geophysical factors. A satellite taken to be the SRS origin moves on a geodesic in GRS.

The absence of dynamical rotation means that the space axes do not rotate with respect to the axes subjected to Fermi–Walker transport along the world line of the origin of the reference system (a gyroscopic triad). Mathematically, this implies the absence of Coriolis terms in the mixed components of the relevant metric tensor (the absence of the term $c^{-1}\epsilon_{ijk}\omega^j x^k$ in g_{0i}). For BRS describing the isolated Solar System the notions of dynamical and kinematic rotation are equivalent and BRS is non-rotating in the kinematic sense as well (constant direction towards fixed distant objects). The GRS space axis rotates kinematically with respect to the BRS space axis (geodesic precession). The TRS and SRS space axis rotate kinematically with respect to the GRS space axis.

Barycentric reference system (BRS). The BRS metric expressed in coordinates $x^0 = ct$, $\boldsymbol{x} = (x^1, x^2, x^3)$,is described by (4.1.2)–(4.1.4) (vanishing a_0, a_k) with values (4.1.11)–(4.1.14). The constants K^i and N^i of integrals (4.1.60) and (4.1.61) should thereby vanish.

Geocentric reference system (GRS). The GRS metric expressed in coordinates $w^0 = ct$, $\boldsymbol{w} = (w^1, w^2, w^3)$ is described by (4.2.2)–(4.2.4) with values (4.2.13)–(4.2.15) and (4.2.27). Mathematically, the coincidence of the GRS origin with the geocentre is dictated by conditions: the absence of the dipole term in the expansion of the geopotential (4.2.5) and the choice of the GRS origin acceleration in form (4.2.15).

The relationship between BRS and GRS is given by expressions (4.2.7), (4.2.8) with (4.2.11), (4.2.17), (4.2.18), (4.2.24)–(4.2.26). As mentioned above, this relationship is established more accurately in Kopejkin (1988).

Topocentric reference system (TRS). Let $z^0 = c\tau$, $\boldsymbol{z} = (z^1, z^2, z^3)$ be the TRS coordinates and $\tilde{g}_{\alpha\beta} = \tilde{g}_{\alpha\beta}(\tau, \boldsymbol{z})$ be the TRS metric tensor. TRS is constructed for an observer located on the Earth's surface. The TRS metric is determined by the gravitational field of external bodies (Earth, Sun, Moon, etc) and has the form (Zhang 1986, Brumberg and Kopejkin 1989a)

$$\tilde{g}_{00}(\tau, \boldsymbol{z}) = 1 - c^{-2}(2E_i z^i + 3E_{ik} z^i z^k + \ldots) + \ldots \tag{6.2.1}$$

$$\tilde{g}_{0i}(\tau, \boldsymbol{z}) = c^{-3} 4\epsilon_{ijk} H_{jm} z^k z^m + \dots \tag{6.2.2}$$

$$\tilde{g}_{ij}(\tau, \boldsymbol{z}) = -\delta_{ij} - c^{-2}(2E_k z^k + 3E_{km} z^k z^m + \dots)\delta_{ij} + \dots . \tag{6.2.3}$$

The TRS origin, i.e. some point on the Earth's surface, does not move along the geodesic. Thus the acceleration E_i is significant. The TRS origin is characterized in GRS by the coordinates $\boldsymbol{w}_T$, velocity $\boldsymbol{v}_T$ and acceleration $\boldsymbol{a}_T$ with

$$\boldsymbol{v}_T = (\hat{\boldsymbol{\omega}}_E \times \boldsymbol{w}_T) + \boldsymbol{v}_{TT} \tag{6.2.4}$$

$$\boldsymbol{a}_T = \left(\frac{\mathrm{d}\hat{\boldsymbol{\omega}}_E}{du} \times \boldsymbol{w}_T \right) + [\hat{\boldsymbol{\omega}}_E \times (\hat{\boldsymbol{\omega}}_E \times \boldsymbol{w}_T)] + 2(\hat{\boldsymbol{\omega}}_E \times \boldsymbol{v}_{TT}) + \boldsymbol{a}_{TT} \tag{6.2.5}$$

$\hat{\boldsymbol{\omega}}_E$ being the GRS vector of the angular velocity of rotation of the Earth. Due to tectonic deformations and other geophysical factors the ground observatory has relative velocity $\boldsymbol{v}_{TT}$ and relative acceleration $\boldsymbol{a}_{TT}$.

The transformation from GRS coordinates u and $\boldsymbol{w}$ to TRS coordinates τ and $\boldsymbol{z}$ is built by analogy with the transformation (4.2.7) and (4.2.8) from BRS to GRS, namely

$$\tau = u - c^{-2}[V(u) + v_T^k(w^k - w_T^k)] + \dots \tag{6.2.6}$$

$$\begin{aligned} z^i &= w^i - w_T^i + c^{-2}[(\tfrac{1}{2} v_T^i v_T^k + \Phi^{ik} + \mathcal{D}^{ik})(w^k - w_T^k) \\ &\quad + \mathcal{D}^{ikm}(w^k - w_T^k)(w^m - w_T^m)] + \dots . \end{aligned} \tag{6.2.7}$$

The matching of the GRS and TRS metric forms enables one to determine the coefficients occurring in (6.2.1)–(6.2.3) and (6.2.6), (6.2.7). One has

$$E_i = -a_T^i + \hat{U}_{E,i}(\boldsymbol{w}_T) + Q_i + 3Q_{ik} w_T^k + \tfrac{15}{2} Q_{ijk} w_T^j w_T^k + \dots \tag{6.2.8}$$

$$E_{ij} = \tfrac{1}{3} \hat{U}_{E,ij}(\boldsymbol{w}_T) + Q_{ij} + 5Q_{ijk} w_T^k + \dots \tag{6.2.9}$$

$$\frac{\mathrm{d}V}{du} = \tfrac{1}{2} v_T^2 + \hat{U}_E(\boldsymbol{w}_T) + Q_k w_T^k + \tfrac{3}{2} Q_{km} w_T^k w_T^m + \tfrac{5}{2} Q_{kmn} w_T^k w_T^m w_T^n + \dots \tag{6.2.10}$$

$$\mathcal{D}^{ij} = [\hat{U}_E(\boldsymbol{w}_T) + Q_k w_T^k + \tfrac{3}{2} Q_{km} w_T^k w_T^m + \tfrac{5}{2} Q_{kmn} w_T^k w_T^m w_T^n + \dots]\delta_{ij} \tag{6.2.11}$$

$$\mathcal{D}^{ijk} = \tfrac{1}{2}(\delta_{ij} a_T^k + \delta_{ik} a_T^j - \delta_{jk} a_T^i) \tag{6.2.12}$$

$$\begin{aligned} \frac{\mathrm{d}\Phi^{ik}}{du} &= \tfrac{3}{2}(v_T^i a_T^k - v_T^k a_T^i) + 2[\hat{U}_{E,i}^k(\boldsymbol{w}_T) - \hat{U}_{E,k}^i(\boldsymbol{w}_T)] + 2(v_T^i E_k - v_T^k E_i) \\ &\quad + 4\epsilon_{ikj} C_{jm} w_T^m + 2(\epsilon_{kjm} C_{ji} - \epsilon_{ijm} C_{jk}) w_T^m + (\dot{Q}_k w_T^i - \dot{Q}_i w_T^k). \end{aligned} \tag{6.2.13}$$

Expressions for H_{jm} and the terms $O(c^{-4})$ in (6.2.6) are not needed below but they may be obtained by analogy with (4.2.21)–(4.2.23). Relation

(6.2.6) with expression (6.2.10) serves to derive the relationship between the scales of the TRS coordinate time τ and the GRS coordinate time u.

Satellite reference system (SRS). SRS, being a coordinate system generated by a satellite, is of the same type as TRS with the difference that its origin, i.e. the satellite, moves in GRS in the geodesic. Therefore, the SRS metric does not contain any terms related to acceleration E_i. Retaining all previous designations and only replacing the index T by S one obtains the SRS metric and its relationship with GRS. The right-hand side of (6.2.8) is therefore equal to zero identically, resulting in an expression for the GRS acceleration of the satellite, i.e. its equations of motion. Expression (6.2.8) is sufficient to derive the Newtonian satellite equations of motion. If the matching of the GRS and SRS metrics is performed with greater accuracy then one obtains the post-Newtonian equations of the satellite motion derived in section 5.1 by a different technique.

6.2.3 Rotating system

Along with GRS, TRS and SRS it is useful to introduce the rotating systems GRS$^+$, TRS$^+$ and SRS$^+$ resulting from the rigid-body rotation of the space axes of the corresponding systems (Brumberg and Kopejkin 1989a). The use of harmonic coordinates is not suitable here because of the complicated transformations involved (Suen 1986).

Remember first of all the basic formulae of theoretical mechanics concerning motion in a rotating system. Let x^i be the coordinates of the fixed system and y^i be the coordinates of the moving system, whose rotation is given by the vector $\boldsymbol{\omega}$. The origins of the systems are assumed to coincide. If $\boldsymbol{r}$ is the position vector of the moving point then its velocity vector consists of two components due to the rotation of the moving system and the change of the point's position relative this moving system:

$$\boldsymbol{v} = \boldsymbol{\omega} \times \boldsymbol{r} + \frac{\tilde{\mathrm{d}}\boldsymbol{r}}{\mathrm{d}t} \tag{6.2.14}$$

where $\tilde{\mathrm{d}}/\mathrm{d}t$ denotes the relative derivative. The transformation from x^i to y^i is given by the orthogonal matrix $P = (P_{ik})$

$$y^i = P_{ik} x^k . \tag{6.2.15}$$

The reciprocal transformation is given by the inverse matrix coinciding with the transposed matrix $P^{-1} = P'$

$$x^i = P_{ki} y^k . \tag{6.2.16}$$

When operating with the orthogonal matrix P one may use the following relations:

$$P_{ik}P_{jk} = \delta_{ij} \qquad \epsilon_{lmn}P_{il}P_{jm}P_{kn} = \epsilon_{ijk}. \tag{6.2.17}$$

Denoting by ω^i and $\tilde{\omega}^i$ the projections of the vector $\boldsymbol{\omega}$ on the moving axes y^i and the fixed axes x^i, respectively, one has from (6.2.15) and (6.2.16)

$$\omega^i = P_{ik}\tilde{\omega}^k \qquad \tilde{\omega}^i = P_{ki}\omega^k. \tag{6.2.18}$$

In projecting onto moving axes y^i relation (6.2.14) yields

$$v^i = P_{ik}\dot{x}^k = \epsilon_{ijk}\omega^j y^k + \dot{y}^i. \tag{6.2.19}$$

The same relation projected onto fixed axes x^i gives

$$P_{ki}v^k = \dot{x}^i = \epsilon_{ijk}\tilde{\omega}^j x^k + P_{ki}\dot{y}^k. \tag{6.2.20}$$

Differentiating (6.2.15) and (6.2.16) and comparing with (6.2.19) and (6.2.20) one obtains

$$\dot{P}_{ik} = \epsilon_{imj}\omega^j P_{mk} \tag{6.2.21}$$

or, in projections on the fixed axes,

$$\dot{P}_{ik} = \epsilon_{kjm}\tilde{\omega}^j P_{im}. \tag{6.2.22}$$

If the metric in coordinates ct, x^i has the standard form with the metric coefficients

$$g_{00} = 1 + h_{00} \qquad g_{0i} = h_{0i} \qquad g_{ij} = -\delta_{ij} + h_{ij} \tag{6.2.23}$$

then in transforming to coordinates ct, y^i

$$\mathrm{d}x^i = c^{-1}\dot{P}_{ki}y^k c\,\mathrm{d}t + P_{ki}\mathrm{d}y^k \tag{6.2.24}$$

one has

$$\tilde{g}_{00} = 1 + h_{00} - c^{-2}\dot{P}_{kj}\dot{P}_{mj}y^k y^m + \dot{P}_{ki}\dot{P}_{mj}h_{ij}y^k y^m + 2c^{-1}h_{0m}\dot{P}_{km}y^k \tag{6.2.25}$$

$$\tilde{g}_{0i} = -c^{-1}P_{ij}\dot{P}_{kj}y^k + h_{0k}P_{ik} + c^{-1}P_{ik}\dot{P}_{nm}h_{km}y^n \tag{6.2.26}$$

$$\tilde{g}_{ij} = -\delta_{ij} + P_{ik}P_{jm}h_{km}. \tag{6.2.27}$$

Using (6.2.21) and identities (4.2.28) we obtain

$$\tilde{g}_{00} = 1 + h_{00} - c^{-2}(\boldsymbol{\omega}\times\boldsymbol{y})^2 + \epsilon_{krm}\epsilon_{lsn}\omega^m\omega^n y^k y^l P_{ri}P_{sj}h_{ij} + 2c^{-1}\epsilon_{kmj}\omega^j P_{mn}y^k h_{0n} \tag{6.2.28}$$

$$\tilde{g}_{0i} = -c^{-1}\epsilon_{ijk}\omega^j y^k + h_{0k}P_{ik} + c^{-1}\epsilon_{ljn}\omega^j y^n P_{ik}P_{lm}h_{km} \tag{6.2.29}$$

$$\tilde{g}_{ij} = -\delta_{ij} + P_{ik}P_{jm}h_{km}. \tag{6.2.30}$$

These relations are used below.

6.2.4 Rotating systems GRS$^+$, TRS$^+$, SRS$^+$

Denoting the GRS$^+$ coordinates by y^α one has

$$y^0 = w^0 \qquad y^i = P_{ik}w^k. \tag{6.2.31}$$

The orthogonal matrix $P = SN\mathsf{P}$ includes the matrix P of precession, the matrix N of nutation and the matrix S of diurnal rotation (Moritz and Mueller 1987). But in doing so, geodesic precession should be eliminated from the precession matrix P since GRS is dynamically non-rotating and geodesic precession is taken into account explicitly in the relationship (4.2.8) between BRS and GRS. The matrix P satisfies the equation

$$\frac{\mathrm{d}P_{ik}}{\mathrm{d}u} = \epsilon_{imj}\hat{\omega}^j_E P_{mk}. \tag{6.2.32}$$

Therefore, the GRS$^+$ metric has the form

$$\begin{aligned}\hat{g}^+_{00}(u,\boldsymbol{y}) = &1 - c^{-2}[2\hat{U}^+_E + (\hat{\omega}_E \times \boldsymbol{y})^2 + 2Q^+_k y^k \\ &+ 3Q^+_{km}y^k y^m + 5Q^+_{kmn}y^k y^m y^n + \ldots]\end{aligned} \tag{6.2.33}$$

$$\begin{aligned}\hat{g}^+_{0i}(u,\boldsymbol{y}) = &- c^{-1}\epsilon_{ijk}\hat{\omega}^j_E y^k + c^{-3}(4\hat{U}^{i+}_E - 2\epsilon_{ijk}\hat{\omega}^j_E y^k \hat{U}^+_E \\ &- 2\epsilon_{ijk}\hat{\omega}^j_E y^k y^m Q^+_m + 4\epsilon_{ijk}C^+_{jm}y^k y^m + \ldots) + \ldots\end{aligned} \tag{6.2.34}$$

$$\begin{aligned}\hat{g}^+_{ij}(u,\boldsymbol{y}) = &- \delta_{ij} - c^{-2}(2\hat{U}^+_E + 2Q^+_k y^k + 3Q^+_{km}y^k y^m \\ &+ 5Q^+_{kmn}y^k y^m y^n + \ldots)\delta_{ij} + \ldots\end{aligned} \tag{6.2.35}$$

$\hat{U}^+_E$ being the potential $\hat{U}_E$ expressed in new variables and

$$\hat{U}^{i+}_E = P_{ik}\hat{U}^k_E \qquad Q^+_{ij} = P_{ik}P_{jm}Q_{km} \qquad Q^+_{ijk} = P_{im}P_{jn}P_{kl}Q_{mnl} \tag{6.2.36}$$

and similarly for Q^+_i and C^+_{ij}. The terms with derivatives of acceleration Q_k are omitted in (6.2.34).

Denoting the TRS$^+$ coordinates by ξ^α one has

$$\xi^0 = z^0 \qquad \xi^i = \tilde{P}_{ik} z^k \qquad\qquad (6.2.37)$$

with the rotation matrix $\tilde{P}_{ik}$ satisfying the equation

$$\frac{\mathrm{d}\tilde{P}_{ik}}{\mathrm{d}\tau} = \epsilon_{imj} \tilde{\omega}_E^j \tilde{P}_{mk}. \qquad\qquad (6.2.38)$$

$\tilde{\omega}_E$ is the vector of the angular velocity of rotation of the Earth in TRS. To Newtonian accuracy it is evident that

$$\tilde{P}_{ik}(\tau) = P_{ik}(u) + \dots \qquad \tilde{\omega}_E^k = \hat{\omega}_E^k + \dots . \qquad\qquad (6.2.39)$$

The TRS$^+$ metric is of the form

$$\tilde{g}_{00}^+(\tau,\xi) = 1 - c^{-2}[(\tilde{\omega}_E \times \xi)^2 + 2E_i^+ \xi^i + 3E_{ij}^+ \xi^i \xi^j + \dots] + \dots \qquad (6.2.40)$$

$$\tilde{g}_{0i}^+(\tau,\xi) = -c^{-1} \epsilon_{ijk} \tilde{\omega}_E^j \xi^k + c^{-3}(4\epsilon_{ijk} H_{jm}^+ \xi^k \xi^m - 2\epsilon_{ijk} \tilde{\omega}_E^j \xi^k \xi^m E_m^+ + \dots) + \dots \qquad (6.2.41)$$

$$\tilde{g}_{ij}^+(\tau,\xi) = -\delta_{ij} - c^{-2}(2E_k^+ \xi^k + 3E_{km}^+ \xi^k \xi^m + \dots)\delta_{ij} + \dots . \qquad (6.2.42)$$

Here

$$E_i^+ = \tilde{P}_{ik} E_k \qquad E_{ij}^+ = \tilde{P}_{ik} \tilde{P}_{jm} E_{km} \qquad\qquad (6.2.43)$$

and similarly for H_{ij}^+.

Finally, the SRS$^+$ metric in retaining the previous designations is described by the same formulae (6.2.37), (6.2.38), (6.2.40)–(6.2.42) with $E_i^+ = 0$ and replacing $\tilde{\omega}_E$ by the rotation velocity $\tilde{\omega}_S$ of the TRS$^+$ space axes with respect to TRS.

6.2.5   Hierarchy of the systems

The hierarchy of the systems described may be schematically represented as follows:

$$\text{BRS} :t, \boldsymbol{x} \rightarrow \text{GRS} :\boldsymbol{u}, \boldsymbol{w} \xrightarrow{y^i = P_{ik} w^k} \text{GRS}^+ :\boldsymbol{u}, \boldsymbol{y}$$

$$\text{SRS}^+ :\tau, \xi \xleftarrow{\xi^i = \tilde{P}_{ik} z^k} \text{SRS} :\tau, \boldsymbol{z} \qquad \text{TRS} :\tau, \boldsymbol{z} \xrightarrow{\xi^i = \tilde{P}_{ik} z^k} \text{TRS}^+ :\tau, \xi$$

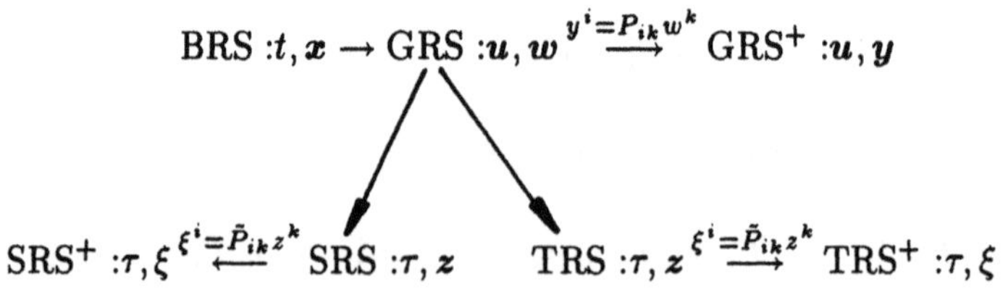

For TRS, TRS$^+$ on the one hand and for SRS, SRS$^+$ on the other hand the same designations are used. It is evident that in simultaneously using these two types of systems one has to introduce different designations for the coordinates and functions associated to these two types.

To convert the reference systems constructed above into reference frames it is necessary to describe the observation procedure in terms of a particular system and to connect the coordinate system with the reference astronomical objects. The relations of the next section may serve as the basis for this.

6.3 RELATIVISTIC REDUCTION OF ASTROMETRIC OBSERVATIONS

6.3.1 Cauchy problem of light propagation in BRS

Let the motion of a photon is given by the conditions

$$\boldsymbol{x}(t_0) = \boldsymbol{x}_0 \qquad \frac{\mathrm{d}\boldsymbol{x}(-\infty)}{\mathrm{d}t} = c\boldsymbol{\sigma} \qquad \boldsymbol{\sigma}^2 = 1 \tag{6.3.1}$$

i.e. its position at some moment t_0 of the coordinate time and its velocity direction at an infinitely far distance from the Solar System in the remote past. Retaining the previous designations $\boldsymbol{r}_A = \boldsymbol{x} - \boldsymbol{x}_A$, $\boldsymbol{r}_{0A} = \boldsymbol{x}_0 - \boldsymbol{x}_A$, $m_A = GM_A/c^2$ and integrating the equations of motion of a photon (2.2.61) one obtains

$$\boldsymbol{x}(t_0) = \boldsymbol{x}_0 + c(t - t_0)\boldsymbol{\sigma} + 2\sum_A m_A \left(\frac{\boldsymbol{\sigma} \times (\boldsymbol{r}_{0A} \times \boldsymbol{\sigma})}{r_{0A} - \boldsymbol{\sigma}\boldsymbol{r}_{0A}} - \frac{\boldsymbol{\sigma} \times (\boldsymbol{r}_A \times \boldsymbol{\sigma})}{r_A - \boldsymbol{\sigma}\boldsymbol{r}_A} \right.$$
$$\left. - \boldsymbol{\sigma} \ln \frac{r_A + \boldsymbol{\sigma}\boldsymbol{r}_A}{r_{0A} + \boldsymbol{\sigma}\boldsymbol{r}_{0A}} \right) \tag{6.3.2}$$

$$c^{-1}\frac{\mathrm{d}\boldsymbol{x}(t)}{\mathrm{d}t} = \boldsymbol{\sigma} - 2\sum_A \frac{m_A}{r_A}\left(\boldsymbol{\sigma} + \frac{\boldsymbol{\sigma} \times (\boldsymbol{r}_A \times \boldsymbol{\sigma})}{r_A - \boldsymbol{\sigma}\boldsymbol{r}_A}\right). \tag{6.3.3}$$

In the process of integration the coordinates of the bodies $\boldsymbol{x}_A$ have been treated as fixed. Therefore, strictly speaking, the results obtained are valid for such a time interval $t - t_0$ during which the change of position of the bodies may be neglected. For a photon moving inside the Solar System this assumption is justified.

Expression (6.3.3) determines the coordinate velocity of the light propagation. To obtain the BRS observed (coordinate-independent) direction $\boldsymbol{p}(t)$ at the point of observation one has by (6.1.19)

$$\boldsymbol{p} = (1 + c^{-2}2U)c^{-1}\frac{\mathrm{d}\boldsymbol{x}(t)}{\mathrm{d}t} \tag{6.3.4}$$

or

$$\boldsymbol{p} = \boldsymbol{\sigma} - 2\sum_A \frac{m_A}{r_A} \frac{\boldsymbol{\sigma} \times (\boldsymbol{r}_A \times \boldsymbol{\sigma})}{r_A - \boldsymbol{\sigma r}_A}. \tag{6.3.5}$$

6.3.2 Boundary value problem of light propagation in BRS

For the problem with boundary conditions

$$\boldsymbol{x}(t_0) = \boldsymbol{x}_0 \qquad \boldsymbol{x}(t) = \boldsymbol{x} \qquad \boldsymbol{R}(t, t_0) = \boldsymbol{x} - \boldsymbol{x}_0 \qquad (t > t_0) \tag{6.3.6}$$

one may use the previous formula (6.3.2) for the light propagation with the value

$$\boldsymbol{\sigma} = \frac{\boldsymbol{R}}{R} + 2\sum_A \frac{m_A}{R} \frac{r_A - r_{0A} + R}{|\boldsymbol{r}_{0A} \times \boldsymbol{r}_A|^2} [\boldsymbol{R} \times (\boldsymbol{r}_{0A} \times \boldsymbol{r}_A)]. \tag{6.3.7}$$

Substitution of (6.3.6) into (6.3.3) and (6.3.4) yields the coordinate and observed directions, respectively, for the moment t of reception of signal:

$$c^{-1}\frac{d\boldsymbol{x}(t)}{dt} = \frac{\boldsymbol{R}}{R} - \frac{2}{R}\sum_A \frac{m_A}{r_A}\left(\boldsymbol{R} + \frac{\boldsymbol{R} \times (\boldsymbol{r}_{0A} \times \boldsymbol{r}_A)}{r_{0A}r_A + \boldsymbol{r}_{0A}\boldsymbol{r}_A}\right) \tag{6.3.8}$$

$$\boldsymbol{p} = \frac{\boldsymbol{R}}{R} - \frac{2}{R}\sum_A \frac{m_A}{r_A} \frac{\boldsymbol{R} \times (\boldsymbol{r}_{0A} \times \boldsymbol{r}_A)}{r_{0A}r_A + \boldsymbol{r}_{0A}\boldsymbol{r}_A}. \tag{6.3.9}$$

The transit time of the light propagation is

$$c(t - t_0) = R + 2\sum_A m_A \ln \frac{r_A + r_{0A} + R}{r_A + r_{0A} - R}. \tag{6.3.10}$$

Consider now the case of a light particle coming from outside the Solar System so that

$$\rho \equiv |\boldsymbol{x}| \ll |x_0| \equiv \rho_0. \tag{6.3.11}$$

Expanding in powers ρ/ρ_0 one has

$$\boldsymbol{\sigma} = -\frac{\boldsymbol{x}_0}{\rho_0} + \frac{1}{\rho_0^3}[\boldsymbol{x}_0 \times (\boldsymbol{x} \times \boldsymbol{x}_0)] + \ldots + \frac{2}{\rho_0^3}\sum_A \frac{m_A}{r_A} \frac{\boldsymbol{x}_0 \times (\boldsymbol{r}_A \times \boldsymbol{x}_0)}{1 + (\boldsymbol{x}_0\boldsymbol{r}_A/\rho_0 r_A)} \tag{6.3.12}$$

$$\boldsymbol{p} = -\frac{\boldsymbol{x}_0}{\rho_0} + \frac{1}{\rho_0^3}[\boldsymbol{x}_0 \times (\boldsymbol{x} \times \boldsymbol{x}_0)] - \frac{2}{\rho_0^2}\sum_A \frac{m_A}{r_A^2} \frac{\boldsymbol{x}_0 \times (\boldsymbol{r}_A \times \boldsymbol{x}_0)}{1 + (\boldsymbol{x}_0\boldsymbol{r}_A/\rho_0 r_A)} \tag{6.3.13}$$

$$c(t - t_0) = \rho_0\left(1 - \frac{\boldsymbol{x}\boldsymbol{x}_0}{\rho_0^2} + \ldots\right) + 2\sum_A m_A \ln \frac{2\rho_0^2}{\rho_0 r_A + \boldsymbol{x}_0\boldsymbol{r}_A}. \tag{6.3.14}$$

It is of interest that in the limit $\rho_0 \to \infty$ the relativistic correction disappears in $\boldsymbol{\sigma}$ and remains in $\boldsymbol{p}$ as may be expected from physical considerations. This relativistic correction in $\boldsymbol{p}$ determines the actual deflection of the light ray emitted by a distant source. The Newtonian correction of order ρ/ρ_0 determines the parallactic displacement caused by the BRS position vector $\boldsymbol{x}$ of the observer.

6.3.3 Angular distance between two sources

The angular distance ψ between two light sources recorded by the BRS observer may be determined by (6.1.12) with (6.1.10), (6.1.11) and (6.3.3) or by (6.1.24) with (6.3.4). The result is

$$\cos\psi = \boldsymbol{p}_1\boldsymbol{p}_2 = \boldsymbol{\sigma}_1\boldsymbol{\sigma}_2 + 2\sum_A \frac{m_A}{r_A}\left(\frac{\boldsymbol{r}_A \times \boldsymbol{\sigma}_1}{r_A - \boldsymbol{\sigma}_1\boldsymbol{r}_A} - \frac{\boldsymbol{r}_A \times \boldsymbol{\sigma}_2}{r_A - \boldsymbol{\sigma}_2\boldsymbol{r}_A}\right)(\boldsymbol{\sigma}_1 \times \boldsymbol{\sigma}_2). \tag{6.3.15}$$

With the aid of (6.3.7) and (6.3.12) one may obtain particular cases when one of the sources (or both) is at a finite or infinite distance from the BRS origin.

6.3.4 Relativistic reduction in BRS

Let us summarize the main statements of relativistic reduction for an observer at rest in BRS. The observer records the unit vector $\boldsymbol{p}$ of the observed propagation of signal from the light source. If this source is at an infinite distance then formula (6.3.4) enables one to take into account the gravitational deflection of the ray in the Solar System and so to obtain $\boldsymbol{\sigma}$. If this source is at a finite distance (a planet) and the transit time is to be taken into account then with the aid of (6.3.9) and (6.3.10) by iterations one finds the moment t_0 of the signal emission (the inclusion of the planetary aberration) and the unit vector $\boldsymbol{R}/R$. Then formula (6.3.7) again yields $\boldsymbol{\sigma}$. Finally, if the source is at a large (but not infinitely large) distance then equations (6.3.13) and (6.3.14) enable one to include the annual parallax and the proper motion of the source (if this is necessary). As a result, the unit vector $\boldsymbol{x}_0/\rho_0$ becomes known yielding by (6.3.12) $\boldsymbol{\sigma}$.

6.3.5 Relativistic reduction in GRS

On the basis of (4.2.7) the differential relationship of GRS time u and BRS time t is determined by the equation

$$\frac{\mathrm{d}u}{\mathrm{d}t} = 1 + c^{-2}\left(-v_E^k\frac{\mathrm{d}x^k}{\mathrm{d}t} + v_E^2 - \dot{S}(t) - a_E^k(x^k - x_E^k)\right). \tag{6.3.16}$$

For the photon the first term in the large brackets is of first order. Differentiating (4.2.8) and using (6.3.16) one gets the coordinate GRS light velocity

$$c^{-1}\frac{dw^i}{du} = c^{-1}\frac{dx^i}{dt} + c^{-1}\left(-v_E^i + v_E^k\frac{dx^k}{cdt}\frac{dx^i}{cdt}\right) + c^{-2}\left[\left(v_E^k\frac{dx^k}{cdt}\right)^2\frac{dx^i}{cdt}\right.$$
$$- \frac{1}{2}v_E^2\frac{dx^i}{cdt} - \frac{1}{2}v_E^i v_E^k\frac{dx^k}{cdt} + [F^{ik} + 2D^{ik} + 2D^{ikm}(x^m - x_E^m)]\frac{dx^k}{cdt}$$
$$\left. + a_E^k(x^k - x_E^k)\frac{dx^i}{cdt}\right]. \qquad (6.3.17)$$

Expression (6.1.19) applied to the GRS metric (4.2.2)–(4.2.4) yields the observed GRS light direction $\hat{p}$

$$\hat{p} = [1 + c^{-2}(2\hat{U}_E + 2Q_k w^k + 3Q_{km}w^k w^m + \ldots)]\frac{dw}{cdu}. \qquad (6.3.18)$$

Using now (4.3.4), (4.3.17) and (4.3.18) one derives the relationship between the observed light directions $\hat{p}$ and p in GRT and BRS, respectively:

$$\hat{p}^{(i)} = p^{(i)} + c^{-1}[p \times (p \times v_E)]^{(i)} + c^{-2}(pv_E)[p \times (p \times v_E)]^{(i)}$$
$$- \tfrac{1}{2}c^{-2}[v_E \times (p \times v_E)]^{(i)} + c^{-2}F^{ik}p^{(k)} + c^{-2}(a_E^k w^i - a_E^i w^k)p^{(k)}$$
$$(6.3.19)$$

$p^{(i)}$ being determined by (6.3.4), (6.3.9) or (6.3.13). This formula of reduction from GRS to BRS includes first-order annual aberration (the second term on the right-hand side) and second-order annual aberration (the third and the fourth terms), geodesic precession (the fifth term) and relativistic contraction (the last term). Corrections for the gravitational deflection of light and (if necessary) for planetary aberration, annual parallax and proper motion are contained in p.

For the cosine of the angular distance between two sources one has, from (6.1.24) and (6.3.15),

$$\cos\hat{\psi} = \hat{p}_1\hat{p}_2 = \cos\psi + c^{-1}(p_1p_2 - 1)(p_1v_E + p_2v_E)$$
$$+ c^{-2}(p_1p_2 - 1)[(p_1v_E)^2 + (p_2v_E)^2 + (p_1v_E)(p_2v_E) - v_E^2]$$
$$(6.3.20)$$

demonstrating that both the geodesic precession and gravitational contraction do not interfere with the results of relative astronomical measurements. This formula may be derived directly from the BRS metric by means of the classical transformation $x = x_E(t) + \rho$ with further application of (6.1.12). The pure kinematic approach of section 2.3 with splitting (2.3.42), (2.3.43) results in expression (6.3.19) without the terms due to the geodesic precession and the gravitational contraction. This corresponds to the kinematic structure of the geocentric system (g) defined by (2.3.41).

6.3.6 Relativistic reduction in TRS (SRS)

Using (6.2.6) one obtains the following differential relationship between the coordinate TRS (SRS) time τ and the coordinate GRS time u:

$$\frac{d\tau}{du} = 1 + c^{-2}\left(-v_T^k \frac{dw^k}{cdu} + v_T^2 - \frac{dV}{du} - a_T^k(w^k - w_T^k)\right). \tag{6.3.21}$$

For the photon the first term in the large brackets is first order. For SRS the index T is to be replaced by S. Differentiating (6.2.7) and using (6.3.21) one has

$$\begin{aligned} c^{-1}\frac{dz^i}{d\tau} = {} & c^{-1}\frac{dw^i}{du} + c^{-1}\left(-v_T^i + v_T^k \frac{dw^k}{cdu}\frac{dw^i}{cdu}\right) + c^{-2}\left[\left(v_T^k \frac{dw^k}{cdu}\right)^2 \frac{dw^i}{cdu}\right. \\ & - \frac{1}{2}v_T^2 \frac{dw^i}{cdu} - \frac{1}{2}v_T^i v_T^k \frac{dw^k}{cdu} + [\Phi^{ik} + 2\mathcal{D}^{ik} \\ & \left. + 2\mathcal{D}^{ikm}(w^m - w_T^m)]\frac{dw^k}{cdu} + a_T^k(w^k - w_T^k)\frac{dw^i}{cdu}\right]. \end{aligned} \tag{6.3.22}$$

Application of (6.1.19) to the metric (6.2.1)–(6.2.3) yields the expression for the observed TRS (SRS) light direction

$$\tilde{\boldsymbol{p}} = [1 + c^{-2}(2E_k z^k + 3E_{km} z^k z^m + \ldots)]\frac{d\boldsymbol{z}}{cd\tau}. \tag{6.3.23}$$

Therefore, the relations (6.3.18), (6.3.22) and (6.3.23) result in the following relationship for the observed directions in TRS (SRS) and GRS:

$$\begin{aligned} \tilde{p}^{(i)} = {} & \hat{p}^{(i)} + c^{-1}[\hat{\boldsymbol{p}} \times (\hat{\boldsymbol{p}} \times \boldsymbol{v}_T)]^{(i)} + c^{-2}(\hat{\boldsymbol{p}}\boldsymbol{v}_T)[\hat{\boldsymbol{p}} \times (\hat{\boldsymbol{p}} \times \boldsymbol{v}_T)]^{(i)} \\ & - \tfrac{1}{2}c^{-2}[\boldsymbol{v}_T \times (\hat{\boldsymbol{p}} \times \boldsymbol{v}_T)]^{(i)} + c^{-2}\Phi^{ik}\hat{p}^{(k)} + c^{-2}(a_T^k z^i - a_T^i z^k)\hat{p}^{(k)}. \end{aligned} \tag{6.3.24}$$

Equation (6.3.24) applied to TRS includes first-order diurnal aberration (the second term) and second-order diurnal aberration (the third and the fourth terms), an analogue of geodesic precession (the fifth term) and a relativistic contraction (the sixth term). All these terms are due to the motion of the TRS origin in GRS. $\hat{p}^{(i)}$ is expressed in terms of $p^{(i)}$, $p^{(i)}$ in turn is expressed in terms of σ^i and one may express σ^i, if needed, similarly to (6.3.12) in terms of the TRS coordinates and take into account the diurnal parallax. When applied to SRS the terms of (6.3.24) may be interpreted in a similar manner.

For the mutual angular distance in TRS (SRS) one has a formula analogous to (6.3.20):

$$\cos\tilde{\psi} = \tilde{\boldsymbol{p}}_1\tilde{\boldsymbol{p}}_2 = \cos\hat{\psi} + c^{-1}(\hat{\boldsymbol{p}}_1\hat{\boldsymbol{p}}_2 - 1)(\hat{\boldsymbol{p}}_1\boldsymbol{v}_T + \hat{\boldsymbol{p}}_2\boldsymbol{v}_T) + c^{-2}(\hat{\boldsymbol{p}}_1\hat{\boldsymbol{p}}_2 - 1) \times [(\hat{\boldsymbol{p}}_1\boldsymbol{v}_T)^2 + (\hat{\boldsymbol{p}}_2\boldsymbol{v}_T)^2 + (\hat{\boldsymbol{p}}_1\boldsymbol{v}_T)(\hat{\boldsymbol{p}}_2\boldsymbol{v}_T) - \boldsymbol{v}_T^2]. \tag{6.3.25}$$

The kinematic approach of section 2.3 involving system s (2.3.46) enables one also to obtain (6.3.25). The observed direction (6.3.24) is therefore obtained without the precession and gravitational contraction terms.

6.3.7 Relativistic reduction in TRS$^+$ (SRS$^+$)

Actual astrometric measurements are performed now from ground observatories (TRS$^+$) or satellites (SRS$^+$) and as a rule the results are taken just in the reference system origin ($\xi = 0$). On the basis of (6.2.37) and (6.2.38) one has

$$c^{-1}\frac{\mathrm{d}\xi^i}{\mathrm{d}\tau} = \tilde{P}_{ik}\frac{\mathrm{d}z^k}{c\mathrm{d}\tau} + c^{-1}\epsilon_{imj}\tilde{\omega}_E^j\tilde{P}_{mk}z^k. \tag{6.3.26}$$

Application of (6.1.19) to the metric (6.2.40)–(6.2.42) gives the expression for the observed light direction in TRS$^+$ (SRS$^+$):

$$\tilde{p}^{(i)+} = [1 + c^{-2}(3E_k^+\xi^k + 3E_{km}^+\xi^k\xi^m + \ldots)]\frac{\mathrm{d}\xi^i}{c\mathrm{d}\tau} + c^{-1}\epsilon_{kjm}\tilde{\omega}_E^j\xi^m\frac{\mathrm{d}\xi^k}{c\mathrm{d}\tau}\frac{\mathrm{d}\xi^i}{c\mathrm{d}\tau} + \ldots. \tag{6.3.27}$$

The terms quadratic in $\tilde{\omega}_E^k$ and ξ^k are omitted here. Using (6.3.24) and (6.3.26) one obtains

$$\tilde{p}^{(i)+} = \tilde{P}_{ik}\tilde{p}^{(k)} + c^{-1}\epsilon_{kjm}\tilde{\omega}_E^j\xi^m\tilde{P}_{in}\tilde{P}_{kl}\tilde{p}^{(n)}\tilde{p}^{(l)} + c^{-1}\epsilon_{imj}\tilde{\omega}_E^j\xi^m + \ldots. \tag{6.3.28}$$

Considering the orthogonality of the matrix $\tilde{P}_{ik}$ it may be easily verified that the right-hand side (6.3.28) represents a unit vector. Using now (6.3.5), (6.3.19), (6.3.24) and (6.3.28) one obtains the expression of $\tilde{p}^{(i)+}$ in terms of the BRS vector $\boldsymbol{\sigma}$. This expression serves as a basis for constructing the astronomical reference frame. In the actual reduction calculation it is not suitable to analytically substitute into (6.3.28) all the previous values. It is more effective to use the sequence of formulae derived here. Let us note only that the terms due to the relativistic precession are combined with the Newtonian precession in the form

$$\tilde{p}^{(i)+} = [\tilde{P}_{im} + c^{-2}\tilde{P}_{ik}(F^{km} + \Phi^{km})]\sigma^m + \ldots \tag{6.3.29}$$

and in practice the relativistic part of precession is not separated from the Newtonian part.

Finally, for the cosine of the mutual angular distance between two sources one has by (6.1.24)

$$\cos\tilde{\psi}^{+} = \cos\tilde{\psi} + O(c^{-2}). \tag{6.3.30}$$

It is evident that the terms $O(c^{-1})$ do not affect the relative angular distances.

6.4 RELATIVISTIC REDUCTION OF RADIO OBSERVATIONS

6.4.1 Basic tools

The basic formula for the relativistic reduction of radio observations is (6.3.10) or as its particular case (6.3.14). This formula is usually applied not to a single measurement but rather to a sequence of measurements performed during a sufficiently long interval of time. If $\boldsymbol{x}_0 = \boldsymbol{x}_0(t)$ is the BRS position vector of the moving light source then in terms of t_0 it may be linear (inclusion of proper motion) or a solution of the two-body problem (a pulsar binary system) or else a still more complicated function of time (the motion of a planet or space probe).If the vector $\boldsymbol{x} = \boldsymbol{x}(t)$ represents the motion of an observer one should perform a reduction due to this motion. This requires the transformation of both the space coordinates and the time. The full theory of transformations BRS→GRS→TRS→TRS$^+$ is not usually needed here and it suffices to employ the simplified formulae. In the linear approximation the transformation BRS→GRS of the space coordinates (4.2.8) is described in the form

$$\boldsymbol{x} = \boldsymbol{x}_E + \boldsymbol{w} - c^{-2}[\tfrac{1}{2}(\boldsymbol{v}_E\boldsymbol{w})\boldsymbol{v}_E + \bar{U}_E(\boldsymbol{x}_E)\boldsymbol{w} + \boldsymbol{w}\times\boldsymbol{F} + O(\boldsymbol{w}^2)] \tag{6.4.1}$$

whereas the transformation GRS→TRS of the space coordinates (6.2.7) reduces to

$$\boldsymbol{w} = \boldsymbol{w}_T + \boldsymbol{z} - c^{-2}[\tfrac{1}{2}(\boldsymbol{v}_T\boldsymbol{z})\boldsymbol{v}_T + \hat{U}_E(\boldsymbol{w}_T)\boldsymbol{z} + \boldsymbol{z}\times\boldsymbol{\Phi} + O(\boldsymbol{w}^2, \boldsymbol{z}^2)]. \tag{6.4.2}$$

The precession terms are represented here in vector form by means of

$$F^i = \tfrac{1}{2}\epsilon_{ikm}F^{km} \qquad \Phi^i = \tfrac{1}{2}\epsilon_{ikm}\Phi^{km}. \tag{6.4.3}$$

Superposition of these transformations and rotations (6.2.31) and (6.2.37) yields the transformations to GRS$^+$ and TRS$^+$ coordinates, respectively

$$x^i = x^i_E + P_{ki}y^k - c^{-2}(\tfrac{1}{2}v^i_E v^k_E P_{mk} + \bar{U}_E(\boldsymbol{x}_E)P_{mi} + F^{ik}P_{mk})y^m + \ldots \tag{6.4.4}$$

$$w^i = w_T^i + \tilde{P}_{ki}\xi^k - c^{-2}(\tfrac{1}{2}v_T^i v_T^k \tilde{P}_{mk} + \hat{U}_E(\boldsymbol{w}_T)\tilde{P}_{mi} + \Phi^{ik}\tilde{P}_{mk})\xi^m + \ldots \quad (6.4.5)$$

For reduction purposes such linear formulae turn out to be quite adequate. They differ from those commonly used (Hellings 1986) by the presence of the precession terms due to the fact that the systems GRS and TRS used here are dynamically non-rotating.

Within an accuracy sufficient for reduction the relationship (4.2.7) of the BRS and GRS timescales has the form

$$t = u + c^{-2}[S(t) + \boldsymbol{v}_E\boldsymbol{w}] \qquad S(t) = \int_0^t [\tfrac{1}{2}v_E^2 + \bar{U}_E(\boldsymbol{x}_E)]\, dt \quad (6.4.6)$$

whereas the relationship (6.2.7) of the GRS and TRS timescales reduces to

$$u = \tau + c^{-2}[V(u) + \boldsymbol{v}_T\boldsymbol{z}] \qquad V(u) = \int_0^u [\tfrac{1}{2}v_T^2 + \hat{U}_E(\boldsymbol{w}_T)]\, du. \quad (6.4.7)$$

The superposition of these transformations results in

$$t = \tau + c^{-2}[S(t) + V(u) + \boldsymbol{v}_E\boldsymbol{w}_T + (\boldsymbol{v}_E + \boldsymbol{v}_T)\boldsymbol{z}]. \quad (6.4.8)$$

One may often use therefore the approximate relation

$$V(u) = (\tfrac{1}{2}v_T^2 + \hat{U}_E)u \quad (6.4.9)$$

since for the observer on the geoid the coefficient in u takes a constant value independent of time and location of the observer on the geoid.

These relations combined with the basic formula (6.3.10) are of great importance for the relativistic reduction of radio measurements.

6.4.2 Radio ranging

In radio ranging a signal is emitted by an observer at some BRS moment t_0, it reaches the planet or space probe at some pre-calculated moment t_1 and then returns to the observer at moment t_2. The observer directly records the moments of emission and reception of the signal in the observer's proper time. The moments t_0 and t_2 are calculated using (6.4.6) and (6.4.7). The value of t_1 can be improved by iteration. If accuracy of the post-Newtonian approximation (6.3.10) is insufficient one may take into account the post-post-Newtonian effect due to the Sun (3.2.51). In doing this the third-order effect caused by the motion of the planets should also be considered (Klioner 1989). From (6.3.10) it follows that

$$\frac{dt}{dt_0} = \left[1 - c^{-1}\frac{\boldsymbol{R}\dot{\boldsymbol{x}}_0}{R} - 4c\sum_A \frac{m_A}{(r_A + r_{0A})^2 - R^2}\right.$$

$$\times \left(\frac{r_A + r_{0A}}{R}\boldsymbol{R}\dot{\boldsymbol{x}}_0 + \frac{\boldsymbol{R}}{r_{0A}}\boldsymbol{r}_{0A}\dot{\boldsymbol{r}}_{0A}\right)\Bigg]$$
$$\times \left[1 - c^{-1}\frac{\boldsymbol{R}\dot{\boldsymbol{x}}}{R} - 4c\sum_A \frac{m_A}{(r_A + r_{0A})^2 - R^2}\right.$$
$$\left.\times \left(\frac{r_A + r_{0A}}{R}\boldsymbol{R}\dot{\boldsymbol{x}} - \frac{\boldsymbol{R}}{r_A}\boldsymbol{r}_A\dot{\boldsymbol{r}}_A\right)\right]^{-1} \tag{6.4.10}$$

generalizing equation (3.2.29) for the Schwarzschild field. In differentiating (6.3.10) it is assumed that

$$\boldsymbol{r}_A = \boldsymbol{x}(t) - \boldsymbol{x}_A(t) \qquad \boldsymbol{r}_{0A} = \boldsymbol{x}_0(t_0) - \boldsymbol{x}_A(t_0) \qquad \boldsymbol{R} = \boldsymbol{x}(t) - \boldsymbol{x}_0(t_0)$$

with the dot over $\boldsymbol{x}$ or $\boldsymbol{x}_0$ denoting a derivative with respect to t or t_0, respectively. Planetary radio ranging is described in detail in Hellings (1986).

In radio ranging the Doppler shift of the signal frequency can also be measured. If ν_0 is the frequency of the light emitter at the point $\boldsymbol{x}_0(t_0)$ and ν is the frequency of the signal received at the point $\boldsymbol{x}(t)$ then considering that the frequency is inversely proportional to the period of the signal in its proper time and generalizing (3.2.28) one has

$$\frac{\nu_0}{\nu} = \frac{\delta\tau}{\delta\tau_0} = \frac{1 + \frac{1}{2}(h_{00})_t - \frac{1}{2}v^2 + c^{-1}(h_{0k})_t v^k}{1 + \frac{1}{2}(h_{00})_{t_0} - \frac{1}{2}v_0^2 + c^{-1}(h_{0k})_{t_0} v_0^k}\frac{\mathrm{d}t}{\mathrm{d}t_0}. \tag{6.4.11}$$

In radio ranging the light signal is emitted by the observer with frequency ν_0. It is received and reflected by the ranging object (planet, space probe, etc) with frequency ν_1 and is received again by the observer with frequency ν_2. The measurable quantity is the ratio ν_0/ν_2 which can be calculated by the repeated application of (6.4.10) and (6.4.11).

6.4.3 Lunar laser ranging

In LLR the starting formula is again (6.3.10). In the relativistic part of this formula one may perform significant simplifications. If t_0 is the moment of the signal emission by a ground observer and t is the moment of the reception of the signal at a point on the surface of the Moon then

$$\boldsymbol{x}_0 = \boldsymbol{x}_E(t_0) + \boldsymbol{\rho}_E(t_0) \qquad \boldsymbol{x} = \boldsymbol{x}_L(t) + \boldsymbol{\rho}_L(t).$$

$\boldsymbol{x}_E$ and $\boldsymbol{x}_L$ are the BRS positions of the centres of mass of the Earth and the Moon, respectively, $\boldsymbol{\rho}_E$ is the geocentric position vector of the ground station, and $\boldsymbol{\rho}_L$ is the selenocentric position vector of the reflector on the Moon. In the relativistic terms the difference between t_0 and t may be ignored. If we consider the case where LLR is performed for the position of the Moon near the meridian of the ground station and denoting the distance between the centres of mass of the Earth and the Moon by r one may use in the term of (6.3.10) containing the factor m_E the approximate relations

$$R = r - \rho_E - \rho_L \qquad r_{0E} = \rho_E \qquad r_E = r - \rho_L. \tag{6.4.12}$$

For the solar term with factor m_S one has $R \approx r$, $r_{0S} \approx |\boldsymbol{x}_E|$,

$$r_S \approx |\boldsymbol{x}_L| = |\boldsymbol{x}_E| + \frac{\boldsymbol{x}_E \boldsymbol{r}}{|\boldsymbol{x}_E|} + \frac{1}{2|\boldsymbol{x}_E|^2}\left(r^2 - \frac{(\boldsymbol{x}_E \boldsymbol{r})^2}{|\boldsymbol{x}_E|^2}\right) + \ldots \tag{6.4.13}$$

and the corresponding logarithmic term may be expanded in powers of $r/|\boldsymbol{x}_E|$. Substitution of these values into (6.3.10) gives the expression for the one-way BRS time transit interval of the signal:

$$c(t - t_0) = R + 2m_S \frac{r}{|\boldsymbol{x}_E|}\left(1 - \frac{1}{2|\boldsymbol{x}_E|^2}\boldsymbol{x}_E \boldsymbol{r} + \ldots\right) + 2m_E \ln \frac{r - \rho_L}{\rho_E}. \tag{6.4.14}$$

The quantity R occurring on the right-hand side is calculated as the magnitude of the BRS vector

$$\boldsymbol{R} = \boldsymbol{x}(t) - \boldsymbol{x}(t_0) = \boldsymbol{x}_E(t) + \boldsymbol{r}(t) + \boldsymbol{\rho}_L(t) - \boldsymbol{x}_E(t_0) - \boldsymbol{\rho}_E(t_0) \tag{6.4.15}$$

$\boldsymbol{r} = \boldsymbol{r}(t)$ being the geocentric BRS position vector of the Moon. In actual discussion the quantity R is determined by numerical iteration. The relativistic contribution to the measurable quantities may be easily evaluated analytically. Neglecting the sizes of the Earth and the Moon one has

$$\boldsymbol{R} = \boldsymbol{x}_E(t) + \boldsymbol{r}(t) - \boldsymbol{x}_E(t_0) = \boldsymbol{r}(t) + \dot{\boldsymbol{x}}_E(t_0)(t - t_0) + \tfrac{1}{2}\ddot{\boldsymbol{x}}_E(t_0)(t - t_0)^2 + \ldots. \tag{6.4.16}$$

From this it follows that

$$R = r\left[1 + \frac{1}{r^2}\boldsymbol{r}\dot{\boldsymbol{x}}_E(t - t_0) + \frac{1}{2r^2}\left(\dot{\boldsymbol{x}}_E^2 + \boldsymbol{r}\ddot{\boldsymbol{x}}_E \right.\right.$$
$$\left.\left. - \frac{1}{r^2}(\boldsymbol{r}\dot{\boldsymbol{x}}_E)^2\right)(t - t_0)^2 + \ldots\right]. \tag{6.4.17}$$

Substituting (6.4.17) into (6.4.14), solving with respect to $t - t_0$ and replacing the acceleration of the geocentre by the main Newtonian term one obtains

$$c(t - t_0) = r\left[1 + \frac{c^{-1}}{r}\boldsymbol{r}\dot{\boldsymbol{x}}_E + \tfrac{1}{2}c^{-2}\left(\dot{\boldsymbol{x}}_E^2 + \frac{1}{r^2}(\boldsymbol{r}\dot{\boldsymbol{x}}_E)^2\right)\right]$$
$$+ m_S \frac{r}{|\boldsymbol{x}_E|}\left(2 - \frac{3}{2|\boldsymbol{x}_E|^2}\boldsymbol{r}\boldsymbol{x}_E\right) + 2m_E \ln \frac{r - \rho_L}{\rho_E} + \ldots. \tag{6.4.18}$$

Multiplied by two this expression gives the BRS time interval of light propagation in LLR (two-way). In actual calculations the Newtonian part of this expression should obviously be extended by (6.4.15). In reality, one measures the quantity $2(\tau - \tau_0)$, τ_0 and τ being the moments of the TRS time corresponding to the moments t_0 and t of the BRS time. Restricting ourselves only to the main terms one has by (6.4.7)

$$\tau_0 = t_0 - c^{-2} \int_0^{t_0} (\tfrac{1}{2}\dot{\boldsymbol{x}}_E^2 + \bar{U}_E(\boldsymbol{x}_E))\, \mathrm{d}t$$

$$\tau = t - c^{-2} \left(\int_0^{t_0} (\tfrac{1}{2}\dot{\boldsymbol{x}}_E^2 + \bar{U}_E(\boldsymbol{x}_E))\, \mathrm{d}t + \dot{\boldsymbol{x}}_E \boldsymbol{r} \right)$$

or neglecting the variation of the integrand over $t - t_0$

$$c(\tau - \tau_0) = c(t - t_0) \left(1 - \tfrac{1}{2} c^{-2} \dot{\boldsymbol{x}}_E^2 - \frac{m_S}{|\boldsymbol{x}_E|} \right) - c^{-1} \dot{\boldsymbol{x}}_E \boldsymbol{r}. \qquad (6.4.19)$$

This expression follows directly from the kinematic relation (2.3.42) as well by identifying variations $\mathrm{d}\xi^i$ with the components of $\boldsymbol{r}$. From this it follows by (6.4.18) that

$$\begin{aligned} 2c(\tau - \tau_0) = {} & 2r \left(1 + \frac{c^{-2}}{2r^2} (\boldsymbol{r} \dot{\boldsymbol{x}}_E)^2 + \frac{m_S}{|\boldsymbol{x}_E|} - \frac{3m_S}{2|\boldsymbol{x}_E|^3} (\boldsymbol{r} \boldsymbol{x}_E) \right) \\ & + 4m_E \ln \frac{r - \rho_L}{\rho_E} + \ldots . \end{aligned} \qquad (6.4.20)$$

Section 5.3 contains the main relativistic perturbations in the BRS coordinates of the Moon. Substitution of $\boldsymbol{r}$ into (6.4.20) gives the real relativistic effect in LLR. It is of importance that this effect turns out to be two orders of magnitude smaller than the BRS relativistic perturbations in the motion of the Moon. In fact, the main relativistic terms in r have the form (5.3.27)

$$r/a_0 = 1 - \tfrac{9}{4}\sigma + \sigma m + \tfrac{1}{4}\sigma \cos 2D + \ldots . \qquad (6.4.21)$$

However, the term in (6.4.20) due to the Lorentz transformation is

$$\frac{c^{-2}}{2r^2} (\boldsymbol{r} \dot{\boldsymbol{x}}_E)^2 = \tfrac{1}{2}\sigma \sin^2 D = \tfrac{1}{4}\sigma(1 - \cos 2D). \qquad (6.4.22)$$

Thus, the trigonometric term in (6.4.21) with amplitude $\sigma/4$ disappears in (6.4.20) and the amplitude of the relativistic terms in the measurable quantity (6.4.20) does not exceed several centimetres. This was first noted by Nordtvedt (1973) and then thoroughly investigated by Soffel *et al* (1986, Soffel 1989). In constructing the GRS theory of the motion of the Moon one would come directly to this conclusion. Indeed, restricting in (4.2.8) by the purely solar terms one has for the GRS coordinates of the Moon w^i

$$w^i = r^i + \tfrac{1}{2}c^{-2}v_E^i v_E^k r^k + c^{-2}F^{ik}r^k + \frac{m_S}{|\boldsymbol{x}_E|}r^i + \frac{m_S}{|\boldsymbol{x}_E|^3}(\tfrac{1}{2}r^2 x_E^i - r^i r^k x_E^k). \tag{6.4.23}$$

From this for the absolute value of $\rho = |\boldsymbol{w}|$ we obtain

$$\rho = r\left(1 + \frac{c^{-2}}{2r^2}(v_E^k r^k)^2 + \frac{m_S}{|\boldsymbol{x}_E|} - \frac{1}{2}\frac{m_S}{|\boldsymbol{x}_E|}(r^k x_E^k)\right). \tag{6.4.24}$$

Thus, in terms of the GRS quantities the right-hand side of (6.4.20) takes the form

$$2c(\tau - \tau_0) = 2\rho\left(1 - \frac{m_S}{|\boldsymbol{x}_E|^3}(\boldsymbol{x}_E\boldsymbol{w})\right) + 4m_E \ln\frac{\rho - \rho_L}{\rho_E}. \tag{6.4.25}$$

In neglecting in the relativistic part by the parallactic ratio $\rho/|\boldsymbol{x}_E|$ it may be seen that the relativistic time delay in LLR reduces to the Schwarzschild delay due to the Earth. Needless to say, this conclusion holds true for artificial Earth satellites as well. Along with this it is evident that the BRS theory of the motion of the Moon has not lost its importance because converting $2r$ to the actually measurable quantity $2c(\tau - \tau_0)$ using (6.4.20) presents no difficulties.

6.4.4 Pulsar timing

It is well known that the timing of millisecond pulsars is at present one of the most accurate methods of obtaining astronomical information (Hellings 1986, Backer and Hellings 1986). A relativistic theory of pulsar timing can be developed on the basis of (6.3.10). If the BRS position vector of a pulsar at moment T_0 is $\boldsymbol{X}_0$ and the pulsar moves with constant velocity $\boldsymbol{V}$ then for the moment T_n of the emission of impulse n its position vector will be

$$\boldsymbol{X}_n = \boldsymbol{X}_0 + \boldsymbol{V}(T_n - T_0). \tag{6.4.26}$$

This impulse n is recorded by the ground station at moment t_n in the BRS position $\boldsymbol{x}_n$. In accordance with (6.3.10)

$$c(t_n - T_n) = |\boldsymbol{x}_n - \boldsymbol{X}_n| + 2\sum_A m_A \ln\frac{|\boldsymbol{x}_n - \boldsymbol{x}_A(t_n)| + |\boldsymbol{X}_n - \boldsymbol{x}_A(T_n)| + |\boldsymbol{x}_n - \boldsymbol{X}_n|}{|\boldsymbol{x}_n - \boldsymbol{x}_A(t_n)| + |\boldsymbol{X}_n - \boldsymbol{x}_A(T_n)| - |\boldsymbol{x}_n - \boldsymbol{X}_n|}. \tag{6.4.27}$$

Denoting now

$$\boldsymbol{k} = \frac{1}{\rho_0}\boldsymbol{X}_0 \qquad \rho_0 = |\boldsymbol{X}_0| \qquad \Delta t_n = T_n - T_0 \tag{6.4.28}$$

and expanding $|\boldsymbol{x}_n - \boldsymbol{X}_n|$ in powers of Δt_n and the ratio $|\boldsymbol{x}_n|/\rho_0$ one obtains

$$\begin{aligned} c(t_n - T_n) = {} & \rho_0 + [(\boldsymbol{k}\boldsymbol{V})\Delta t_n - \boldsymbol{k}\boldsymbol{x}_n] + \frac{1}{2\rho_0}[\boldsymbol{x}_n^2 - (\boldsymbol{k}\boldsymbol{x}_n)^2] \\ & - \frac{1}{\rho_0}[\boldsymbol{x}_n\boldsymbol{V} - (\boldsymbol{k}\boldsymbol{V})(\boldsymbol{k}\boldsymbol{x}_n)]\Delta t_n + \frac{1}{2\rho_0}[\boldsymbol{V}^2 - (\boldsymbol{k}\boldsymbol{V})^2](\Delta t_n)^2 \\ & - 2\sum_A m_A \ln\frac{|\boldsymbol{r}_A(t_n) + \boldsymbol{k}\boldsymbol{r}_A(t_n)|}{2\rho_0}. \end{aligned} \tag{6.4.29}$$

Let ct_0 be the quantity

$$ct_0 = cT_0 + \rho_0 + 2\sum_A m_A \ln(2\rho_0). \tag{6.4.30}$$

Then, one has finally

$$\begin{aligned} c(t - t_0) = {} & c(T_n - T_0) + [(\boldsymbol{k}\boldsymbol{V})\Delta t_n - \boldsymbol{k}\boldsymbol{x}_n] + \frac{1}{2\rho_0}[\boldsymbol{x}_n^2 - (\boldsymbol{k}\boldsymbol{x}_n)^2] \\ & - \frac{1}{\rho_0}[\boldsymbol{x}_n\boldsymbol{V} - (\boldsymbol{k}\boldsymbol{V})(\boldsymbol{k}\boldsymbol{x}_n)]\Delta t_n + \frac{1}{2\rho_0}[\boldsymbol{V}^2 - (\boldsymbol{k}\boldsymbol{V})^2](\Delta t_n)^2 \\ & - 2\sum_A m_A \ln|\boldsymbol{r}_A(t_n) + \boldsymbol{k}\boldsymbol{r}_A(t_n)|. \end{aligned} \tag{6.4.31}$$

Each term in this equation admits a simple physical interpretation. The second term represents the first-order Doppler effect. The third term is caused by parallax. The fourth term describes the proper motion of the pulsar. The fifth term is the second-order Doppler effect. Finally, the sixth term represents the Shapiro effect, i.e. the relativistic time delay in light propagation. It remains for us to convert in (6.4.31) from t_n to the corresponding moment of proper time τ_n by means of (6.4.8) and to replace $\boldsymbol{x}_n$ by the GRS+ coordinates $\boldsymbol{y}$ of the point of observation, assuming $\boldsymbol{z} = 0$.

In the case of a binary pulsar, i.e. a pulsar in a binary system, the timing formula becomes more complicated, due, first of all, to replacing (6.4.26) by the more cumbersome formula

$$\boldsymbol{X}_n = \boldsymbol{X}_0 + \boldsymbol{V}\Delta t_n + \boldsymbol{X}_{1n}$$

where $\boldsymbol{X}_0$ and $\boldsymbol{V}$ now represent the BRS position and velocity of the centre of mass of the binary and $\boldsymbol{X}_{1n}$ is the pulsar's position relative to the barycentre of the binary. This latter quantity can be calculated using the formulae of the relativistic two-body problem. A detailed derivation of the timing formula for the binary pulsar is given by Backer and Hellings (1986) and Damour and Deruelle (1986).

6.4.5 VLBI observations

Let us now apply equations (6.4.31) to the reduction of the VLBI observations. Let impulse n from a radio source be recorded by two ground stations with BRS position vectors $\boldsymbol{x}_1$ and $\boldsymbol{x}_2$ at BRS moments t_1 and t_2, respectively. Taking the difference of equations (6.4.31) for both stations and ignoring the parallactic term and the terms due to the proper motion of the source during the interval $t_2 - t_1$ one has

$$c(t_2 - t_1) = -\boldsymbol{k}(\boldsymbol{x}_2 - \boldsymbol{x}_1) - 2\sum_A m_A \ln\left(\frac{r_A^{(2)} + \boldsymbol{k}\boldsymbol{r}_A^{(2)}}{r_A^{(1)} + \boldsymbol{k}\boldsymbol{r}_A^{(1)}}\right). \qquad (6.4.32)$$

In the relativistic part it suffices to take into account only the solar term, which in GRS coordinates takes the form

$$-2m_S \ln\left(\frac{\boldsymbol{k}(\boldsymbol{x}_E + \boldsymbol{w}_2) + |\boldsymbol{x}_E| + \boldsymbol{n}\boldsymbol{w}_2}{\boldsymbol{k}(\boldsymbol{x}_E + \boldsymbol{w}_1) + |\boldsymbol{x}_E| + \boldsymbol{n}\boldsymbol{w}_1}\right) = -\frac{2m_S}{|\boldsymbol{x}_E|}\frac{(\boldsymbol{k} + \boldsymbol{n})(\boldsymbol{w}_2 - \boldsymbol{w}_1)}{1 + \boldsymbol{n}\boldsymbol{k}} \qquad (6.4.33)$$

with $\boldsymbol{n} = \boldsymbol{x}_E/|\boldsymbol{x}_E|$. The value of $\boldsymbol{x}_E$ may be taken either at moment t_1 or at t_2. The Newtonian term $\boldsymbol{x}_2 - \boldsymbol{x}_1$ is also transformed to GRS coordinates by means of (6.4.1). Using the approximate relations

$$\boldsymbol{x}_E(t_2) - \boldsymbol{x}_E(t_1) \approx \boldsymbol{v}_E(t_2 - t_1)$$

$$\boldsymbol{w}_2(t_2) - \boldsymbol{w}_1(t_1) \approx \boldsymbol{w}_2(t_1) - \boldsymbol{w}_1(t_1) + \boldsymbol{v}_T^{(2)}(t_2 - t_1)$$

with $\boldsymbol{v}_T^{(2)}$ being the GRS velocity of the second station at moment t_1 one obtains

$$\begin{aligned} c(t_2 - t_1) = & -\boldsymbol{k}(\boldsymbol{v}_E + \boldsymbol{v}_T^{(2)})(t_2 - t_1) - \boldsymbol{k}(\boldsymbol{w}_2(t_1) - \boldsymbol{w}_1(t_1)) \\ & + c^{-2}\{\tfrac{1}{2}(\boldsymbol{k}\boldsymbol{v}_E)[\boldsymbol{v}_E(\boldsymbol{w}_2 - \boldsymbol{w}_1)] + \boldsymbol{k}(\boldsymbol{w}_2 - \boldsymbol{w}_1)\bar{U}_E(\boldsymbol{x}_E) \\ & + \boldsymbol{k}[(\boldsymbol{w}_2 - \boldsymbol{w}_1) \times \boldsymbol{F}]\} - \frac{2m_S}{|\boldsymbol{x}_E|}\frac{(\boldsymbol{k} + \boldsymbol{n})(\boldsymbol{w}_2 - \boldsymbol{w}_1)}{1 + \boldsymbol{n}\boldsymbol{k}}. \end{aligned} \qquad (6.4.34)$$

As the actually measurable quantity one may consider the interval $\tau_2 - \tau_1$ expressed in the proper time of the first station. Because of the smallness of the interval $t_2 - t_1$ one finds by (6.4.8)

$$\tau_2 - \tau_1 = (t_2 - t_1)[1 - c^{-2}(\tfrac{1}{2}v_E^2 + \tfrac{1}{2}v_T^2 + U)] - c^{-2}(\boldsymbol{v}_E + \boldsymbol{v}_T)(\boldsymbol{w}_2 - \boldsymbol{w}_1). \quad (6.4.35)$$

If VLBI observations are used for improving the GRS network of ground stations then it remains only to substitute (6.4.35) into (6.4.34)

$$\begin{aligned} c(\tau_2 - \tau_1) = & - \boldsymbol{k}(\boldsymbol{w}_2 - \boldsymbol{w}_1)[1 + c^{-1}\boldsymbol{k}(\boldsymbol{v}_E + \boldsymbol{v}_T^{(2)})]^{-1} \\ & \times [1 - c^{-2}(\tfrac{1}{2}v_E^2 + \tfrac{1}{2}v_T^2 + 2U - U_E)] \\ & - c^{-1}(\boldsymbol{v}_E + \boldsymbol{v}_T)(\boldsymbol{w}_2 - \boldsymbol{w}_1) + \tfrac{1}{2}c^{-2}(\boldsymbol{k}\boldsymbol{v}_E)[\boldsymbol{v}_E(\boldsymbol{w}_2 - \boldsymbol{w}_1)] \\ & + c^{-2}\boldsymbol{k}[(\boldsymbol{w}_2 - \boldsymbol{w}_1) \times \boldsymbol{F}] - \frac{2m_S}{|\boldsymbol{x}_E|}\frac{(\boldsymbol{k} + \boldsymbol{n})(\boldsymbol{w}_2 - \boldsymbol{w}_1}{1 + \boldsymbol{n}\boldsymbol{k}} \end{aligned} \tag{6.4.36}$$

and to introduce the GRS$^+$ space coordinates by means of (6.2.31). On the other hand, if VLBI observations are applied to construct the local TRS network then in (6.4.36) one should introduce the TRS coordinates $\boldsymbol{z}$ of the second station relative the first station. Then it follows that

$$\begin{aligned} c(\tau_2 - \tau_1) = & - (\boldsymbol{k}\boldsymbol{z})[1 + c^{-1}\boldsymbol{k}(\boldsymbol{v}_E + \boldsymbol{v}_T^{(2)})]^{-1}[1 - c^{-2}(\tfrac{1}{2}v_E^2 + \tfrac{1}{2}v_T^2 + 2U)] \\ & - c^{-1}(\boldsymbol{v}_E + \boldsymbol{v}_T)\boldsymbol{z} + \tfrac{1}{2}c^{-2}(\boldsymbol{k}\boldsymbol{v}_E)(\boldsymbol{v}_E\boldsymbol{z}) + \tfrac{1}{2}c^{-2}(\boldsymbol{k}\boldsymbol{v}_T)(\boldsymbol{v}_T\boldsymbol{z}) \\ & + c^{-2}\boldsymbol{k}[\boldsymbol{z} \times (\boldsymbol{F} + \boldsymbol{\Phi})] - \frac{2m_S}{|\boldsymbol{x}_E|}\frac{(\boldsymbol{k} + \boldsymbol{n})\boldsymbol{z}}{1 + \boldsymbol{n}\boldsymbol{k}}. \end{aligned} \tag{6.4.37}$$

It remains to transform to the TRS$^+$ coordinates by means of (6.2.37). In a result, the relativistic precession Φ^i will enter into the matrix of the diurnal rotation of the Earth. Equation (6.4.37) is identical to the reduction formula of VLBI observations obtained by Hellings (1986) except for the presence of precession terms. When using the technique of chronometric invariant time intervals and distances (section 2.3) one obtains the same formulae with the exception of the precession terms.

Among other algorithms of the relativistic reduction of VLBI observations one may note those of Cannon *et al* (1986), Zeller *et al* (1986) and Zhu and Groten (1988).

7

Relativistic Effects in Geodynamics

7.1 TIMESCALES ON THE EARTH

7.1.1 Timescale requirements

In contemporary celestial mechanics and ephemeris astronomy the most important timescales are TDB, TDT and TAI (Japanese Ephemeris 1985, Guinot 1986, Guinot and Seidelmann 1988). Without going into detail, TDB may be characterized as the time associated with the Solar System barycentre and serves as the argument of the theory of motion of the planets; TDT may be regarded as the time associated with the geocentre and is able to serve as the best argument in constructing the geocentric theories of motion of the Moon and Earth satellites; TAI manifests itself as the time most closely related to the time directly recorded on an observer's clock. The papers indicated above present definitions of these timescales which are adequate for real use. However, precision of time measurement increases from year to year, and this calls for more rigorous definitions of the timescales to be consistent with the statements of GRT. The self-consistent theory of timescales may be developed on the basis of the relativistic theory of astronomical reference systems. The hierarchy of BRS, GRS and TRS systems described in chapter 6 has a quite definite meaning. It is to be noted that the astronomical directions of the space axes are not significant in the problem of timescales. If BRS, GRS and TRS are uniquely defined, their scales of coordinate time are also uniquely defined. For rigorous relativistic definitions of TB, TT and TAI (TB and TT are used here instead of TDB and TDT, respectively, as suggested by Guinot and Seidelmann) the following are necessary:

(1) to link TB and TT with the coordinate timescales of BRS and GRS, respectively;
(2) to describe the operational procedure relating TAI with the coordinate timescale of arbitrary TRS;

(3) to choose units of measurement of TB, TT and TAI so as to reduce as much as possible the secular (non-periodic) difference between these scales (the IAU recommendation of 1976 implies the absence of such a secular trend but, as seen below, it may be realized only within a limited accuracy).

Since the BRS coordinate time and TB on the one hand and the GRS coordinate time and TT on the other hand differ from one another by constant factors then to compensate these differences all linear sizes in BRS and GRS should be multiplied by corresponding factors.

7.1.2 Relation between TB and TT

TB is a timescale differing only by a constant factor from the BRS coordinate time t,

$$TB = k_B t \tag{7.1.1}$$

while TT is a timescale differing only by a constant factor from the GRS coordinate time u,

$$TT = k_G u. \tag{7.1.2}$$

The interrelation between u and t is defined by (4.2.7):

$$u = t - c^{-2}[S(t) + v_E^k(x^k - x_E^k)] + \ldots \tag{7.1.3}$$

where x_E^k and v_E^k are the BRS coordinates and velocity components of the geocentre and where function $S = S(t)$ satisfies the differential equation

$$\frac{\mathrm{d}S}{\mathrm{d}t} = \tfrac{1}{2}v_E^2 + \bar{U}_E(\boldsymbol{x}_E). \tag{7.1.4}$$

$\bar{U}_E(\boldsymbol{x}_E)$ is the external mass potential evaluated in the geocentre

$$\bar{U}_E(\boldsymbol{x}_E) = \sum_{A \neq E} \frac{GM_A}{r_{EA}} + \ldots. \tag{7.1.5}$$

For the geocentre the rate of the time u is determined only by equation (7.1.4).

This equation has been investigated by many authors. The first detailed solution was given by Moyer (1981). This solution was based on the substitution into (7.1.4) of Keplerian motion for Solar System bodies and provided an accuracy of about 20 microseconds. As mentioned above, the most accurate analytical theories of motion of the major planets and the Moon are the VSOP82/ELP2000 theories. Fairhead *et al* (1988) and Hirayama *et al* (1988) substituted these theories into (7.1.4) and then integrated. The BRS velocity of the Earth v_E^k corresponding to these theories may be found in Sôma *et al* (1987). Therefore, it is not necessary anymore to transform the barycentric motion of the Earth to heliocentric motion of the Earth, geocentric motion of the Moon and heliocentric motion of the Earth–Moon centre of mass, as was done by Moyer. In the VSOP82 theory the rectangular coordinates and the velocity components of the planets are represented by trigonometric series in the mean longitudes of the planets, the coefficients of the series being slowly changing polynomial functions of time. Hence, the solution of (7.1.4) is represented in the same form, i.e.

$$S = A_0+B_0t+C_0t^2+D_0t^3+\ldots+\sum_i(A_i+B_it+C_it^2+D_it^3+\ldots)\sin(\omega_it+\varphi_i). \tag{7.1.6}$$

The numerical values of the coefficients are reproduced in the papers cited above. The coefficients C_0, D_0, ..., B_i, C_i, D_i, ... are caused by the secular terms of the planetary theories and vanish for pure Keplerian motion of the planets. These coefficients would also vanish in using a purely trigonometric planetary theory, but such a theory leads to a drastic extension of the number of periodic terms so that its actual use inevitably results in loss of accuracy. For astronomical purposes it is customary to have the timescales differing from one another only by periodic terms. The linear function $A_0 + B_0t$ in (7.1.6) does not interfere with this since the constant A_0 may be removed with a suitable choice of initial moment (which will be assumed in the following) and the constant B_0 is removed by adopting different units of measurement for different timescales. This is allowed for by the IAU 1976 recommendation.

Equation (7.1.4) may be also integrated numerically by using in the right-hand side the numerical theories DE200/LE200. Such integration is described, for instance, in Hellings (1986), and Backer and Hellings (1986). In numerical integration, the analytical structure (7.1.6) is latent but the numerically determined secular trend of function $S(t)$ depends on the interval of integration. A comparison of analytical and numerical approaches to solving equation (7.1.4) is performed in detail in Chotimskaya (1989).

In all cases the function S is represented in the form

$$S(t) = S^*t + S_p(t) \tag{7.1.7}$$

where S^* is constant and $S_p(t)$ includes both periodic and non-periodic terms caused by the secular evolution of the planetary orbits. Substitution of (7.1.7) into (7.1.3) results in

$$u = (1 - c^{-2}S^*)t - c^{-2}[S_p(t) + v_E^k(x^k - x_E^k)]. \tag{7.1.8}$$

Using (7.1.1) and (7.1.2) one finds the interrelation between TB and TT in the form

$$k_G^{-1}TT = k_B^{-1}(1 - c^{-2}S^*)TB - c^{-2}[S_p(t) + v_E^k(x^k - x_E^k)]. \tag{7.1.9}$$

The choice of constants k_B and k_G is arbitrary. Without determining them completely, it is suitable to set their ratio as

$$k_B/k_G = 1 - c^{-2}S^*. \tag{7.1.10}$$

In this case one obtains

$$TB = TT + c^{-2}[S_p(t) + v_E^k(x^k - x_E^k)]. \tag{7.1.11}$$

The constant S^* is not have to be equal to the coefficient B_0 in (7.1.6). The representation (7.1.6) itself is rather conventional. In using (7.1.7) over a restricted interval of time the very-long-period terms in $S_p(t)$ manifest themselves as secular terms and are included in the secular part when averaging numerically. Therefore, S^* should be chosen so as to minimize the influence of $S_p(t)$ on the secular rate of the difference $TB - TT$ over some definite time interval. After completing this interval the value of S^* may be changed so as to compensate for the accumulated secular rate in $TB - TT$. To an order of magnitude one has

$$c^{-2}S^* = 1.48 \times 10^{-8}.$$

7.1.3 TRS timescale

For an arbitrary point with GRS coordinates u, w^k the TRS coordinate time τ is given by

$$\tau = u - c^{-2}[V(u) + v_T^k(w^k - w_T^k)] \tag{7.1.12}$$

where w_T^k and v_T^k are the GRS coordinates and velocity components of the TRS origin with

$$\frac{\mathrm{d}\boldsymbol{w}_T}{\mathrm{d}u} = \boldsymbol{v}_T = (\hat{\boldsymbol{\omega}}_E \times \boldsymbol{w}_T) + \boldsymbol{v}_{TT}. \tag{7.1.13}$$

$\hat{\boldsymbol{\omega}}_E$ is the GRS vector of the angular velocity of rotation of the Earth, $\boldsymbol{v}_{TT}$ is the relative velocity of the TRS origin due to geophysical factors (deviation of the rotation of the ground station around the centre of mass of the Earth from the law of rigid-body rotation). The function $V(u)$ satisfies the differential equation (6.2.10) or

$$\frac{\mathrm{d}V}{\mathrm{d}u} = \tfrac{1}{2}v_T^2 + \hat{U}_E(\boldsymbol{w}_T) + Q_k w_T^k + \tfrac{1}{2}\bar{U}_{E,km}(\boldsymbol{x}_E)w_T^k w_T^m + \tfrac{1}{6}\bar{U}_{E,kmn}(\boldsymbol{x}_E)w_T^k w_T^m w_T^n + \dots. \tag{7.1.14}$$

$\hat{U}_E(\boldsymbol{w}_T)$ is the value of the Newtonian geopotential in the GRS origin

$$\hat{U}_E(\boldsymbol{w}_T) = \frac{G\hat{M}_E}{d_T} + \frac{1}{2d_T^3}\left(-\delta_{km} + \frac{3}{d_T^2}w_T^k w_T^m\right)G\hat{I}_E^{km} + \dots \tag{7.1.15}$$

where $d_T = (w^k_T w^k_T)^{1/2}$ and $\hat{I}^{km}_E$ are the quadrupole moments

$$\hat{I}^{km}_E = \int_{(E)} \hat{\rho} w^k w^m \, \mathrm{d}^3 w \qquad (7.1.16)$$

and $\hat{\rho}$ and $\hat{M}_E$ are the GRS mass density and mass of the Earth, respectively. For an observer at rest in the TRS origin, with $w^k = w^k_T, \tau$ is its proper time.

Introducing now the GRS$^+$ space coordinates y^k, using (6.2.19) for the velocity of the ground station and neglecting the acceleration Q_k, the right-hand side of (7.1.14) takes the form

$$\frac{\mathrm{d}V}{\mathrm{d}u} = \epsilon_{ijk}\hat{\omega}^j_E y^k_T \dot{y}^i_T + \tfrac{1}{2}\dot{y}^k_T \dot{y}^k_T + W_E(\boldsymbol{y}_T) + \tfrac{1}{2}\bar{U}_{E,km}(\boldsymbol{x}_E) P_{ik} P_{jm} y^i_T y^j_T + \ldots. \qquad (7.1.17)$$

$\hat{\omega}^j_E$ are the GRS$^+$ components of the angular velocity of rotation of the Earth. $W_E(\boldsymbol{y}_T)$ is the potential (of opposite sign) of the force of gravity due to the Earth at the point $\boldsymbol{y}_T$. The last term in (7.1.17) describes the tidal external perturbation of the potential of the force of gravity in the quadrupole approximation. This tidal perturbation from the Sun and the Moon gives a contribution to $\mathrm{d}\tau/\mathrm{d}u$ of the order of 10^{-17}. It is necessary to reveal in the right-hand side of (7.1.17) the constant part independent of time and coordinates $\boldsymbol{y}_T$ of the ground station. This calls for more detailed examination of the force gravity potential using the technique of Zhongolovich (1957).

7.1.4 Geoid

The potential of the force of gravity generated by the Earth alone at any point $\boldsymbol{y}$ on the surface of the Earth is made up of the centrifugal potential and the Earth's attractive potential

$$W_E(\boldsymbol{y}) = \tfrac{1}{2}(\epsilon_{ijk}\omega^j y^k)^2 + \hat{U}_E(\boldsymbol{y}). \qquad (7.1.18)$$

The GRS$^+$ components of the angular velocity of rotation of the Earth are denoted here, for simplicity, as ω^j. Let us introduce the geocentric spherical coordinates r, ψ and l, choosing the direction of the y^3 axis to be along the axis of rotation of the Earth. Then,

$$y^1 = r\cos\psi\cos l \qquad y^2 = r\cos\psi\sin l \qquad y^3 = r\sin\psi \qquad (7.1.19)$$

and $\omega^1 = \omega^2 = 0$, $\omega^3 = \omega$. One obtains

$$W_E(r, l, \psi) = \tfrac{1}{2}\omega^2 r^2 \cos^2\psi + \hat{U}_E(\boldsymbol{y}). \qquad (7.1.20)$$

In spherical coordinates (7.1.19) the quadrupole external tidal perturbation occurring in (7.1.17) may be presented in the form (Moritz and Mueller 1987)

$$Q_2 = Q_{20} + Q_{21} + Q_{22} \tag{7.1.21}$$

where the zonal, tesseral and sectorial parts are, respectively,

$$Q_{20} = r^2 P_{20}(\sin\psi) \sum_{A\neq E} \frac{GM_A}{r_{EA}^3} P_{20}(\sin\psi_A) \tag{7.1.22}$$

$$Q_{21} = \tfrac{1}{3} r^2 P_{21}(\sin\psi) \sum_{A\neq E} \frac{GM_A}{r_{EA}^3} P_{21}(\sin\psi_A)\cos(l - l_A) \tag{7.1.23}$$

$$Q_{22} = \tfrac{1}{12} r^2 P_{22}(\sin\psi) \sum_{A\neq E} \frac{GM_A}{r_{EA}^3} P_{22}(\sin\psi_A)\cos 2(l - l_A). \tag{7.1.24}$$

Here $P_{nm}(x)$ are associated Legendre functions, l_A and ψ_A are the geocentric longitude and latitude of body A, respectively. With the use of the theories of motion for Solar System bodies the right-hand sides of these expressions may be expanded in trigonometric series of type

$$Q_{2m} = r^2 P_{2m}(\sin\psi) \sum_j C_{mj} \cos(\omega_{mj} t + ml + \beta_{mj}). \tag{7.1.25}$$

The zonal terms ($m = 0$) contain long-period harmonics (half a month for the Moon and half a year for the Sun), the tesseral terms ($m = 1$) include harmonics with a diurnal period and the sectorial terms ($m = 2$) involve harmonics with a semidiurnal period. Moreover, the zonal part contains the time-independent term

$$\bar{Q}_{20} = -r^2 P_{20}(\sin\psi)\tfrac{1}{2} \sum_{A\neq E} \frac{GM_A}{\bar{r}_{EA}^3} \tag{7.1.26}$$

with $\bar{r}_{EA}$ denoting some mean value of r_{EA}. It is suitable to add this value to the potential of the force of gravity due to the Earth and to consider the potential

$$W(r,l,\psi) = W_E(r,l,\psi) + \bar{Q}_{20}(r,\psi) \tag{7.1.27}$$

although the term $\bar{Q}_{20}$ is usually treated as a perturbation and is not included in the potential for gravitational force.

The surface $r = r(l,\psi)$ closely approximating the mean sea level and providing a constant value for the gravity potential

$$W = W_0 = \text{constant} \tag{7.1.28}$$

is called the geoid. Nowadays, there are attempts to define the geoid within the framework of GRT (Soffel *et al* 1987b) but for our purposes the Newtonian definition (7.1.28) is quite satisfactory. Introduce now the quantities

$$r_0 = \frac{GM_E}{W_0} \qquad g_0 = \frac{GM_E}{r_0^2} \qquad \sigma = \frac{\omega^2}{2}\frac{r_0}{g_0} \qquad \epsilon = \frac{1}{2}\frac{r_0}{g_0}\sum_{A\neq E}\frac{GM_A}{\bar{r}_{EA}^3} \tag{7.1.29}$$

with σ and ϵ being small dimensionless parameters. The dimensionless ratio $\rho = r_0/r$ is closed to 1 for the points of the geoid. Expansion of W in spherical harmonics yields

$$W(r, l, \psi) = W_0\left(\rho + \sum_{k=2}^{\infty} A_k\rho^{k+1} + B_2\rho^{-2}\right) \tag{7.1.30}$$

with

$$B_2 = (\sigma + \tfrac{3}{2}\epsilon)\cos^2\psi - \epsilon. \tag{7.1.31}$$

The coefficients A_k admit the expansion

$$A_k = \sum_{m=0}^{k}(c_{km}\cos ml + d_{km}\sin ml)P_{km}(\sin\psi). \tag{7.1.32}$$

The coefficients c_{km}, d_{km} are expressed in terms of the moments of inertia of the Earth. If one ignores the small angle (less than 1″) between the axis of rotation and axis of inertia then among the second-order harmonics only the coefficients

$$c_{20} = \frac{1}{2M_E r_0^2}(A + B - 2C) \qquad c_{22} = \frac{1}{4M_E r_0^2}(B - A) \qquad d_{22} = \frac{1}{2M_E r_0^2}D \tag{7.1.33}$$

are not equal to zero. Here $A = \hat{I}_E^{22} + \hat{I}_E^{33}$, $B = \hat{I}_E^{33} + \hat{I}_E^{11}$, $C = \hat{I}_E^{11} + \hat{I}_E^{22}$, $D = \hat{I}_E^{12}$ are the quadrupole moments of inertia of the Earth. In accordance with (7.1.28) and (7.1.30) the equation of the geoid has the form

$$\rho + \sum_{k=2}^{\infty} A_k\rho^{k+1} + B_2\rho^{-2} = 1. \tag{7.1.34}$$

Solving this equation with respect to ρ one finds the initial terms

$$\rho^{-1} = r/r_0 = 1 + A_2 + B_2 + A_3 + A_4 - 2A_2^2 + 3B_2^2 + A_2B_2 + \ldots \tag{7.1.35}$$

or in a more general form

$$r = r_0\left(p_{00} + \sum_{k=2}^{\infty}\sum_{m=0}^{k}(p_{km}\cos ml + q_{km}\sin ml)P_{km}(\sin\psi)\right) \tag{7.1.36}$$

starting with

$$p_{00} = 1 + \tfrac{2}{3}\sigma + \ldots \qquad p_{20} = c_{20} - \tfrac{2}{3}\sigma - \epsilon$$
$$p_{22} = c_{22} + \ldots \qquad q_{22} = d_{22} + \ldots. \tag{7.1.37}$$

At the point on the surface of the Earth considered one may introduce the triad of unit vectors

$$e_1 = \begin{pmatrix} \cos l \cos\psi \\ \sin l \cos\psi \\ \sin\psi \end{pmatrix} \qquad e_2 = \begin{pmatrix} -\cos l \sin\psi \\ -\sin l \sin\psi \\ \cos\psi \end{pmatrix} \qquad e_3 = \begin{pmatrix} -\sin l \\ \cos l \\ 0 \end{pmatrix} \tag{7.1.38}$$

directed along the radius vector, the tangent to the parallel of latitude and the tangent to the meridian of the given point, respectively. The components of the force of gravity $g = \mathrm{grad}(W)$ onto these directions are

$$g_1 = g e_1 = \frac{\partial W}{\partial r} = -g_0\left(\rho^2 + \sum_{k=2}^{\infty}(k+1)A_k\rho^{k+2} - 2B_2\rho^{-1}\right) \tag{7.1.39}$$

$$g_2 = g e_1 = \frac{1}{r}\frac{\partial W}{\partial \psi} = g_0\left(\sum_{k=2}^{\infty}\frac{\partial A_k}{\partial\psi}\rho^{k+2} + \frac{\partial B_2}{\partial\psi}\rho^{-1}\right) \tag{7.1.40}$$

$$g_3 = g e_3 = \frac{1}{r\cos\psi}\frac{\partial W}{\partial l} = g_0 \sec\psi \sum_{k=2}^{\infty}\frac{\partial A_k}{\partial l}\rho^{k+2}. \tag{7.1.41}$$

Using (7.1.27) one finds for the initial terms of the force of gravity components on the surface of the geoid

$$g_1 = -g_0(1 + A_2 - 4B_2 + \ldots) \qquad g_2 = g_0\left(\frac{\partial A_2}{\partial\psi} + \frac{\partial B_2}{\partial\psi} + \ldots\right)$$
$$g_3 = g_0 \sec\psi\left(\frac{\partial A_2}{\partial l} + \ldots\right). \tag{7.1.42}$$

The magnitude of the force of gravity on the geoid is

$$g = |g| = g_0(1 + A_2 - 4B_2 + \ldots). \tag{7.1.43}$$

The coefficients A_k and B_2 are expressed in terms of geocentric spherical coordinates l and ψ. In astronomical measurements one uses the astronomical longitude λ and latitude φ which determine the direction of the external normal n to the geoid at the point of observation

$$n = \begin{pmatrix} \cos\lambda\cos\varphi \\ \sin\lambda\cos\varphi \\ \sin\varphi \end{pmatrix}. \tag{7.1.44}$$

Since the force of gravity is directed along the interior normal to the geoid implying $\boldsymbol{g} = -g\boldsymbol{n}$ one may derive from (7.1.39)–(7.1.43) the components of the normal

$$\boldsymbol{n}\boldsymbol{e}_1 = -g_1/g \qquad \boldsymbol{n}\boldsymbol{e}_2 = -g_2/g \qquad \boldsymbol{n}\boldsymbol{e}_3 = -g_3/g. \tag{7.1.45}$$

On the other hand, using (7.1.38) and (7.1.44) one finds

$$\sin\varphi - \sin\psi = (\boldsymbol{n}\boldsymbol{e}_1 - 1)\sin\psi + (\boldsymbol{n}\boldsymbol{e}_2)\cos\psi \tag{7.1.46}$$

$$\sin(\lambda - l) = (\boldsymbol{n}\boldsymbol{e}_3)\sec\varphi. \tag{7.1.47}$$

From this, it follows that

$$\varphi - \psi = -\frac{\partial A_2}{\partial\psi} - \frac{\partial B_2}{\partial\psi} + \ldots \qquad \lambda - l = -\frac{\partial A_2}{\partial l}\sec^2\psi + \ldots. \tag{7.1.48}$$

These relations enable us to express A_k and B_2 in terms of λ and φ and to obtain the final expansion of the force of gravity on the geoid in the form

$$g = g_0\left(a_{00} + \sum_{k=2}^{\infty}\sum_{m=0}^{k}(a_{km}\cos m\lambda + b_{km}\sin m\lambda)P_{km}(\sin\varphi)\right) \tag{7.1.49}$$

with initial values

$$a_{00} = 1 - \tfrac{8}{3}\sigma + \ldots \qquad a_{20} = c_{20} + \tfrac{8}{3}\sigma + 4\epsilon + \ldots$$

$$a_{22} = c_{22} + \ldots \qquad b_{22} = d_{22} + \ldots. \tag{7.1.50}$$

The potential of the force of gravity at any point on the surface of the Earth may be now presented in the form

$$W(h, \lambda, \varphi) = W_0 - g(\varphi, \lambda)h + \ldots \tag{7.1.51}$$

h being the height of the observer above the geoid. More detailed results (with $\epsilon = 0$) may be found in Zhongolovich (1957).

7.1.5 Relation between GRS and TRS timescales

Returning to (7.1.17) and using (7.1.51) one obtains

$$\frac{\mathrm{d}V}{\mathrm{d}u} = \epsilon_{ijk}\hat{\omega}^j_E y^k_T \dot{y}^i_T + \tfrac{1}{2}\dot{y}^k_T\dot{y}^k_T + W_0 + Q_2 - \bar{Q}_{20} - g(\varphi, \lambda)h + \ldots \tag{7.1.52}$$

where φ, λ, h and y^k_T are the coordinates of the ground station. The solution of this equation is of the form

$$V(u) = V^* u + V_p(u). \tag{7.1.53}$$

V^* is a constant, being the same for all possible TRS and, hence, independent of location of the ground station. Thus, one should put

$$V^* = W_0. \tag{7.1.54}$$

$V_p(u)$ contains all the remaining terms resulting from integration of equation (7.1.52). The term $-g(\varphi, \lambda)h$ leads to the linear function of u dependent on the ground station coordinates. The contribution from $Q_2 - \bar{Q}_{20}$ contains periodic tidal terms caused by the Moon and the Sun. Finally, the terms with $\dot{y}_T^k$ give the contribution due to geophysical factors. As a result, equation (7.1.12) gives the following relationship between u and τ:

$$\tau = (1 - c^{-2}V^*)u - c^{-2}[V_p(u) + v_T^k(w^k - w_T^k)] \tag{7.1.55}$$

or

$$(1 - c^{-2}V^*)u = \tau + c^{-2}[V_p(u) + v_T^k(w^k - w_T^k)]. \tag{7.1.56}$$

7.1.6 International atomic time TAI

Along the world line of the TRS origin ($w^k = w_T^k$) τ is proper time of the ground station and may be measured. The quantity V^* is the same for all systems TRS, and the functions $V_p(u)$ specific to each TRS are known from theoretical calculations. International atomic time TAI is formed by averaging the clock readings of many ground observatories (Guinot 1986). This may be interpreted as averaging over many TRS so that

$$TAI = \mathrm{mean}(\tau + c^{-2}V_p(u)). \tag{7.1.57}$$

From (7.1.56) TAI is related to TT by means of

$$TAI = k_G^{-1}(1 - c^{-2}V^*)TT. \tag{7.1.58}$$

Thus, the TRS coordinate time τ at an arbitrary point is expressed in terms of TAI, TT and TB, respectively, as follows:

$$\tau = TAI - c^{-2}[V_p(u) + v_T^k(w^k - w_T^k)] \tag{7.1.59}$$

$$\tau = k_G^{-1}(1 - c^{-2}V^*)TT - c^{-2}[V_p(u) + v_T^k(w^k - w_T^k)] \tag{7.1.60}$$

$$\begin{aligned}\tau = {} & k_G^{-1}(1 - c^{-2}V^*)TB - c^{-2}[S_p(t) + v_E^k(x^k - x_E^k)] \\ & - c^{-2}[V_p(u) + v_T^k(w^k - w_T^k)].\end{aligned} \tag{7.1.61}$$

Putting now

$$k_G = 1 - c^{-2}V^* \approx 1 - 0.7 \times 10^{-9} \tag{7.1.62}$$

one has finally

$$TB = TT + c^{-2}[S_p(TB) + v_E^k(x^k - x_E^k)] \tag{7.1.63}$$

$$TT = TAI + 32.184s \tag{7.1.64}$$

$$\tau = TAI - c^{-2}[V_p(TAI) + v_T^k(w^k - w_T^k)] \tag{7.1.65}$$

or

$$\tau = TT - 32.184s - c^{-2}[V_p(TT) + v_T^k(w^k - w_T^k)] \tag{7.1.66}$$

or else

$$\begin{aligned}\tau = &TB - 32.184s - c^{-2}[S_p(TB) + v_E^k(x^k - x_E^k)] \\ &- c^{-2}[V_p(TT) + v_T^k(w^k - w_T^k)].\end{aligned} \tag{7.1.67}$$

An additive constant in (7.1.63) may be specified by the condition of coincidence of TB and TT on the world line of the geocentre ($x^k = x_E^k$) at some fundamental epoch. The constant shift 32.184s used in (7.1.64) is due, to historical reasons, to prevent the discontinuity between atomic and ephemeris timescales. The right-hand sides of (7.1.65)–(7.1.67) may in addition involve constants dependent on a specific clock.

These relations make it evident that TAI is the physical realization of the coordinate timescale TT. All specific clocks involved in forming TAI are assumed to be synchronized with respect to coordinate time TT (coordinate synchronization in contrast to Einstein synchronization in proper time). Due to (7.1.64) TAI is defined theoretically in the same domain as TT but the practical realization of this scale has so far been performed only at selected points on the surface of the Earth. It may be possible in future to include clocks on Earth satellites into the procedure for forming TAI. Then the domain of practical realization of TAI will be significantly extended. The unit of measurement of TAI is the SI second determined on the surface of the geoid in rotation.

7.1.7 Units of length

The problem of units of length is closely related to timescales. Indeed, linear distances in BRS and GRS are not directly measurable quantities in astronomy. Therefore, the units of length do not interfere directly in astronomical measurements and their definition is dictated mainly by the considerations of convenience in reduction calculations. Such considerations are, for instance, as follows:

(1) to compensate for the rate difference between TB and the BRS coordinate time all linear sizes and relevant quantities in BRS should be multiplied by a constant factor;

(2) to compensate for the rate difference between TT and the GRS coordinate time all linear sizes and relevant quantities in GRS should be multiplied by a constant factor.

Indeed, the BRS equations of motion are described in terms of t, x_A^k and M_A. In the reduction of observational data the scale TB is used. In keeping the BRS equations invariable one actually deals with the quantities

$$x_A^{*k} = k_B x_A^k \qquad (GM_A)^* = k_B GM_A. \tag{7.1.68}$$

Similarly, the GRS equations of motion are described in terms of u, w_A^k and $\hat{M}_A$. In the reduction of observational data the scale TT is used. In keeping the GRS equations invariable one actually deals with the quantities

$$w'^k_A = k_G w_A^k \qquad (GM_A)' = k_G G\hat{M}_A. \tag{7.1.69}$$

This approach is based on Hellings (1986) and partly on Fukushima *et al* (1986b). The latter also suggest an alternative approach without transforming the spatial BRS and GRS coordinates. It leads to the same results but implies different values of the gravitational constant and the velocity of light in different reference frames and for astronomical problems with the pure auxiliary role of units of length seems to be logically more complicated.

In any case relations (7.1.68) and (7.1.69) involve definite inconveniences. For instance, the BRS planetary equations involve the mass of the Moon and its geocentric coordinates. These quantities should be expressed in BRS units and will differ from the normal GRS quantities for the Moon. Similarly, the GRS equations of motion of the Moon or an Earth satellite involve the mass of the Sun and its geocentric coordinates (with masses and the coordinates of the planets for the lunar equations). These quantities should be expressed in GRS units and will differ from the normal BRS values for the Sun and the planets.

The question of timescales and units of measurement still remains under discussion, involving conflicting considerations of the practical convenience of possible definitions. The paper by Guinot and Seidelmann (1988) reflects the history of different approaches to this question and does not claim to solve the problem (see, for instance, Huang *et al* (1989)). The considerations discussed here and resulting in (7.1.63)–(7.1.65) are dictated by the requirement to retain as far as possible the astronomical tradition to use timescales having no mutual secular trend. But the disadvantage of this tradition is the inconvenience related to the necessity to re-determine the length-dependent quantities in different reference systems.

These questions are discussed in more detail in Brumberg and Kopejkin (1990).

7.2 CLOCKS AND SATELLITES IN CIRCUMTERRESTRIAL SPACE

7.2.1 Doppler effect and gravitational displacement of frequency

In solving problems related to the measurement of time in circumterrestrial space it is sufficient to use only the first terms of the coefficients of the GRS metric (4.2.2)–(4.2.4)

$$h_{00} = -2c^{-2}\Phi(u, \boldsymbol{w}) \qquad h_{0i} = 0 \qquad h_{ij} = 0 \tag{7.2.1}$$

with $\boldsymbol{w} = (w^1, w^2, w^3)$ being the geocentric position of a current point at the GRS moment u and

$$\Phi(u, \boldsymbol{w}) = \hat{U}_E(u, \boldsymbol{w}) + \tfrac{1}{2}\bar{U}_{E,km}(\boldsymbol{x}_E)w^k w^m + \dots . \tag{7.2.2}$$

To obtain results within an accuracy of 1 ns (or, equivalently, an accuracy of 10^{-14} in frequency) one may neglect in (7.2.2) the lunar and solar tidal terms and retain in the geopotential only the second zonal harmonic. In this case expression (7.2.2) in equatorial coordinates is replaced by

$$\Phi(\boldsymbol{w}) = \frac{GM_E}{d}\left[1 + \frac{1}{2}\left(\frac{A_E}{d}\right)^2 J_2(1 - 3\sin^2\varphi)\right] \tag{7.2.3}$$

where A_E is the equatorial radius of the Earth, J_2 is the coefficient in the second zonal harmonic, and d and φ are the radius vector and latitude of the given point, respectively.

Consider the Doppler effect in the gravitational field of the Earth. Let a light signal of duration δu be emitted from the moving point 1 at the GRS moment u. This signal reaches the moving point 2 at moment u' as the impulse of duration $\delta u'$. The corresponding intervals of the proper time at points 1 and 2 are

$$\delta\tau = (1 - 2\Phi_1 c^{-2} - \dot{\boldsymbol{w}}_1^2 c^{-2})^{1/2}\delta u \tag{7.2.4}$$

$$\delta\tau' = (1 - 2\Phi_2 c^{-2} - \dot{\boldsymbol{w}}_2^2 c^{-2})^{1/2}\delta u' \tag{7.2.5}$$

with

$$\Phi_1 = \Phi(\boldsymbol{w}_1) \qquad \Phi_2 = \Phi(\boldsymbol{w}_2). \tag{7.2.6}$$

The moment u' of reception of the light signal is a function of the moment u of its emission. Therefore,

$$\delta u' = \frac{\mathrm{d}u'}{\mathrm{d}u}\delta u \tag{7.2.7}$$

and

$$\frac{\delta\tau'}{\delta\tau} = \frac{(1 - 2\Phi_2 c^{-2} - \dot{w}_2^2 c^{-2})^{1/2}}{(1 - 2\Phi_1 c^{-2} - \dot{w}_1^2 c^{-2})^{1/2}} \frac{du'}{du}. \tag{7.2.8}$$

In the Newtonian approximation

$$u' - u = c^{-1}|w_1(u) - w_2(u')|. \tag{7.2.9}$$

Introducing the unit vector

$$k = \frac{w_1(u) - w_2(u')}{|w_1(u) - w_2(u')|} \tag{7.2.10}$$

and differentiating (7.2.9) one finds

$$\frac{du'}{du} = 1 + c^{-1}k\dot{w}_1(u) - c^{-1}k\dot{w}_2(u')\frac{du'}{du}$$

or

$$\frac{du'}{du} = \frac{1 + c^{-1}k\dot{w}_1(u)}{1 + c^{-1}k\dot{w}_2(u')}. \tag{7.2.11}$$

Finally,

$$\frac{\delta\tau'}{\delta\tau} = \frac{(1 - 2\Phi_2 c^{-2} - \dot{w}_2^2 c^{-2})^{1/2}}{(1 - 2\Phi_1 c^{-2} - \dot{w}_1^2 c^{-2})^{1/2}} \frac{1 + c^{-1}k\dot{w}_1(u)}{1 + c^{-1}k\dot{w}_2(u')} \tag{7.2.12}$$

or after expanding in series

$$\frac{\delta\tau'}{\delta\tau} = 1 + c^{-2}(\Phi_1 - \Phi_2) + c^{-1}k\dot{w}_1(u) - c^{-1}k\dot{w}_2(u') + \tfrac{1}{2}c^{-2}(\dot{w}_1^2 - \dot{w}_2^2) + c^{-2}(k\dot{w}_2)^2 - c^{-2}(k\dot{w}_1)(k\dot{w}_2) + \ldots. \tag{7.2.13}$$

Small third-order quantities are omitted in (7.2.12) and (7.2.13). In the second-order terms one may not distinguish between the arguments u and u'. If f_1 is the proper frequency of the emitted signal and f_{12} is the frequency of this signal recorded at point 2 then

$$\frac{f_1}{f_{12}} = \frac{\delta\tau'}{\delta\tau}. \tag{7.2.14}$$

This relation, together with (7.2.12) or (7.2.13), determines the frequency displacement due to the motions of emitter and receiver (intrinsic Doppler effect) and the difference in the gravitation potentials at the points of emission and reception of the signal (gravitational displacement of frequency).

The first-order terms in (7.2.13) contain two instants, the moment u of emission and the moment u' of reception. Sometimes it may be useful to rewrite this relation as the function of one single argument. Let us choose first the moment of emission u as the basic argument. Starting from (7.2.9) and denoting

$$\boldsymbol{R}(u) = \boldsymbol{w}_1(u) - \boldsymbol{w}_2(u) \tag{7.2.15}$$

one obtains

$$\begin{aligned} u' - u &= c^{-1}|\boldsymbol{R}(u) - (u' - u)\dot{\boldsymbol{w}}_2(u)| = c^{-1}R(u) - c^{-1}(u' - u)\frac{\boldsymbol{R}\dot{\boldsymbol{w}}_2}{R} \\ &= c^{-1}R(u) - c^{-2}\boldsymbol{R}\dot{\boldsymbol{w}}_2(u) + \dots . \end{aligned} \tag{7.2.16}$$

After some manipulation we obtain

$$\boldsymbol{k} = \frac{\boldsymbol{R}(u)}{R(u)} - c^{-1}\dot{\boldsymbol{w}}_2 + c^{-1}\frac{(\boldsymbol{R}\dot{\boldsymbol{w}}_2)}{R^2}\boldsymbol{R} \tag{7.2.17}$$

and

$$\dot{\boldsymbol{w}}_2(u') = \dot{\boldsymbol{w}}_2(u) + c^{-1}R\ddot{\boldsymbol{w}}_2(u). \tag{7.2.18}$$

Substituting these expressions into (7.2.13) one gets

$$\frac{\delta\tau'}{\delta\tau} = 1 + c^{-2}(\Phi_1 - \Phi_2) + c^{-1}\dot{R}(u) + \tfrac{1}{2}c^{-2}\dot{\boldsymbol{R}}^2 - c^{-2}\boldsymbol{R}\ddot{\boldsymbol{w}}_2 + \dots . \tag{7.2.19}$$

In neglecting the acceleration of the light receiver this formula reduces to the Doppler shift of special relativity

$$\frac{1 + c^{-1}\dot{R}(u)}{(1 - c^{-2}\dot{\boldsymbol{R}}^2)^{1/2}}$$

in combination with the additive gravitational displacement $c^{-2}(\Phi_1 - \Phi_2)$. Equation (7.2.19) may be derived from (7.2.8) more easily by using the relation resulting from (7.2.16)

$$\frac{du'}{du} = 1 + c^{-1}\dot{R}(u) - c^{-2}\dot{\boldsymbol{R}}\dot{\boldsymbol{w}}_2 - c^{-2}\boldsymbol{R}\ddot{\boldsymbol{w}}_2. \tag{7.2.20}$$

Let us transform (7.2.13) to the argument u'. From

$$u' - u = c^{-1}R(u') - c^{-2}\boldsymbol{R}\dot{\boldsymbol{w}}_1 \tag{7.2.21}$$

and

$$\frac{du}{du'} = 1 - c^{-1}\dot{R}(u') + c^{-2}\dot{\boldsymbol{R}}\dot{\boldsymbol{w}}_1 + c^{-2}\boldsymbol{R}\ddot{\boldsymbol{w}}_1 \tag{7.2.22}$$

one has from (7.2.8)

$$\frac{\delta\tau}{\delta\tau'} = 1 + c^{-2}(\Phi_2 - \Phi_1) - c^{-1}\dot{R}(u') + \tfrac{1}{2}c^{-2}\dot{\boldsymbol{R}}^2 + c^{-2}\boldsymbol{R}\ddot{\boldsymbol{w}}_1. \tag{7.2.23}$$

Needless to say, equations (7.2.12), (7.2.19) and (7.2.23) are quite equivalent.

7.2.2 Integrated Doppler effect

Consideration of the integrated Doppler effect is performed here by a technique used by Boucher (1978) in a slightly different form. A clock with proper frequency f_1 located at moving point 1 (a satellite) emits light signals during the interval (τ_1, τ_2) of the proper time τ of this point. These signals are received with frequency f_{12} by the moving point 2 (a ground station) in the interval (τ'_1, τ'_2) of the proper time τ' of this point and are compared with the signals with frequency f_2 of the clock located at point 2. The measurable quantity over the interval (τ'_1, τ'_2) is the difference between the number of proper impulses generated by the clock at point 2 and the number of impulses received at point 2 from the clock at point 1. Mathematically, this quantity equals

$$N = \int_{\tau'_1}^{\tau'_2} (f_2 - f_{12})\,\mathrm{d}\tau'. \tag{7.2.24}$$

The frequency f_2 is constant with respect to proper time τ' and may be put outside the integral sign. The number of impulses received at point 2 is equal to the number of impulses emitted at point 1. Hence,

$$\int_{\tau'_1}^{\tau'_2} f_{12}\,\mathrm{d}\tau' = \int_{\tau_1}^{\tau_2} f_1\,\mathrm{d}\tau \tag{7.2.25}$$

and this time the frequency f_1 is constant with respect to τ and may be put outside the integral sign. The remaining integral is transformed again to the proper time τ' with the use of (7.2.8)

$$\begin{aligned}\int_{\tau_1}^{\tau_2} \mathrm{d}\tau &= \int_{\tau'_1}^{\tau'_2} \frac{\mathrm{d}\tau}{\mathrm{d}\tau'}\,\mathrm{d}\tau' \\ &= \int_{\tau'_1}^{\tau'_2} \left[1 + c^{-2}(\Phi_2 - \Phi_1) + \tfrac{1}{2}c^{-2}(\dot{w}_2^2 - \dot{w}_1^2) + \left(\frac{\mathrm{d}u}{\mathrm{d}u'} - 1\right)\right] \mathrm{d}\tau'.\end{aligned} \tag{7.2.26}$$

It is convenient to express $\mathrm{d}u/\mathrm{d}u'$ here by (7.2.22) in terms of the moment of reception u', as in integrating within the adopted accuracy the difference between u' and τ' may be neglected. Therefore,

$$\int_{\tau_1}^{\tau_2} \mathrm{d}\tau = \tau'_2 - \tau'_1 - c^{-1}[R(\tau'_2) - R(\tau'_1)] + c^{-2} \int_{\tau'_1}^{\tau'_2} (\Phi_2 - \Phi_1 + \tfrac{1}{2}\dot{\boldsymbol{R}}^2 + \boldsymbol{R}\ddot{\boldsymbol{w}}_1)\,\mathrm{d}\tau'. \tag{7.2.27}$$

Using all intermediate results, one finds finally

$$N = (f_2 - f_1)(\tau_2' - \tau_1') + c^{-1} f_1 [R(\tau_2') - R(\tau_1')] \\ - c^{-2} f_1 \int_{\tau_1'}^{\tau_2'} (\Phi_2 - \Phi_1 + \tfrac{1}{2}\dot{\boldsymbol{R}}^2 + \boldsymbol{R}\ddot{\boldsymbol{w}}_1)\, \mathrm{d}\tau'. \qquad (7.2.28)$$

In the limit $\tau_2' \to \tau_1'$ taking account of (7.2.14) one returns again to (7.2.23).

7.2.3 Synchronization of ground and satellite clocks

As suggested by Ashby and Allan (1979), to compare the rate of clocks at different points of circumterrestrial space it is suitable to take as the basis the coordinate time of a geocentric reference frame. In particular, this leads to a significant simplification of the procedure of clock synchronization. This enables us to avoid difficulties related to non-transitivity and the dependence on the traversed path in the clock synchronization in proper time (Einstein synchronization).

The satellite clock may be synchronized with the ground station clock by (indirectly) recording at station A the coordinate time interval T_A between emission of signal to the satellite and its return to the station after reflection from the satellite. Due to the Earth's rotation, with angular velocity $\boldsymbol{\omega}$, station A during the interval T_A changes its geocentric location from $\boldsymbol{w}_A = \boldsymbol{w}_A(u_A)$ at moment u_A of the signal emission to

$$\boldsymbol{w}_A(u_A + T_A) = \boldsymbol{w}_A(u_A) + T_A(\boldsymbol{\omega} \times \boldsymbol{w}_A). \qquad (7.2.29)$$

Clock synchronization implies that after converting to coordinate time the satellite clock at the moment of signal arrival at the satellite has a reading

$$u_A' = u_A + \tfrac{1}{2} T_A + c^{-2} \boldsymbol{w}_A'(\boldsymbol{\omega} \times \boldsymbol{w}_A) \qquad (7.2.30)$$

$\boldsymbol{w}_A' = \boldsymbol{w}'(u_A')$ being the geocentric position of the satellite at moment u_A'. Indeed, for a signal propagating from the ground station to the satellite one has

$$u_A' = u_A + c^{-1} D_A \qquad D_A = |\boldsymbol{w}'(u_A') - \boldsymbol{w}_A(u_A)| \qquad (7.2.31)$$

and for the return journey

$$u_A + T_A = u_A' + c^{-1} |\boldsymbol{w}_A(u_A + T_A) - \boldsymbol{w}'(u_A')| \\ = u_A' + c^{-1} D_A - c^{-1} \frac{T_A}{D_A} (\boldsymbol{\omega} \times \boldsymbol{w}_A) \boldsymbol{w}_A'. \qquad (7.2.32)$$

The difference between these expressions yields (7.2.30).

If the satellite clock is chosen as the reference clock then the ground clock may be synchronized by the emission of a signal from the satellite to the ground station and its further reception again at the satellite. During the two-way transit time T'_A the satellite, moving with velocity $\boldsymbol{v}'$, changes its position from $\boldsymbol{w}'_A = \boldsymbol{w}'(u'_A)$ at initial moment u'_A to

$$\boldsymbol{w}'(u'_A + T'_A) = \boldsymbol{w}'(u'_A) + T'_A \boldsymbol{v}'. \tag{7.2.33}$$

For clock synchronization the ground clock at the moment of reception of the signal should have a reading which corresponds to the coordinate time

$$u_A = u'_A + \tfrac{1}{2}T'_A - c^{-2}\boldsymbol{v}'(\boldsymbol{w}'_A - \boldsymbol{w}_A). \tag{7.2.34}$$

Indeed, for the path from the satellite to the ground station one has

$$u_A = u'_A + c^{-1}D_A \tag{7.2.35}$$

and for the way back from the ground station to the satellite

$$\begin{aligned} u'_A + T'_A &= u_A + c^{-1}|\boldsymbol{w}'(u'_A + T'_A) - \boldsymbol{w}_A(u_A)| \\ &= u_A + c^{-1}D_A + c^{-1}\frac{T'_A}{D_A}\boldsymbol{v}'(\boldsymbol{w}'_A - \boldsymbol{w}_A). \end{aligned} \tag{7.2.36}$$

The difference between these expressions gives (7.2.34). Relations (7.2.30) and (7.2.34) enable one to synchronize the clocks provided that the readings of one clock and the two-way transit time of the signal are known. The coordinates and velocities of the satellite and the ground station occur in these relations only in the post-Newtonian terms and, hence, may be known to an accuracy of the first (Newtonian) approximation. For a satellite in a circular orbit one has $\boldsymbol{v}'\boldsymbol{w}'_A = 0$, resulting in some simplification of (7.2.34).

7.2.4 Synchronization of two ground clocks via satellite

The basic relations (7.2.30) and (7.2.34) underlie the analysis by Ashby and Allan (1979) of different real cases of interest in the rate comparison of two ground clocks via satellite. Because of their practical importance some results are reproduced below. In the first case the stations are connected with one another by light signals via satellite. In all other cases each station interacts separately with the satellite.

(1) Ground station A at moment u_A transmits the signal to the satellite. Being reflected from the satellite this signal is received by ground station B and is re-transmitted to the satellite. The signal is again reflected from the satellite and returns to station A at moment $u_A + T_A$. Measurement of u_A and T_A(including the necessary conversion from proper to

coordinate time) enables us to calculate the moment of the signal's arrival at B and to obtain the synchronization formula

$$u_B = u_A + \tfrac{1}{2}T_A + \boldsymbol{v}'\left(\boldsymbol{w}_B - \boldsymbol{w}'_B + \frac{D_B}{D_A}(\boldsymbol{w}_A - \boldsymbol{w}'_A)\right)c^{-2} + c^{-2}\left(1 + \frac{D_B}{D_A}\right)(\boldsymbol{\omega} \times \boldsymbol{w}_A)\boldsymbol{w}'_A \quad (7.2.37)$$

where D_A and D_B are the distances travelled by the photon in propagating from stations A and B, respectively, to the satellite. All other designations are the same as used in (7.2.30) and (7.2.34). In fact, denoting by u'_A, u'_B the moments of arrival of signals from A and B, respectively, at the satellite S, one has for the four segments (A, S), (S, B), (B, S) and (S, A)

$$u'_A = u_A + c^{-1}D_A \qquad D_A = |\boldsymbol{w}'(u'_A) - \boldsymbol{w}_A(u_A)| \quad (7.2.38)$$

$$\begin{aligned} u_B &= u'_A + c^{-1}|\boldsymbol{w}'(u'_A) - \boldsymbol{w}_B(u_B)| \\ &= u'_A + c^{-1}D_B + c^{-1}\frac{u'_B - u'_A}{D_B}\boldsymbol{v}'[\boldsymbol{w}_B(u_B) - \boldsymbol{w}'(u'_B)] \end{aligned} \quad (7.2.39)$$

$$u'_B = u_B + c^{-1}D_B \qquad D_B = |\boldsymbol{w}'(u'_B) - \boldsymbol{w}_B(u_B)| \quad (7.2.40)$$

$$\begin{aligned} u_A + T_A &= u'_B + c^{-1}|\boldsymbol{w}'(u'_B) - \boldsymbol{w}_A(u_A + T_A)| \\ &= u'_B + c^{-1}D_A - \frac{c^1}{D_A}[T_A(\boldsymbol{\omega} \times \boldsymbol{w}_A)\boldsymbol{w}'(u'_A) \\ &\quad + (u'_B - u'_A)\boldsymbol{v}'(\boldsymbol{w}_A(u_A) - \boldsymbol{w}'(u'_A))]. \end{aligned} \quad (7.2.41)$$

Excluding from this the moments u'_A, u'_B and the Newtonian terms with distances D_A, D_B one obtains the synchronization formula (7.2.37). To post-Newtonian accuracy one may replace $\boldsymbol{w}'_B$ by $\boldsymbol{w}'_A$ in (7.2.37).

(2) Ground station A transmits a signal to the satellite at moment u_A. This signal is reflected from the satellite and returns to A at moment $u_A + T_A$. Similarly, station B transmits a second signal at moment u_B ($> u_A$), returning to it at moment $u_B + T_B$ after reflection from the satellite. The known quantities are T_A, T_B and the time interval $T' = u'_B - u'_A$ between recording the two signals on the satellite. The synchronization formula for clocks A and B has the form

$$u_B = u_A + \tfrac{1}{2}(T_A - T_B) + T' + c^{-2}[(\boldsymbol{\omega} \times \boldsymbol{w}_A)\boldsymbol{w}'_A - (\boldsymbol{\omega} \times \boldsymbol{w}_B)\boldsymbol{w}'_B]. \quad (7.2.42)$$

In fact, applying (7.2.30) to clocks A and B and forming the difference of the corresponding expressions, one obtains (7.2.42).

(3) Ground station A transmits a signal to the satellite at moment u_A which returns to it at moment $u_A + T_A$ after reflection from the satellite. In addition, the second signal is transmitted from the satellite to station B and then returns to the satellite. Let the round-trip transit time for this signal be T'_B. The known quantities are T_A, T'_B and the interval $T' = u'_B - u'_A$ recorded on the satellite between the arrival of the signal from A and the departure of signal to station B. The synchronization formula at moment u_A of the first signal emission at A and moment u_B of the second signal's reception at B takes the form

$$u_B = u_A + \tfrac{1}{2}(T_A + T'_B) + T' + c^{-2}[\boldsymbol{w}'_A(\boldsymbol{\omega} \times \boldsymbol{w}_A) - \boldsymbol{v}'(\boldsymbol{w}'_B - \boldsymbol{w}_B)]. \quad (7.2.43)$$

Indeed, by applying (7.2.30) and (7.2.34) to A and B, respectively, and adding the results one obtains relation (7.2.43).

(4) Two signals are transmitted with an interval T' from the satellite to stations A and B, respectively, and after their return to the satellite their transit times T'_A and T'_B are recorded. The synchronization formula for the moments of arrival of these signals at A and B is of the form

$$u_B = u_A + \tfrac{1}{2}(T'_B - T'_A) + T' - c^{-2}[\boldsymbol{v}'_B(\boldsymbol{w}'_B - \boldsymbol{w}_B) - \boldsymbol{v}'_A(\boldsymbol{w}'_A - \boldsymbol{w}_A)]. \quad (7.2.44)$$

This follows from application of (7.2.34) to A and B. If the time interval is sufficiently small one may make the values $\boldsymbol{v}'_A$ and $\boldsymbol{v}'_B$ of the satellite velocity identical.

All these clock synchronization techniques involve the geocentric time u. Actual measurements give the proper time of the ground station (the TRS origin) or the proper time of the satellite (the SRS origin). The transformation from u to these proper times has been considered in section 6.2 and in more detail for the ground station in section 7.1.

7.2.5 Use of navigation satellites

The problem of navigation with the use of the systems of navigation satellites like NAVSTAR is closely related to the problem of clock synchronization on the surface of the Earth and in circumterrestrial space. Consider three moving objects: a navigation satellite 1, a user 2 (ship or aircraft) and a ground station 3. All three objects are supplied with clocks indicating their individual proper time τ, τ' and τ'', respectively. For the clock on the satellite the conversion from proper time τ to GRS coordinate time u is reduced to the integral

$$u = \int (1 - c^{-2}\Phi_1 + \tfrac{1}{2}c^{-2}\dot{\boldsymbol{w}}_1^2)\, \mathrm{d}\tau \quad (7.2.45)$$

to be taken along the satellite trajectory. If in the geopotential (7.2.2) one may be restricted by the approximation (7.2.3) then using the energy integral

$$\tfrac{1}{2}\dot{\boldsymbol{w}}_1^2 + \Phi_1 = H \tag{7.2.46}$$

one obtains

$$u = \int \left(1 + c^{-2}H + \frac{2m}{d_1} + \frac{mA_E^2}{d_1^3} J_2 (1 - 3\sin^2\varphi_1)\right) \mathrm{d}\tau \tag{7.2.47}$$

$m = GM_E/c^2$ being the gravitational Earth parameter. For the reference clock of the ground station one has, by analogy with (7.2.45),

$$u = \int (1 - c^{-2}\Phi_3 + \tfrac{1}{2}c^{-2}\dot{\boldsymbol{w}}_3^2)\,\mathrm{d}\tau''. \tag{7.2.48}$$

From (7.1.52)–(7.1.56) this relation takes the form

$$u = (1 + c^{-2}V^*)\tau'' + c^{-2}V_p(u). \tag{7.2.49}$$

If one may again be restricted here by the approximation (7.2.3) then the relations of section 7.1.4. are replaced by the approximate relations

$$d_3 = A_E - A_E f \sin^2\varphi + h \tag{7.2.50}$$

$$f = \frac{A_E - A_p}{A_E} = \frac{\omega^2 A_E^3}{2GM_E} + \tfrac{3}{2}J_2 \tag{7.2.51}$$

$$\dot{\boldsymbol{w}}_3 = \boldsymbol{\omega} \times \boldsymbol{w}_3 \tag{7.2.52}$$

$$W_0 = \frac{GM_E}{A_E}(1 + \tfrac{1}{2}J_2) + \tfrac{1}{2}\omega^2 A_E^2 \tag{7.2.53}$$

$$g(\varphi) = \frac{GM_E}{A_E^2} + \frac{3GM_E}{2A_E^2}J_2 - \omega^2 A_E + \left(2\omega^2 A_E - \frac{3GM_E}{2A_E^2}J_2\right)\sin^2\varphi. \tag{7.2.54}$$

A_p is the polar radius of the Earth, f is the Earth's oblateness, W_0 is, as previously, the value of the potential of the force of gravity on the surface of the geoid. Integral (7.2.48) is transformed to

$$u = \int [1 + c^{-2}W_0 - c^{-2}g(\varphi)h]\,\mathrm{d}\tau''. \tag{7.2.55}$$

Numerical analysis of expressions (7.2.47) and (7.2.55) has been performed by Ashby and Allan (1979).

The user of navigation satellites can derive, on the basis of approximately known values of his own position and velocity, the coordinate time from proper time τ' in a similar fashion to (7.2.45) and (7.2.48)

$$u = \int (1 - c^{-2}\Phi_2 + \tfrac{1}{2}c^{-2}\dot{w}_2^2)\, d\tau'. \qquad (7.2.56)$$

In principle, the determination of the position of user 2 by means of a system of navigation satellites like NAVSTAR may be realized as follows. At the epoch u_0 of GRS time let clocks 1, 2 and 3 be synchronized, implying that their readings τ_1, τ_1' and τ_1'' in individual proper time correspond to the moment u_0 of GRS coordinate time. Satellite 1 transmits signals periodic in its proper time τ. Consider a signal emitted by the satellite at moment τ_2 in its own timescale τ and received by the user at moment τ_2' in the user's timescale τ'. The moments τ_2 and τ_2' correspond to the epochs u_1 and u_2 of GRS time, respectively. Then by (7.2.45) and (7.2.56) one has

$$u_1 - u_0 = \tau_2 - \tau_1 + c^{-2} \int_{\tau_1}^{\tau_2} (-\Phi_1 + \tfrac{1}{2}\dot{w}_1^2)\, d\tau \qquad (7.2.57)$$

$$u_2 - u_0 = \tau_2' - \tau_1' + c^{-2} \int_{\tau_1'}^{\tau_2'} (-\Phi_2 + \tfrac{1}{2}\dot{w}_2^2)\, d\tau'. \qquad (7.2.58)$$

The moments u_1 and u_2 are related by (7.2.21):

$$u_2 - u_1 = c^{-1} R_{12}(u_2) - c^{-2} \boldsymbol{R}_{12} \dot{\boldsymbol{w}}_1 \qquad (7.2.59)$$

with $\boldsymbol{R}_{12}$ being determined by (7.2.15). Using (7.2.57)–(7.2.59) one finds

$$R_{12}(u_2) = c(\tau_2' - \tau_1') - c(\tau_2 - \tau_1) + c^{-1} \int_{\tau_1'}^{\tau_2'} (-\Phi_2 + \tfrac{1}{2}\dot{w}_2^2)\, d\tau' - c^{-1} \int_{\tau_1}^{\tau_2} (-\Phi_1 + \tfrac{1}{2}\dot{w}_1^2)\, d\tau + c^{-1} \boldsymbol{R}_{12} \dot{\boldsymbol{w}}_1. \qquad (7.2.60)$$

The epochs τ_1, τ_2, τ_1' and τ_2' are measurable quantities and the integrals, as well as the last term in (7.2.60), may be calculated with approximate known data. Thus, this relation enables one to find the exact value of the distance between user 2 and satellite 1 at moment u_2.

The navigation system envisages that the user receives the signals simultaneously from at least four navigation satellites. Three relations of type (7.2.60) allow the determination of the geocentric coordinates of the user on the basis of three distance values. The user's clock as a rule is not sufficiently accurate and one actually determines pseudodistances involving the unknown correction to the user's clock. To determine this correction a fourth equation of the type (7.2.60) is needed. The technique of calculating the geocentric position of the user and the correction of his clock with the aid of navigation satellites is described in detail in a set of papers published in *Navigation* **25** no 2 1978. The consideration of the relativity effects developed here differs a little from the technique used in GPS (Spilker 1978).

Ground station 3 also receives signals from satellite 1. The time interval from the epoch of the synchronization to the moment u_3 of signal reception is equal to

$$u_3 - u_0 = \tau_2'' - \tau_1'' + c^{-2} \int_{\tau_1''}^{\tau_2''} (-\Phi_3 + \tfrac{1}{2}\dot{w}_3^2)\, \mathrm{d}\tau'' \qquad (7.2.61)$$

with τ_2'' being the moment of signal reception in proper time τ'' of the ground station. The moments u_1 and u_3 are related, by analogy with (7.2.21), by

$$u_3 - u_1 = c^{-1} R_{13}(u_3) - c^{-2} \boldsymbol{R}_{13}\dot{\boldsymbol{w}}_1 \qquad (7.2.62)$$

or similarly to (7.2.16)

$$u_3 - u_1 = c^{-1} R_{13}(u_1) - c^{-2} \boldsymbol{R}_{13}\dot{\boldsymbol{w}}_3 = c^{-1} R_{13}(u_1) - c^{-2} \boldsymbol{w}_1(\boldsymbol{\omega} \times \boldsymbol{w}_3). \qquad (7.2.63)$$

The use of (7.2.57), (7.2.61) and (7.2.62) or (7.2.63) permits us in general to relate the readings τ of the satellite clock with the readings τ'' of the reference clock of the ground station and to refer all measurements of users to the timescale τ''. But the use of GRS time u seems to be a simpler and more convenient method.

7.2.6 Synchronization by clock transportation or electromagnetic signal transmission

From the GRS$^+$ metric (6.1.16)–(6.1.18) or (7.1.52) it is seen that the proper time of a clock moving on the surface of the Earth is

$$\mathrm{d}\tau = \{1 - c^{-2}[W_0 + Q_2 - \bar{Q}_{20} - g(\varphi, \lambda)h + \epsilon_{ijk}\omega_E^j y^k v^i + \tfrac{1}{2}v^k v^k + \ldots]\}\, \mathrm{d}u \qquad (7.2.64)$$

where v^k is the velocity of the moving clock in GRS$^+$ with φ and λ being its latitude and longitude, respectively. Introducing TT instead of u and neglecting the lunar and solar tidal terms one obtains

$$\mathrm{d}(TT) = [1 - c^{-2} g(\varphi, \lambda)h + \tfrac{1}{2}c^{-2}v^2 + c^{-2}(\boldsymbol{\omega} \times \boldsymbol{y})\boldsymbol{v}]\, \mathrm{d}\tau. \qquad (7.2.65)$$

In transporting the clock along some path on the surface of the Earth the change of time TT is determined by the curvilinear integral of the right-hand side of (7.2.65). The last term of this expression leads to the integral

$$\int(\boldsymbol{\omega}\times\boldsymbol{y})\boldsymbol{v}\,\mathrm{d}\tau=\boldsymbol{\omega}\int(\boldsymbol{y}\times\boldsymbol{v})\,\mathrm{d}\tau=\boldsymbol{\omega}\int(\boldsymbol{y}\times\mathrm{d}\boldsymbol{y})=2\boldsymbol{\omega}\boldsymbol{S}=2\omega S_E. \quad (7.2.66)$$

$\mathrm{d}\boldsymbol{S}=(\boldsymbol{y}\times\mathrm{d}\boldsymbol{y})/2$ is the area vector element swept out by the geocentric position vector of the clock in its transfer over the Earth's surface, $\boldsymbol{S}$ is the integral area and S_E is the projection of this area onto the equatorial plane. This projection is taken to be positive for eastward motion of the clock. This term, due completely to the rotation of the Earth, is called the Sagnac effect. In clock movement over a closed circuit its readings will differ by a quantity proportional to the area covered.

Clock synchronization may be performed by transmitting electromagnetic signals as well. The determining relation is obtained from applying (2.3.25) to the GRS$^+$ metric

$$\mathrm{d}(TT)=c^{-1}[1-c^{-2}g(\varphi,\lambda)h+c^{-2}(\boldsymbol{\omega}\times\boldsymbol{y})\boldsymbol{V}]\,\mathrm{d}\ell \quad (7.2.67)$$

where $\boldsymbol{V}$ is the coordinate velocity of light in GRS$^+$ and $\mathrm{d}\ell$ is the proper distance element along the light trajectory. One has approximately

$$\mathrm{d}\ell^2=(1+2c^{-2}\hat{U}_E)\,\mathrm{d}y^k\mathrm{d}y^k. \quad (7.2.68)$$

On integration, the third term in (7.2.67) again gives the Sagnac effect in the form

$$c^{-3}\int(\boldsymbol{\omega}\times\boldsymbol{y})\boldsymbol{V}\,\mathrm{d}\ell=c^{-3}\boldsymbol{\omega}\int(\boldsymbol{y}\times\boldsymbol{V})\,\mathrm{d}\ell=c^{-2}\boldsymbol{\omega}\int(\boldsymbol{y}\times\boldsymbol{V})\,\mathrm{d}\tau$$
$$=c^{-2}\boldsymbol{\omega}\int(\boldsymbol{y}\times\mathrm{d}\boldsymbol{y})=2c^{-2}\omega S_E. \quad (7.2.69)$$

At present, all effects related to light propagation and clock rate comparison on the Earth and in circumterrestrial space have been well tested experimentally. The numerical estimates and other relevant data may be found in Allan and Ashby (1986) and Alley (1983).

Relativistic geodynamics is still taking its first steps. Exposition here has been restricted only to methodological questions with limited accuracy. A review of other relativistic problems of geodynamics may be found in Boucher (1986).

Postscript

Since the main part of this book was written, IAU Colloquium 127 on reference systems has taken place at Virginia Beach (14–20 October 1990). One of the principal aims of this Colloquium was to discuss the draft recommendations for the forthcoming 21st IAU General Assembly (Buenos Aires, 23 July–1 August 1991). It is of importance that in the draft recommendation on time there appeared for the first time the coordinate timescales t and u of BRS and GRS, under the names TCB (*temps coordonné barycentrique*) and TCG (*temps coordonné géocentrique*), respectively (section 7.1). This does not prevent us from using the scales TB and TT, but in operating with t and u one can avoid the necessity of introducing scaling factors in the values for the masses and coordinates of the bodies (cf section 7.1.7).

In the draft recommendations specifying the main reference systems is the fact that, for the first time, these systems are defined as GRT four-dimensional systems to be postulated using corresponding metric forms. In so doing only the values of $h_{00}^{(2)}$ and $h_{ij}^{(2)}$ are fixed in the expansions (4.1.1) for the metric tensor components, with the condition $h_{ij}^{(2)} = h_{00}^{(2)}\delta_{ij}$. $h_{00}^{(2)}$ is supposed to include the Newtonian potential of the internal masses and the tidal potential due to external masses. Such a choice of coordinates involves, in particular, the vanishing of the coordinate functions a_i in the description (2.2.38)–(2.2.40) of BRS, enabling one to use both harmonic and PPN type (2.2.41) coordinates, with the coordinate function a_0 not being fixed. Moreover, without indicating the form of $h_{0i}^{(3)}$ one cannot distinguish between dynamically non-rotating and kinematically non-rotating systems (applied to GRS the first case is specified by (4.2.3) and (4.2.8) whereas the second case implies the absence of the term with geodesic precession F^{ik} in (4.2.8) and the addition of a term $-c^{-3}\dot{F}^{ik}w^k$ on the right-hand side of (4.2.3)). This question, as well as the possibility of representing GRS in closed form with respect to the geocentric coordinates w^k, are described in detail in a paper submitted by the author to the Colloquium proceedings. In addition, there is also a paper in the proceedings by Damour, Soffel and Xu dealing with the construction of relativistic reference systems in closed form in ADM coordinates (of the type of PPN frame coordinates). These two papers complement the content of the present book.

References

Allan D W and Ashby N 1986 Coordinate Time in the Vicinity of the Earth *RCMA* p 299

Alley C 1983 Proper Time Experiments in Gravitational Fields with Atomic Clocks, Aircraft and Laser Light Pulses *Quantum Optics, Experimental Gravity and Measurement Theory* ed P Meystre and M O Scully (New York: Plenum) p 363

Anderson J D 1974 Lectures on Physical and Technical Problems Posed by Precision Radio Tracking *Experimental Gravitation* ed B Bertotti (New York: Academic) p 163

Ashby N 1986 Planetary Perturbation Equations Based on Relativistic Keplerian Motion *RCMA* p 41

Ashby N and Allan D W 1979 Practical Implications of Relativity for a Global Coordinate Time Scale *Radio Sci.* **14** 649

Ashby N and Bertotti B 1984 Relativistic Perturbations of an Earth Satellite *Phys. Rev. Lett.* **52**, 485

—— 1986 Relativistic Effects in Local Inertial Frames *Phys. Rev.* D **34** 2246

Backer D C and Hellings R W 1986 Pulsar Timing and General Relativity *Ann. Rev. Astron. Astrophys.* **24** 537

Barker B M and O'Connell R F 1976 Lagrangian–Hamiltonian Formalism for the Gravitational Two-Body Problem with Spin and Parametrized Post-Newtonian Parameters γ and β *Phys. Rev.* D **14** 861

Barker B M, Byrd G G and O'Connell R F 1986 Relativistic Kepler's Third Law *Astrophys. J.* **305** 623

Bertotti B 1954 On the Two-Body Problem in General Relativity *Nuovo Cimento* **12** 226

—— 1986 Local Frames *RCMA* p 233

Bertotti B, Ciufolini I and Bender P 1987 New Test of General Relativity: Measurement of de Sitter Geodetic Precession Rate for Lunar Perigee *Phys. Rev. Lett.* **58** 1062

Bogdan M and Plebanski J 1962 A Study of Geodesic Motion in the Field of Schwarzschild's Solution *Acta Phys. Polon.* **21** 239

Bogorodsky A F 1962 *Einstein Field Equations and Their Application in Astronomy* (Kiev: Kiev University) (in Russian)

Bordovitsyna T V, Bykova L E, Tamarov V A and Sharkovsky N A 1985 Structure Analysis of Orbital Perturbations of NAVSTAR-Type Earth Satellites *Kosmicheskie Issledovanya* **23** 713

Boucher C 1978 Relativistic Correction to Satellite Doppler Observation *Proc. IAGSSG 445 Meeting. The Mathematical Structure of the Gravity Field* Lagonissi, Greece (Athens: National Technivcal University)

—— 1986 Relativistic Effects in Geodynamics *RCMA* p 241

Boyer R H and Lindquist R W 1967 Maximal Analytic Extension of the Kerr Metric *J. Math. Phys.* **8** 265

Bretagnon P 1982 Théorie du mouvement de l'ensemble des planètes Solution VSOP82 *Astron. Astrophys.* **114** 278

Brumberg V A 1972 *Relativistic Celestial Mechanics* (Moscow: Nauka) (in Russian)

—— 1981 Relativistic Effects in Radar, Optical and Radiointerferometric Measurements *Astron. Zh.* **58** 181

—— 1986 Relativistic Reduction of Astrometric Observations *Astrometric Techniques* ed H K Eichhorn and R J Leacock (Dordrecht: Reidel) p 19

—— 1987a Post-Post Newtonian Propagation of Light in the Schwarzschild Field *Kinematika i Fisika Nebesn. Tel* **3** 8

—— 1987b Contemporary Problems of Relativistic Celestial Mechanics and Astrometry *Proc. 10th ERAM of the IAU* **3** 3

Brumberg V A and Ivanona T V 1985 Relativistic Effects in the Earth-Moon Dynamics *Trans ITA (Leningrad)* **19** 3

Brumberg V A and Kopejkin S M 1989a Relativistic Theory of Celestial Reference Frames *Reference Frames* ed J Kovalevsky, I I Mueller and B Kolaczek (Dordrecht: Kluwer) p 115

—— 1989b Relativistic Reference Systems and Motion of Test Bodies in the Vicinity of the Earth *Nuovo Cimento* B **103** 63

—— 1990 Relativistic Time Scales in the Solar System *Cel. Mech.* **48** 23

Brumberg V A and Tarasevich S V 1983 Application of System GRATOS to Determining Perturbations of the Spherically Symmetric Metric *Analytical Calculations by Computer and Their Applications in Theoretical Physics* (Dubna: Joint Institute for Nuclear Research) p 149

Cannon W H, Lisewski D, Finkelstein A M and Kreinovich V Ya 1986 Relativistic Effects in Earth Based and Cosmic Long Baseline Interferometry *RCMA* p 255

Carter B 1966 Complete Analytic Extension of the Symmetry Axis of Kerr's Solution of Einstein's Equations *Phys. Rev.* **141** 1242

Chapront J and Chapront-Touzé M 1981 Comparaison de ELP-2000 à une integration numerique du JPL *Astron. Astrophys.* **103** 295

Chapront-Touzé M and Chapront J 1983 The Lunar Ephemeris ELP-2000 *Astron. Astrophys.* **124** 50

Chazy J 1928 *La théorie de la relativité et la mécanique céleste* vol 1 (Paris: Gauthier-Villars)

—— 1930 *La théorie de la relativité et la mécanique céleste* vol 2 (Paris: Gauthier-Villars)

Cherny S D 1949 Motion of Material Point under Action of Force Producing the Acceleration $-m_1r^{-2} - 3m_2r^{-4}$ *Bull. ITA (Leningrad)* **4** 287

Chotimskaya E Z 1989 The Relation between TDB and TDT: The Numerical Integration and the Comparison of the Result Obtained with the Analytical Solutions *Preprint Institute of Applied Astronomy (Leningrad)* no 10

Cugusi L and Proverbio E 1978 Relativistic Effects on the Motion of Earth's Artificial Satellites *Astron. Astrophys.* **69** 321

Damour T 1983 Gravitational Radiation and the Motion of Compact Bodies *Gravitational Radiation* ed N Deruelle and T Piran (Amsterdam: North-Holland) p 59

—— 1984 The Motion of Compact Bodies and Gravitational Radiation *General Relativity and Gravitation* ed B Bertotti *et al* (Dordrecht: Reidel) p 89

—— 1987a An Introduction to the Theory of Gravitational Radiation *Gravitation in Astrophysics* ed B Carter and J B Hartle (New York: Plenum) p 3

—— 1987b The Problem of Motion in Newtonian and Einsteinian Gravity *300 Years of Gravitation* ed S W Hawking and W Israel (Cambridge: Cambridge University Press) p 128

Damour T and Deruelle N 1985 General Relativistic Celestial Mechanics of Binary Systems: I. The Post-Newtonian Motion *Ann. Inst. H. Poincaré* **43** 107

—— 1986 General Relativistic Celestial Mechanics of Binary Systems: II. The Post-Newtonian Timing Formula *Ann. Inst. H. Poincaré* **44** 263

Damour T, Grishchuk L P, Kopejkin S M and Schäfer G 1989 Higher-Order Relativistic Dynamics of Binary Systems *Proc. 5th Marcel Grossmann mtg on General Relativity* ed D Blair and M J Buckingham (Singapore: World Scientific) part B, pp 1311–13

Damour T and Schäfer G 1988 Higher-Order Relativistic Periastron Advances and Binary Pulsars *Nuovo Cimento* B **101** 127

Eddington A S and Clark G L 1938 The Problem of n Bodies in General Relativity Theory *Proc. R. Soc.* A **166** 465

Einstein A and Grommer J 1927 Allgemeine Relativitätstheorie und Bewegungsgesetz *Sitzber. Preuss. Akad. Wiss.* **1** 2

Einstein A and Infeld L 1949 On the Motion of Particles in General Relativity Theory *Can. J. Math.* **1** 209

Einstein A, Infeld L and Hoffman B 1938 The Gravitational Equations and the Problem of Motion *Ann. Math.* **39** 65

Estabrook F B 1969 Post-Newtonian n-Body Equations of the Brans-Dicke Theory *Astrophys. J.* **158** 81

Fairhead L, Bretagnon P and Lestrade J-F 1988 The Time Transformation TDB–TDT: an Analytical Formula and Related Problem of Convention *The Earth's Rotation and Reference Frames for Geodesy and Geodynamics* ed G Wilkins and A Babcock (Dordrecht: Kluwer) p 419

Fock V A 1955 *The Theory of Space, Time and Gravitation* (Moscow: State Technical Publications) (in Russian)

Fujimoto M and Grafarend E 1986 Spacetime Coordinates in the Geocentric Reference Frame *RCMA* p 269

Fukushima T 1988 The Fermi Coordinate System in the Post-Newtonian Framework *Cel. Mech.* **44** 61

Fukushima T, Fujimoto M K, Kinoshita H and Aoki S 1986a Coordinate Systems in the General Relativistic Framework *RCMA* p 145

—— 1986b A System of Astronomical Constants in the Relativistic Framework *Cel. Mech.* **38** 215

Ganea I-M 1973 The Exact Solution of Schwarzschild Problem in Painlevé Coordinates *Studii cerc. astr. (Bucuresti)* **18** 213

Gaposchkin E M and Kolaczek B (ed) 1981 *Reference Coordinate Systems for Earth Dynamics* (Dordrecht: Reidel)

Green R M 1985 *Spherical Astronomy* (Cambridge: Cambridge University Press)

Grishchuk L P and Kopejkin S M 1983 Motion of Gravitating Bodies with Consideration of Radiation Damping Force *Astron. Zh. Lett.* **9** 436

—— 1986 Equations of Motion for Isolated Bodies with Relativistic Corrections Including the Radiation Reaction Force *RCMA* p 19

Guinot B 1986 Is the International Atomic Time TAI a Coordinate Time or a Proper Time? *Cel. Mech.* **38** 155

Guinot B and Seidelmann PK 1988 Time Scales—Their History, Definition and Interpretation *Astron. Astrophys.* **194** 304

Heimberger J, Soffel M and Ruder H 1990 Relativistic Effects in the Motion of Artificial Satellites: II. The Oblateness of the Central Body *Cel. Mech.* **47** 205

Hellings R W 1986 Relativistic Effects in Astronomical Timing Measurements *Astron. J.* **91** 650, **92** 1446

Hirayama Th, Kinoshita H, Fujimoto M K and Fukushima T 1988 Analytical Expression of TDB-TDT_0 *19th IUGG Gen. Assembly, Vancouver, t1, Relativistic Effects in Geodesy* p 91

Huang T-Y, Zhu J, Xu B-X and Zhang H 1989 The Concept of TAI and TDT *Astron. Astrophys.* **220** 329

Infeld L 1954 On the Motion of Bodies in General Relativity Theory *Acta Phys. Polon.* **13** 187

—— 1957 Equations of Motion in General Relativity Theory and the Action Principle *Rev. Mod. Phys.* **29** 398

Infeld L and Plebanski J 1960 *Motion and Relativity* (Oxford: Pergamon)

Infeld L and Schild A 1949 On the Motion of Test Particles in General Relativity *Rev. Mod. Phys.* **21** 408

Irvine W M 1965 Local Irregularities in an Expanding Universe *Ann. Phys.* **32** 322

Ivanitskaya O S 1979 *Lorentzian Basis and Gravitation Effects in Einstein Theory of Gravitation* (Minsk: Science and Technics) (in Russian)

Japanese Ephemeris 1985 *Basis of the New Japanese Ephemeris*

Kislik M D 1985 Doppler Shift Frequency of Distant Light Source in the Schwarzschild Field *Astron. Zh.* **62** 905

Klioner S A 1989 Light Propagation in the Barycentric Reference System Considering the Motion of the Gravitating Masses *Preprint Institute of Applied Astronomy (Leningrad)* no 6

Kolaczek B and Weiffenbach G (ed) 1975 *On Reference Coordinate Systems for Earth Dynamics* (Warsaw: Technical University)

Kopejkin S M 1985 The Equations of Motion of Extended Bodies in General Relativity with Conservative Corrections and Radiation Damping Taken into Account *Astron. Zh.* **62** 889

—— 1987 On the Method of Solution of the External and Internal Problem of Motion of Bodies in General Relativity *Trans. Sternberg State Astron. Inst. (Moscow)* **59** 53

—— 1988 Celestial Coordinate Reference Systems in Curved Space-Time *Cel. Mech.* **44** 87

Kottler F 1922 Gravitation und Relativitätstheorie *Encykl. Math. Wiss.* **6** (2) no 22a

Kovalevsky J 1985 Systèmes de référence terrestres et célestes *Bull. Astron. Obs. R. de Belgique* **10** 87

Kovalevsky J and Mueller I I 1981 Comments on Conventional Terrestrial and Quasi-Inertial Reference Systems *Reference Coordinate Systems for Earth Dynamics* ed E M Gaposchkin and B Kolaczek (Dordrecht: Reidel) p 375

Krause H G L 1963 Relativistic Perturbation Theory of an Artificial Satellite in an Arbitrary Orbit about the Rotating Oblated Earth Spheroid and the Time Dilatation Effect for this Satellite *The Use of Artificial Satellites for Geodesy* ed G Veis (Amsterdam: North-Holland) p 69

Kreinovich V A 1975 Influence of Possible Violation of Principle of Equivalence on the Motion of the Moon *Proc. Leningrad State Univ. ser. math. mech. astron.* no 4 144

Krivov A V 1988 Relativistic Effects in Motion and Observations of Earth Satellites: I. Relativistic Perturbations in the Motion of Satellite. II. Calculation of Measurable Quantities. III. Geocentric Technique and Estimates of the Magnitude of Relativistic Effects *Proc. Leningrad State. Univ. ser.1* no 1 84, no 3 83, no 4 78

Landau L D and Lifshitz E M 1962 *The Theory of Fields* (Moscow: Phys. Math. Publ.) (in Russian)

Lestrade J-F and Bretagnon P 1982 Perturbations relativistes pour l'ensemble des planètes *Astron. Astrophys.* **105** 42

Lestrade J-F and Chapront-Touzé M 1982 Relativistic Perturbations of the Moon in ELP2000 *Astron. Astrophys.* **116** 75

Lightman A P, Press W H, Price R H and Teukolsky S A 1975 *Problem Book on Relativity and Gravitation* (Princeton: Princeton University Press)

Martin C F, Torrence M H and Misner C W 1985 Relativistic Effects on an Earth Orbiting Satellite in the Barycentric Coordinate System *J. Geophys. Res.* **90** 9403

Mast C B and Strathdee J 1959 On the Relativistic Interpretation of Astronomical Observations *Proc. R. Soc.* A **252** 476

Møller C 1972 *The Theory of Relativity* (Oxford: Clarendon)

Moyer T D 1981 Transformation from Proper Time on Earth to Coordinate Time in Solar System Barycentric Space-Time Frame of Reference *Cel. Mech.* **23** 33, 57

Moritz H and Mueller I I 1987 *Earth Rotation: Theory and Observation* (New York: Ungar)

Murray C A 1983 *Vectorial Astrometry* (Bristol: Adam Hilger)

Ni W-T and Zimmermann M 1978 Inertial and Gravitational Effects in the Proper Reference Frame of an Accelerated, Rotating Observer *Phys. Rev.* D **17** 1473

Nobili A M and Will C M 1986 The Real Value of Mercury's Perihelion Advance *Nature* **320** 39

Nordtvedt K 1973 Post-Newtonian Gravitational Effects in Lunar Laser Ranging *Phys. Rev.* D **7** 2347

—— 1988 Gravitomagnetic Interaction and Laser Ranging to Earth Satellites *Phys. Rev. Lett.* **61** 2647

Pavlov N V 1984a On Relativistic Theory of Astrometric Observations: I. Determination of Positions of Distant Sources from the Surfaces of the Solar System Bodies *Astron. Zh.* **61** 385

—— 1984b On Relativistic Theory of Astrometric Observations: II. Determination of Positions of Distant Sources from the Surface of the Earth *Astron. Zh.* **61** 600

—— 1985 On Relativistic Theory of Astrometric Observations: III. Radiointerferometry of Distant Sources *Astron. Zh.* **62** 172

Peters P C 1966 Perturbations in the Schwarzschild Metric *Phys. Rev.* **146** 938

Rashevsky P K 1953 *Riemannian Geometry and Tensor Analysis* (Moscow: State Technical Publications) (in Russian)

RCMA 1986 *Relativity in Celestial Mechanics and Astrometry* ed J Kovalevsky and V A Brumberg (Dordrecht: Reidel)

Reasenberg R D and Shapiro I I 1986 Prospects for Observations of Relativistic Effects in the Solar System *RCMA* 1986 p 383

Richter G W and Matzner R A 1982 Second-Order Contributions to Gravitational Deflection of Light in the Parametrized Post-Newtonian Formalism I, II *Phys. Rev.* D **26** 1219, 2549

—— 1983 Second-Order Contributions to Relativistic Time-Delay in the Parametrized Post-Newtonian Formalism *Phys. Rev.* D **28** 3007

Ries J C, Huanghand K and Watkins M 1988 Effect of General Relativity on a Near-Earth Satellite in the Geocentric and Barycentric Reference Frames *Phys. Rev. Lett.* **61** 903

Shapiro I I, Counselman C C and King R W 1976 Verification of the Principle of Equivalence for Massive Bodies *Phys. Rev. Lett.* **36** 555

Shapiro I I, Reasenberg R D, Chandler J F and Babock R W 1988 Measurement of the de Sitter Precession of the Moon: A Relativistic Three-Body Effect *Phys. Rev. Lett.* **61** 2643

de Sitter W 1916 On Einstein's Theory of Gravitation and Its Astronomical Consequences *Mon. Not. R. Astron. Soc.* **76** 699, **77** 155, 481

Sobolev S L 1950 *Some Applications of Functional Analysis in Mathematical Physics* (Leningrad: Leningrad University) (in Russian)

Soffel M 1989 *Relativity in Astrometry, Celestial Mechanics and Geodesy* (Berlin: Springer-Verlag)

Soffel M, Herold H, Ruder H and Schneider M 1987b Relativistic Geodesy: Gravimetric Measurements and the Definition of the Geoid *Manuscripta Geodaetica* **13** 143

Soffel M, Ruder H and Schneider M 1986 The Dominant Relativistic Terms in the Lunar Theory *Astron. Astrophys.* **157** 357

—— 1987a The Two-Body Problem in the (Truncated) PPN-Theory *Cel. Mech* **40** 77

Soffel M, Wirrer R, Schastok J, Ruder H and Schneider M 1988 Relativistic Effects in the Motion of Artificial Satellites: I. The Oblateness of the Central Body *Cel. Mech* **42** 81

Sôma M, Hirayama Th and Kinoshita H 1987 Analytical Expressions of the Earth's Position and Velocity for the Calculation of Apparent Positions *Cel. Mech* **41** 389

Spilker J I 1978 GPS Signal and Performance Characteristics *J. Inst. Navigation* **25** 121

Standish E M, Keesey M S W and Newhall X X 1976 JPL Development Ephemeris Number 96 *JPL TR 32-1603* (Pasadena: JPL)

Suen W M 1986 Multipole Moments for Stationary, Non-Asymptotically-Flat System in General Relativity *Phys. Rev.* D **34** 3617

Synge J L 1960 *Relativity: The General Theory* (Amsterdam: North-Holland)

Tarasevich S V, Krivov A V and Titov V B 1987 System of Analytical Manipulation GRATOS *Algorithms of Celestial Mechanics* nos 53, 54 (Leningrad: Institute of Theoretical Astronomy)

Tausner M J 1966 General Relativity and Its Effects on Planetary Orbits and Interplanetary Observations *Tech. Rep. 425* Lincoln Lab, MIT

Thorne K S and Hartle J B 1985 Laws of Motion and Precession for Black Holes and Other Bodies *Phys. Rev.* D **31** 1815

Vincent M A 1986 The Relativistic Equations of Motion for a Satellite in Orbit About a Finite-Size, Rotating Earth *Cel. Mech* **39** 15

Vladimirov Yu S 1982 *Reference Frames in the Theory of Gravitation* (Moscow: Energy) (in Russian)

Voinov A V 1988 Motion and Rotation of Celestial Bodies in the Post-Newtonian Approximation *Cel. Mech* **42** 293

—— 1990 Relativistic Equations of Earth Satellite Motion *Manuscripta Geodaetica* **15** 65

Will C M 1971 Theoretical Framework for Testing Relativistic Gravity: II. Parametrized Post-Newtonian Hydrodynamics and the Nordtvedt Effect *Astrophys. J.* **163** 611

—— 1974 The Theoretical Tools of Experimental Gravitation *Experimental Gravitation* ed B Bertotti (New York: Academic) p 1

—— 1985 *Theory and Experiment in Gravitational Physics* (Cambridge: Cambridge University Press)

—— 1986 General Relativity Confronts Experiments *RCMA* p 355

Will C M and Nordtvedt K 1972 Conservation Laws and Preferred Frames in Relativistic Gravity: I. Preferred-Frame Theories and an Extended PPN Formalism *Astrophys. J.* **177** 757

Williams J G, Dicke R H, Bender P L *et al* 1976 New Test of the Equivalence Principle from Lunar Laser Ranging *Phys. Rev. Lett.* **36** 551

Yao M, Zhu S Y, Pan R, Cheng Z, Yan H and Zhu W 1988 Relativistic Effects on an Earth-Orbiting Satellite in the Geocentric Coordinate System *Acta Astron. Sin.* **29** 181

Young J H and Coulter C A 1969 Exact Metric for a Nonrotating Mass with a Quadrupole Moment *Phys. Rev.* **184** 1313

Zeller G, Soffel M, Ruder H and Schneider M 1986 Relativistische Effecte bei der Laufzeitdifferenz der VLBI *Veröffentlichungen der Bayer. Komm. für die Intern. Erdmessung* **48** 218

Zel'dovich Ja B and Novikov I D 1967 *Relativistic Astrophysics* (Moscow: Nauka) (in Russian)

Zel'manov A L 1956 Chronometric Invariants and Co-Moving Coordinates in General Relativity Theory *DAN USSR* **107** 815

Zhang X-H 1986 Multipole Expansion of the General Relativistic Gravitational Field of the External Universe *Phys. Rev.* D **34** 991

Zhongolovich I D 1957 Earth Gravity Potential *Bull. ITA (Leningrad)* **6** 505

Zhu S Y and Groten E 1988 Relativistic Effect in VLBI Time Delay Measurement *Manuscripta Geodaetica* **13** 33

Zhu S Y, Groten E, Pan R S, Yan H J, Cheng Z Y, Zhu W Y, Huang C and Yao M 1988 Motion of Satellite—The Choice of Reference Frames *The Few Body Problem* ed M J Valtonen (Dordrecht: Kluwer) p 207

Index

Milton Keynes UK
Ingram Content Group UK Ltd.
UKHW040109071024
449327UK00019B/938

9 780367 403065